The Pleasure of Inquiry:
Readings in Sociology

Ed Ksenych
George Brown College

David Liu
Harrisburg Area Community College

THOMSON

NELSON

Australia   Canada   Mexico   Singapore   Spain   United Kingdom   United States

THOMSON

NELSON

The Pleasure of Inquiry: Readings in Society
by Ed Ksenych and David Liu

Associate Vice President, Editorial Director:
Evelyn Veitch

Editor-in-Chief, Higher Education:
Anne Williams

Executive Editor:
Cara Yarzab

Marketing Manager:
Heather Leach

Publisher's Representative:
Claudine O'Donnell

Developmental Editors:
Lesley Mann and
Kamilah Reid Burrell

Permissions Coordinator:
Nicola Winstanley

Content Production Manager:
Jaime Smith

Production Service:
ICC Macmillan Inc.

Copy Editor:
Susan Fitzgerald

Proofreader:
Dianne Fowlie

Manufacturing Coordinator:
Loretta Lee

Design Director:
Ken Phipps

Cover Design:
Dianna Little

Cover Image:
Astrid Ho

Printer:
Thomson West

Library and Archives Canada Cataloguing in Publication Data

Ksenych, Edward, 1950-
  The pleasure of inquiry / Ed Ksenych, David Liu.

ISBN 13: 978-0-17-610474-0
ISBN 10: 0-17-610474-7

  1. Sociology—Textbooks.
I. Liu, David, 1957-II. Title.
HM586.K84 2007   301
C2006-906510-1

To Alja and Sherri

Ed Ksenych and David Liu

Table of Contents

The following two tables of contents represent different ways of organizing the articles in the reader. The first table organizes the readings in terms of some fundamental problems of sociological inquiry and reflects a general orientation to teaching inductively from everyday examples to social issues and the sociological questions which underlie them. The particular sociological problems here are:

- making problematic the taken-for-granted, everyday social world;
- understanding the social world sociologically;
- being and becoming social through social action and interaction;
- the nature of social order and its constituent social institutions;
- conflict arising from socially constructed differences, inequality, and control; and
- the ongoing relevance of methodically researching the social world in a valid, reliable and ethical way in order to deepen our understanding of it.

The second table of contents organizes the same collection of articles in terms of the topic areas usually presented in introductory textbooks. Some articles are relevant to more than one topic area.

The decision to provide two tables of contents for the same articles reflects an essential aspect of sociological consciousness: That which is could always be organized in different, sensible ways. So the question arises: What rule(s) or principle(s) is (are) governing how something has come to be customarily organized, and why is it usually organized that way rather than in some other possible way?

I.

II.

III.

BEING AND BECOMING SOCIAL: SELF, SOCIAL ACTION, AND CULTURE   **116**

IV.

SOCIAL INSTITUTIONS: ORDERING COLLECTIVE LIFE   **203**

V.

VI.

Table of Contents
by Topic Area

I.

SOCIOLOGICAL PERSPECTIVE/SOCIOLOGICAL THEORY

II.

CULTURE

III.

MEDIA ANALYSIS

IV.

SELF AND SOCIAL INTERACTION

V.

SOCIALIZATION AND EDUCATION

VI.

VII.

VIII.

SOCIAL INEQUALITY

IX.

RACE AND ETHNICITY

X.

SEX AND GENDER

XIV.

ISSUES IN SOCIOLOGICAL RESEARCH

Preface

Introductory sociology texts generally provide a useful overview of research, concepts, and theories within sociology, but they tend to describe sociology rather than exemplify the practice of sociological inquiry. As a complement to introductory texts, or as the primary course text itself, this reader offers students an engaging and lively sense of the practice of sociology.

Sociology may be generally described as a field of study with society as its object of inquiry. But most sociologists would agree that, more accurately, it is a practice, a distinctive way of inquiring into human conduct and collective life. By engaging in the practice students can discover the excitement, the discipline, and the power of sociology.

The process of studying something sociologically reflects a commitment to examining this world in a particular way. Sociology brings a critical and vital spirit to researching, theorizing, and making problematic the social world in order to understand it both in itself and for the sake of the greater human good. Echoing Socrates some two thousand years ago, "What we are engaged in here is not a chance conversation, but a dialogue about the very way we ought to live our lives." These general principles guide the selection of readings in this book.

The organization of the book is also reflective of our particular characterization of sociological inquiry. The first thing you may have noticed is the inclusion of two tables of contents. The first is organized in terms of key areas characteristic of sociological inquiry.

The first section, entitled "The Strange World of the Familiar," consists of articles that challenge and make problematic the everyday, taken-for-granted world we ordinarily inhabit. The second section, "The Sociological Tradition," offers a sample of articles from classical to more contemporary authors who discuss, and have contributed to, sociology's distinctive mode of theorizing. The third section on "Being and Becoming Social" brings together readings which address a puzzling phenomenon that appears when we begin probing into our social existence. While we ordinarily take for granted that humans are social, how we are and how we become social is actually a question which requires attention. How do we learn to be social and to become a member of a group through acting and interacting when these activities are already social in nature? Section 4 on "Social Institutions" focuses on how sociologists research and theorize some of the key social institutions—the family, the economy, sports, the media—which structure our social world and establish the opportunities and constraints we experience within it. The fifth section, "Contested Terrains," extends a phrase first put forward by Richard Edwards in his critical analysis of the world of work and organizations. Highlighting the critical, dialogic, and analytic temperament which sociologists bring to a variety of issues, the readings in this section stress conflicts which arise from socially constructed differences, inequalities, and struggles for control. The final section, "Doing Sociology," returns us to the theme of sociology as a practice.

A key dimension of this practice is critiquing, interpreting, and theorizing. Many sociology courses highlight this by drawing students' attention to, and then challenging, prevailing ideas, views, and explanations about the social world by using sociological research and theory. But there is usually little done today to encourage students to try their own hand, even in a rudimentary way, at gathering their own findings as part of reacquainting themselves with the social world they inhabit, developing their own formulations about what's going on, or applying the formulations of those they study.

Of course, this isn't simply professorial negligence. Two major external constraints on students and teachers alike are limited time and increasingly larger classes. Yet, we would contend there are some very basic field projects that can be undertaken and evaluated at an introductory level even in time-pressured courses with large enrolments.

The final section provides an orientation to two general types of projects—observation and content analysis—which can be adapted in a variety of ways to even a one-semester introductory sociology course. Here students can be provided with a taste of what it is to systematically study the social world and their involvement in it. The observational projects can range from observing more or less familiar social settings to simply observing when one doesn't tell the truth throughout a day.

Content analyses may range from tabulating types of content in local newspapers to documenting how women are portrayed in TV soaps or how murders are depicted in crime dramas. The section also offers a reading which sensitizes students to the ongoing importance of methodically researching the social world in an ethical way as part of deepening their understanding of it. A number of articles in the reader have been chosen because they serve as good examples of field projects, or because they provide theoretical background important to analyzing the material gathered in the field project.

In addition to this organization of the articles, the book also provides a second table of contents that organizes the same collection of articles in terms of topic areas usually presented in introductory textbooks. Some articles are relevant to more than one topic area. The decision to provide two tables of contents reflects an essential aspect of sociological consciousness or awareness: That which is could always be organized in different, albeit equally sensible, ways. Questions then arise: What rule(s) and/or principle(s) govern how something has come to be customarily organized, and why it is usually organized that way rather than some other possible way?

The first table of contents organizes the readings in terms of some of the fundamental problems of sociological inquiry and reflects a general orientation to teaching students inductively, moving from everyday examples to social issues and the sociological questions which underlie them. The second table of contents organizes the readings in accordance with the convention of presenting sociology as a social scientific field of study.

We have already discussed one of the key organizing principles behind the selection of articles for this reader, that is, emphasizing sociology as a pleasurable, lively, and engaging activity. Presenting sociology as a practice is of particular importance given the nature of students at many colleges and universities today. Emphasizing sociology as a practice ties in well with students' practical educational concerns, and gives them an opportunity, through assignments, to apply sociology to areas of interest which are relevant to their everyday lives.

We also want to discuss four other organizing principles which have guided the selection of articles:

- providing a grounding in the theoretical bases of sociology as part of introducing students to the practice of sociological inquiry,
- addressing both the critical and the phenomenological/social constructionist approaches to teaching introductory sociology,
- raising questions regarding the emergence of a postmodern society and the growth of a media/consumer driven society,
- emphasizing global content and research which may be discussed within a particular national context but which make a significant contribution to understanding the topic under discussion in a more transnational context.

An important dimension of introducing students to sociology as a form of inquiry is providing them with some grounding in the theoretical bases of the discipline. To that end we have provided a variety of accessible articles from both classical and contemporary sociological theorists, which orient students to significant themes, concepts, and problems in sociological inquiry. The section on the sociological tradition explicitly offers a sample of theoretic works relevant to conflict theory, structural-functionalism, symbolic interactionism, feminist theory, and social phenomenology, as well as other contemporary examples of theorizing.

However, our concern with orienting students to the theoretical basis of sociology extends beyond providing samples of particular theoretical approaches. It grounds the selection of articles, section introductions, and commentaries on the articles. Driving this interest in theory and theorizing is the sense of wondering about and understanding a phenomenon while describing and explaining it using the methods of science. Central to this process is the presentation and use of familiar topic areas as occasions to raise sociological questions and to articulate sociological problems. This begins with questioning aspects of the familiar topic area which usually go unasked or unnoticed, and then bringing sociology's particular disciplined approach to exploring the question or problem being posed.

As Peter Berger stated it, while the "social problem" may be conventionally identified as, say, divorce or crime, the "sociological problem" is marriage and the family, or law and justice. For example, one cannot begin to grasp the problem of juvenile delinquency or youth's lack of participation in civic society without asking about the socio-cultural construction of "youth" as a meaningfully lived age category. One needs to explore how this was established politically about a century ago through the social institutions of law (e.g., child labour laws, voting laws, and the juvenile delinquency act), education (e.g., universal schooling), and the social science professions (e.g., the biological doctrine of adolescence advanced by psychologists which naturalizes youth as a life stage).

Nor can one understand either delinquency or lack of political participation without exploring the experience of young adults being marginalized from full participation in the political economy of a society and simultaneously being subjected to more surveillance and control than those in any other age category. So while the social problem may be delinquency or lack of civic participation, the sociological problem is this peculiar condition of being juvenile, an age category virtually unknown in other, more traditional, forms of society.

It is with the above sense that we have selected articles written by authors who are moved by, and exhibit, the theoretic impulse to wonder about and understand the phenomenon they're investigating even if they are not explicitly making reference to sociological theory.

Another organizing principle of this reader concerns the attention given to the two prevailing approaches used in presenting sociology at an introductory level—the critical approach and the social constructionist/ phenomenological approach. In highlighting these two approaches our aim is not to exclude any of the diverse ways of teaching introductory sociology, but to emphasize the attention this selection of articles gives to the two most popular approaches used to orient students to sociology.

One of the approaches frequently used in introductory texts and courses is based on an understanding of sociology articulated by C. Wright Mills, which is centred around the development of "the sociological imagination." Mills formulates a difference between problems we may experience as personal troubles and those that are public issues. He characterizes personal troubles as problems that are primarily rooted in the character of the individual and/or his or her local environment.

By contrast, public issues, although they may affect us as individuals, are primarily rooted in the nature of the social arrangements which prevail in a society. The sociological imagination enables one to grasp the relationship between the experiences an individual may be undergoing, which are often defined as personal troubles, and the broader public issues and institutional arrangements of which they are a part.

For example, an unemployed individual will, of course, experience his or her unemployment directly on a personal level and contend with it practically as a personal trouble. He or she may also be vaguely aware that it has some connection to the broader problem of unemployment rates in a nation, the current state of the economy, business trends, government policies, and so on. But the sociological imagination enables us to clarify this connection by first examining how relevant social institutions like an advanced capitalist market economy, corporations, government systems, etc., are structured and operate. Then, by seeing the relation of current economic trends, work opportunities, income levels, the distribution of wealth, the effects of technology, and so forth to one's own and others' unemployment, one begins to grasp the phenomenon as a public issue. Understanding the influence of these structures and the impact of these trends enables individuals to more knowledgeably and effectively conduct themselves in their private lives and as members of communities and nations.

Mills' work is often treated as the modern foundation of the critical approach to sociology. Courses emphasizing this approach focus primarily on the nature and impact of social institutions, are organized around the problem of inequality and equity issues, and usually conceptualize the individual in basic social psychological terms. This approach is compatible with modes of sociological inquiry that describe themselves in terms of conflict theory, Marxism, feminism, or politically-oriented styles of post-modernist inquiry.

A second general approach also used in introducing sociology is based on an understanding of sociology articulated by Peter Berger, and is organized around developing a sociological "form of consciousness."

While attending to the relationship between the individual self and broader social arrangements, this approach places more emphasis on *meaning* in the individual's conscious experience of social life and understanding society. Whether addressing problems as personal troubles or public issues, or even addressing ordinary everyday experiences which may not be seen as especially problematic by individuals, the subjective meaning of the phenomenon is accented.

The initial move is to "bracket" an experience or event, that is, pull it, and our ordinary understandings of it, out of the everyday flow of life. This enables us to carefully examine and reflect on it. Following the earlier example, this approach would consider the objective fact of being unemployed (i.e., someone doesn't have a job, can't find work they can or want to do, and doesn't have money coming in), but would also inquire into what being unemployed *means* subjectively and intersubjectively to individuals and collectives.

We often have, and use, short form descriptions of an experience or occurrence like being unemployed. These are generally sufficient for addressing the problem practically. But, through what we might call a more disciplined reflection on the experience, we find that the phenomenon of unemployment is actually more complex and multilayered than first appears. For instance, what's the primary worry in not having money coming in? How do individuals feel about themselves? What happens to their status in the community? How does this status influence their interactions with family members and friends? What does the social fact of one's gender, race, or religion contribute to the experience of being unemployed? More generally, what does the condition of being "unemployed" mean exactly? In terms of culture and social structure, what makes this condition possible or even probable? We discover that we have more to comprehend and account for than simply not having a job and needing money.

Second, through a particular form of questioning, this kind of investigation enables us to discover the "problematic" character of everyday social experiences and occurrences even when they are typically treated as routine and ordinary. It is in this sense that sociological inquiry makes the familiar strange. For instance, going to work every day is just as problematic from a sociological perspective as being unemployed. How do we frame the question of why we do it? What general rules constitute how we usually go about doing it? Why do we go about doing it this way? What does it to work? After all, "working" might be construed as a strange kind of thing to be spending so much time at, especially given that we are now quite technologically capable of doing far less of it and still having a good quality life. How do we understand the irony that hunting and gathering societies spend much less time toiling than modern ones do and that, throughout most of Western history, we have regarded work as an unhappy necessity or a curse, yet we continue to increase the amount and pace of work that we do.

The second approach we've outlined is often described as phenomenological because of its emphasis on describing and reflecting upon life experience, examining the meaning of the experiences for the participants involved, and inquiring into the social foundations of their possibility. The aim is to develop and expand a sociological form of consciousness, that is, an awareness and understanding of ourselves and others as social beings who inhabit meaningful worlds which are collectively created, re-created, and changed through our conscious experience of them. Within such an approach, the individual is conceptualized in more existentialist terms as a social actor. This approach is intended to characterize the approach of practioners of sociological inquiry who may describe themselves in terms of social phenomenology, interpretive sociology, social constructionism, or cultural studies.

As we've outlined them, these two general approaches aren't intended to be mutually exclusive. They represent a difference in emphasis and theoretic style which usually has implications for course design, curriculum, pedagogy, assignments, and evaluative practices. We have tried to be responsive to both.

Finally, our reader is organized around the principle of offering articles that explore the possible transformation of modern society into what is often referred to as a postmodern society, and the increasingly influential global context in which this is occurring. Postmodern refers here to an emerging form of society which valorizes individuals' subjective experiences and choices; where our role as consumer rivals that of citizen in defining our participation as social members; where diversity, whether cultural, sexual, or even social class, becomes reconstructed as lifestyle; where nations become resituated within transnational social formations; and where the stability of all social institutions are in flux as they are subject to accelerated change. All of this both reflects and requires the development of new forms of social cohesion, inequality, authority, social control, social action, and even the self—which is part of what makes the current changes so intriguing to contemporary sociological inquiry.

To address this we have included articles which represent a postmodern style of inquiry or which topicalize aspects of postmodernity itself. A significant aspect of postmodernism is the growth of a media-driven, consumer society. So a number of articles address the significant influence of the media in contemporary life and encourage the development of media literacy as a strategy for questioning, doubting, and resisting the sophistry which is used to persuade us to consume everything from goods and services to political ideas, as well as to package and disguise so much of the world and what is occurring within it.

In addition to exploring the question of post-modern society by addressing the growth of the media and consumerism, we have also provided articles which address issues within a more global context. This is taken up less by providing comparative studies than by selecting articles which topicalize emerging issues regarding globalization, or which address a topic in a particular national context but in a manner that is relevant transnationally. Regarding the latter, articles which may be researched or discussed within a Canadian, American, or other national context are included usually because they make a significant contribution to understanding issues or topics transnationally. For example, Reiter's study of working in a fast-food restaurant in a Canadian suburb, Blauner's discussion of black and white languages of race in America, Sartre's examination of anti-Semitism in mid-century Europe, Gregg's discussion of diversity and multiculturalism in Canada, or Foucault's examination of the shift in penal style within France all successfully identify and address issues which extend beyond national contexts.

That said, while articles have been drawn from scholarship in many countries, most of the selections are North American and explicitly address Canadian and/or American society. If an essential characteristic of sociology is "debunking," as Peter Berger pointed out, then we have found that in understanding and challenging the normative order one needs to consider the distinction between Canada and the United States, and this is reflected in this reader.

A key difference between the two nations concerns the viability of real and legitimate alternative voices in public discourse about one's society as a political community. Despite a strong tradition of freedom of speech in the United States, there actually are few significant departures from the two dominant political narratives of Democrat and Republican political parties. By contrast, in Canada there is a greater variety of viable regional and national political movements. Labour, separatist, and Native self-government movements are examples which continue to powerfully influence Canada's political and cultural landscape.

This difference is not inconsequential for teaching and practising sociology. In the United States challenging the normative order sociologically has meant increasingly emphasizing the "scientific" aspect of sociology, without necessarily investigating the many issues surrounding sociology as a science. While this is an important part of introducing students to sociology in Canada, we contend that populist social and political movements establish a basis for the expression of alternative viewpoints. They also provide for a clearer articulation of rival theoretic understandings of the modern world. You will find selections throughout the book which reflect such viewpoints.

In conclusion, our overall aim has been to have various sociological concepts, theories, and research methods become vivid in addressing enduring questions of human society such as domination and group conflict, social order and change, and human action and shared meanings, within the context of our current society and historical age. The selections vary in style and complexity. We continue to find that students are capable of enjoying and benefiting from challenging material if they have been appropriately prepared or are assisted. Each selection has been prefaced by remarks which serve to highlight important concerns, identify basic questions, or suggest a way to approach the reading.

The materials within this book do not claim to comprehensively cover the field of sociology. They stand as a resource for introducing students to sociological inquiry. As such, they serve as an opportunity for both students and teachers to undertake a shared re-examination of their own experiences, conduct, and embeddedness in the distinctive web of social relations and meanings we call modern society.

ANCILLARIES

Web Site: http://www.pleasureofinquiry.nelson.com. Student resources include information on degrees and careers, study resources, search engines, and the Online Dictionary of the Social Sciences.

Acknowledgements

To begin we'd like to thank our families. There's a reason why family members, both immediate and extended, are thanked so often in acknowledgements. Writing and working on a book pulls your attention away from your ongoing life with others, and for a longer period of time than you usually anticipate. This includes those whom you deeply care about and who care about you. Acknowledging is a way of asking them to forgive you under cover of thanking them.

Second, we want to thank our many friends and colleagues who offered advice or served as sounding boards for different parts of this project. Schooled as we both were in a collaborative approach to academic work, it is difficult to go about citing the names of so many individuals whose teaching and academic work are also an ongoing part of our work. However, we do want to highlight the contributions of Dianne Acey, John Baker, Brenda Bennett, Sherri Kimmel, and Alja Pirosok for their comments on drafts of the written work, Diane Trullinger for her helpful administrative assistance, as well as Jim Cosgrave for his many thoughtful comments from the first to the last stage of this particular project.

Many authors and publishers gave us permission to freely reproduce their work, or in some cases reduced their initial permissions costs and requirements. We appreciate their consideration in making the articles in this sociology reader available at an affordable price to college and university students.

There's also a reason why particular people in publishing companies are so often thanked in acknowledgements. It just wouldn't have happened without them. Specifically, we want to thank Claudine O'Donnell, Lesley Mann, and Cara Yarzab for their ongoing support, suggestions, and talented work on this project. As well we'd like to acknowledge the following Nelson personnel: Jaime Smith (content production manager), Kelly Smyth (marketing manager), and Nicola Winstanley (permissions researcher).

Finally, we want to thank the five reviewers who took the time to assess the initial prospectus and provide some very valuable feedback: Joan Allen, York University; Barry Scott Green, University of Toronto-Erindale; Raymond Foui, University of Manitoba; Patricia Cormack, St. Francis Xavier University; and, Oliver Stoetzer, Fanshawe College.

Ed Ksenych and David Liu
Toronto, Ontario, Canada
Harrisburg, Pennsylvania, U. S. A.
February 2007

About the Authors

Ed Ksenych (pronounced "senich") did his graduate work in social theory at York University and has been a professor at George Brown College for over twenty years. He teaches various courses in sociology and philosophy, but especially enjoys introducing students to the pleasure of sociological inquiry. He has won the Board of Governors' Award of Excellence in Teaching and the Community Services and Health Sciences Divisional Teaching Award, both at George Brown College, as well as the National Institute for Staff and Organizational Development (NISOD) Teaching Excellence Award, University of Texas at Austin. Ed co-edited *Conflict, Order and Action,* 3rd edition (Toronto: Canadian Scholars' Press, 2001), with his colleague David Liu, and edited *Forbidden Desires: Deviance and Social Control* (Toronto: Canadian Scholars' Press, 2003). In addition, he has written and presented papers regarding education in the college system which reflect his abiding interest in classical and contemporary approaches to pedagogy and curriculum development. Extracurricular interests include playing tai chi, improvising on the piano, camping in Algonquin Park, but most of all, having conversations with his friends.

David Liu grew up in Baltimore, Maryland, and received his B.A. from Dickinson College, Carlisle, Pennsylvania. He did his graduate work at York University, Toronto. Having taught at a number of colleges and universities in Canada and the United States, he has found a home at Harrisburg Area Community College in Harrisburg, Pennsylvania. His teaching and research interests include the sociology of art and culture, mass media, health and nutrition, and social problems. His personal interests correspond to his professional ones and include cooking, playing soccer, and trying to teach his old cats new tricks. David has published poetry, essays, and reviews on diverse topics, and he co-edited the first three editions of *Conflict, Order and Action* (Toronto: Canadian Scholars' Press, 2001) with his colleague Ed Ksenych.

PART ONE

The Strange World of the Familiar: Everyday Life and the Taken-for-Granted

Sociology is often described as the scientific study of human society and social action. However, it is essentially a practice, a distinctive way of inquiring into the social interactions, organizations, and institutional arrangements that constitute our social world. But where and how does such an inquiry begin? By looking at the orderly and disorderly character of everyday life.

Historically, sociology emerged amid transformative events that occurred in the wake of the Industrial Revolution. Traditional forms of social life were giving way to modern forms. People were being forced to move off farms and seek unhealthy work in factories located in expanding cities. Technological innovation was challenging traditional craft industries. And the living conditions and family life of most people were becoming radically transformed.

Contemporary sociological inquiry is often prompted by disturbing events and emerging trends as well. But disturbances in themselves don't generate sociological questions. People buffeted by such events typically focus on coping with, or adjusting to, the resulting turmoil rather than trying to understand the nature of the events themselves. Millions of people have undergone the disruption of modernization, but relatively few have been prompted to ask sociological questions about the process of modernization itself. The instability and disorder resulting from change is often cast as unavoidable and inevitable.

Cataclysmic events do not engender a sociological consciousness, though they can provide the context for its emergence. It is often not until a system or an institution is tested, or something goes wrong, that we begin to think about how a system is organized and is expected to function.

This is true on a macro level—what would happen if suddenly there were no police?—as well as on a micro level—what if you said hello to someone and she ignored you? The organization and meaning of these systems, be they official systems of authority such as the government and police, or more informal systems of expectations that govern interaction such as etiquette and manners, remain in the background, typically unnoticed and taken for granted in everyday life.

The first step in thinking about society is to bring the background into the foreground. We begin our inquiry by making the everyday world, or even extraordinary events occurring within it, "problematic." That is, much of sociological inquiry originates, not with dramatic events, but by simply asking how and why everyday life is organized in the way that it is. That is, we begin by making what is familiar strange.

Sociological inquiry, like all inquiry, commences with looking around. But, in this case, what we see leads to an unsettled feeling. These observations and feelings lead to questions about the social world around us. Why is there inequality among people in society? Why is there misery in the world? Can an individual ordinarily make any difference to what is occurring in the world? Are human beings really free?

But the questions can also be playful and more locally focused exercises of the imagination. Why are we being asked to learn what we're supposed to be learning in school? Why are most decorative images for children cartooned and coloured in solid blocks of pigment when children don't actually see the world this way? Why do we often ask about how people are when we really aren't that interested, and routinely reply that we're fine when we often are not?

There is yet another element to sociological inquiry. Inquiry is preceded by a nagging suspicion, if not a certainty, that there is more to the familiar world that appears before us than our customary stories about it reveal. Moreover, it is accompanied by a skepticism regarding the commonsense and official explanations of why things are as they are. Usual and customary explanations may offer relief from uncertainty and indecision, but they also curb wonder, the source and inspiration for all inquiry. As Collins and Makowsky point out, concealed above all by customary explanations is that the social world and human society itself are mysteries, and human sociality is as much a puzzle as it is a resource we rely upon in order to live together in the way that we do.[1]

Each of the articles in this section is animated by the desire to probe into and ask questions about everyday life, and each invites us, in its own way, to look at and think about what we typically take for granted. Each directs us to a distinctive way of bringing the background into the foreground, and presents us with stepping-off points into the strange world of the familiar.

NOTE

1. Randall Collins and Michael Makowsky, *The Discovery of Society*, 6th edition, McGraw-Hill, 1998, pp. 1–2.

Body Ritual Among the Nacirema

Horace Miner

Here, anthropologist Horace Miner observes and reports on the Nacirema, a people whose culture is riddled with magic and whose everyday lives are organized around superstition and the supernatural. But who are these people? In what ways are they similar to, and different from, us? How does Miner suggest that we understand their particular way of life? In reporting his findings Miner is also sensitizing readers to their own ethnocentricity, and presses us to recognize that the superstitious and magical are less a matter of specific activities than they are of the assumptions that we draw upon to interpret those activities. We are left with the question: When does a practice become a ritual, and a ritual an exercise of superstition?

The anthropologist has become so familiar with the diversity of ways in which different peoples behave in similar situations that he is not apt to be surprised by even the most exotic customs. In fact, if all of the logically possible combinations of behaviour have not been found somewhere in the world, he is apt to suspect that they must be present in some yet undescribed tribe. This point has, in fact, been expressed with respect to clan organization by Murdock (1949: 71). In this light, the magical beliefs and practices of the Nacirema present such unusual aspects that it seems desirable to describe them as an example of the extremes to which human behaviour can go.

Professor Linton first brought the ritual of the Nacirema to the attention of anthropologists twenty years ago (1936: 326), but the culture of this people is still very poorly understood. They are a North American group living in the territory between the Canadian Cree, the Yaqui and Tarahumare of Mexico, and the Carib and Arawak of the Antilles. Little is known of their origin, although tradition states that they came from the east. According to Nacirema mythology, their nation was originated by a culture hero, Notgnihsaw, who is otherwise known for two great feats of

Source: Horace Miner, "Body Ritual Among the Nacirema," *American Anthropologist* 50(3): 503–7. Reprinted with permission from the American Anthropological Association.

strength—the throwing of a piece of wampum across the river Pa-To-Mac and the chopping down of a cherry tree in which the Spirit of Truth resided.

Nacirema culture is characterized by a highly developed market economy which has evolved in a rich natural habitat. While much of the people's time is devoted to economic pursuits, a large part of the fruits of these labours and a considerable portion of the day are spent in ritual activity. The focus of this activity is the human body, the appearance and health of which loom as a dominant concern in the ethos of the people. While such a concern is certainly not unusual, its ceremonial aspects and associated philosophy are unique.

The fundamental belief underlying the whole system appears to be that the human body is ugly and that its natural tendency is to debility and disease. Incarcerated in such a body, man's only hope is to avert these characteristics through the use of the powerful influences of ritual and ceremony. Every household has one or more shrines devoted to this purpose. The more powerful individuals in the society have several shrines in their houses and, in fact, the opulence of a house if often referred to in terms of the number of such ritual centres it possesses. Most houses are of wattle and daub construction, but the shrine rooms of the more wealthy are walled with stone. Poorer families imitate the rich by applying pottery plaques to their shrine walls.

While each family has at least one such shrine, the rituals associated with it are not family ceremonies but are private and secret. The rites are normally only discussed with children, and then only during the period when they are being initiated into these mysteries. I was able, however, to establish sufficient rapport with the natives to examine these shrines and to have the rituals described to me.

The focal point of the shrine is a box or chest which is built into the wall. In this chest are kept the many charms and magical potions without which no native believes he could live. These preparations are secured from a variety of specialized practitioners. The most powerful of these are the medicine men, whose assistance must be rewarded with substantial gifts. However, the medicine men do not provide the curative potions for their clients, but decide what the ingredients should be and then write them down in an ancient and secret language. This writing is understood only by the medicine men and by the herbalists who, for another gift, provide the required charm.

The charm is not disposed of after it has served its purpose, but is placed in the charm-box of the household shrine. As these magical materials are specific for certain ills, and the real or imagined maladies of the people are many, the charm-box is usually full to overflowing. The magical packets are so numerous that people forget what their purposes were and fear to use them again. While the natives are very vague on this point, we can only assume that the idea in retaining all the old magical materials is that their presence in the charm-box, before which the body rituals are conducted, will in some way protect the worshipper.

Beneath the charm-box is a small font. Each day every member of the family, in succession, enters the shrine room, bows his head before the charm-box, mingles different sorts of holy water in the font, and proceeds with a brief rite of ablution. The holy waters are secured from the Water Temple of the community, where the priests conduct elaborate ceremonies to make the liquid ritually pure.

In the hierarchy of magical practitioners, and below the medicine men in prestige, are specialists whose designation is best translated "holy-mouth-men." The Nacirema have an almost pathological horror of and fascination with the mouth, the condition of which is believed to have a supernatural influence on all social relationships. Were it not for the rituals of the mouth, they believe that their teeth would fall out, their gums bleed, their jaws shrink, their friends desert them, and their lovers reject them. They also believe that a strong relationship exists between oral and moral characteristics. For example, there is a ritual ablution of the mouth for children which is supposed to improve their moral fibre.

The daily body ritual performed by everyone includes a mouth-rite. Despite the fact that these people are so punctilious about care of the mouth, this rite involves a practice which strikes the uninitiated stranger as revolting. It was reported to me that the ritual consists of inserting a small bundle of hog hairs into the mouth, along with certain magical powders, and them moving the bundle in a highly formalized series of gestures.

In addition to the private mouth-rite, the people seek out a holy-mouth-man once or twice a year. These practitioners have an impressive set of paraphernalia, consisting of a variety of augers, awls, probes, and prods. The use of these objects in the exorcism of the evils of the mouth involves almost unbelievable ritual torture of the client. The holy-mouth-man opens the client's mouth and, using the above mentioned tools, enlarges any holes which decay may have created in the teeth. Magical materials are put into these holes. If there are no naturally occurring holes in the teeth, large sections of one or more teeth are gouged out so that the supernatural substance can be applied. In the client's view, the purpose of these ministrations is to arrest decay and to draw new friends. The extremely sacred and traditional character of the rite is evident in the fact that the natives return to the holy-mouth-men year after year, despite the fact that their teeth continue to decay.

It is to be hoped that, when a thorough study of the Nacirema is made, there will be careful inquiry into the personality structure of these people. One has but to watch the gleam in the eye of a holy-mouth-man, as he jabs an awl into an exposed nerve, to suspect that a certain amount of sadism is involved. If this can be established, a very interesting pattern emerges, for most of the population shows definite masochistic tendencies. It was to these that Professor Linton referred in discussing a distinctive part of the daily body ritual which is performed only by men. This part of the rite involves scraping and lacerating the surface of the face with a sharp instrument. Special women's rites are performed only four times during each lunar month, but what they lack in frequency is made up in barbarity. As part of this ceremony, women bake their heads in small ovens for about an hour. The theoretically interesting point is that what seems to be a preponderantly masochistic people have developed sadistic specialties.

The medicine men have an imposing temple, or *latipso,* in every community of any size. The more elaborate ceremonies required to treat very sick patients can only be performed at this temple. These ceremonies involve not only the thaumaturge but a permanent group of vestal maidens who move sedately about the temple chambers in distinctive costume and headdress.

The *latipso* ceremonies are so harsh that it is phenomenal that a fair proportion of the really sick natives who enter the temple ever recover. Small children whose indoctrination is still incomplete have been known to resist attempts to take them to the temple because "that is where you go to die." Despite this fact, sick adults are not only willing but eager to undergo the protracted ritual purification, if they can afford to do so. No matter how ill the supplicant or how grave the emergency, the guardians of many temples will not admit a client if he cannot give a rich gift to the custodian. Even after one has gained admission and survived the cere-monies, the guardians will not permit the neophyte to leave until he makes another gift.

The supplicant entering the temple is first stripped of all his or her clothes. In every-day life the Nacirema avoids exposure of his body and its natural functions. Bathing and excretory acts are performed only in the secrecy of the household shrine, where they are ritualized as part of the body-rites. Psychological shock results from the fact that body secrecy is suddenly lost upon entry into the *latipso.* A man, whose own wife has never seen him in an excretory act, suddenly finds himself naked and assisted by a vestal maiden while he performs his natural functions into a sacred vessel. This sort of ceremonial treatment is necessitated by the fact that the excreta are used by a diviner to ascertain the course and nature of the client's sickness. Female clients, on the other hand, find their naked bodies are subjected to the scrutiny, manipulation and prodding of the medicine men.

Few supplicants in the temple are well enough to do anything but lie on their hard beds. The daily ceremonies, like the rites of the holy-mouth-men, involve discomfort and torture. With ritual precision, the vestals awaken their miserable charges each dawn and roll them about on their beds of pain while performing ablutions, in the formal movements of which the maidens are highly trained. At other times they insert magic wands in the supplicant's mouth or force him to eat substances which are supposed to be healing. From time to time the medicine men come to their clients and jab magically treated needles into their flesh. The fact that these temple ceremonies may not cure, and may even kill the neophyte, in no way decreases the people's faith in the medicine men.

There remains one other kind of practitioner, known as a "listener." This witch-doctor has the power to exorcise the devils that lodge in the heads of people who have been bewitched. The Nacirema believe that parents bewitch their own children. Mothers are particularly suspected of putting a curse on children while teaching them the secret body rituals. The counter-magic of the witch-doctor is unusual in its lack of ritual. The patient simply tells the "listener" all his troubles and fears, beginning with the earliest difficulties he can remember. The memory displayed by the Nacirema in these exorcism sessions is truly remarkable. It is not uncommon for the patient to bemoan the rejection he felt upon being weaned as a babe, and a few individuals even see their troubles as going back to the traumatic effects of their own birth.

In conclusion, mention must be made of certain practices which have their base in native aesthetics but which depend upon the pervasive aversion to the natural body and its functions. There are ritual fasts to make fat people thin and ceremonial feasts to make thin people fat. Still other rites are used to make women's breasts larger if they are small, and smaller if they are large. General dissatisfaction with breast shape is symbolized in the fact that the ideal form is virtually outside the range of human variation. A few women afflicted with almost inhuman hyper-mammary development are so idolized that they make a handsome living simply going from village to village and permitting the natives to stare at them for a fee.

Reference has already been made to the fact that excretory functions are ritualized, routinized, and relegated to secrecy. Natural reproductive functions are similarly distorted. Intercourse is a taboo as a topic and scheduled as an act. Efforts are made to avoid pregnancy by the use of magical materials or by limiting intercourse to certain phases of the moon. Conception is actually very infrequent. When pregnant, women dress so as to hide their condition. Parturition takes place in secret, without friends or relatives to assist, and the majority of women do not nurse their infants.

Our review of the ritual life of the Nacirema has certainly shown them to be a magic-ridden people. It is hard to understand how they have managed to exist so long under the burdens which they have imposed upon themselves. But even such exotic customs as these take on real meaning when they are viewed with the insight provided by Malinowski when he wrote (1948: 70):

> Looking from far and above, from our high places of safety in the developed civilization, it is easy to see all the crudity and irrelevance of magic. But without its power and guidance early man could not have mastered his practical difficulties as he has done, nor could man have advanced to the higher stages of civilization.

References

Linton, Ralph. (1936). *The Study of Man*. New York, D. Appleton-Century Co.
Malinowski, Bronislaw. (1948). *Magic, Science, and Religion*. Glencoe, The Free Press.
Murdock, George P. (1949). *Social Structure*. New York, The Macmillan Co.

Life in a Fast-Food Factory

Ester Reiter

In this selection Reiter participates as a worker in a fast-food restaurant. She scrupulously observes and records details of the physical setting, rout-ines, and relationships. While the worker may experience the organization of the workplace as constraining or monotonous, the structure and culture of the workplace reflect something broader and more deeply seated, namely, the pervasive historical process that Max Weber called the rationalization of social life and which George Ritzer has termed "McDonaldization."

This transformation occurs when the social practices and social organization of a group become formalized, and then subjected to standards of rationality. Everyday social relationships, expectations, practices, and so on are replaced by explicit rules, codes, regulations, and procedures. Finding the most efficient means to achieve previously calculated ends becomes the standard. Are Reiter's experiences of work and the results of her participant observation study similar to your experiences of work? What are some longer-term consequences of the McDonaldization of institutions as diverse as the funeral industry, the travel industry, or the health care system? Is the process of rationalization, or McDonaldization, inevitable?

The growth of large multinational corporations in the service industries in the post-World War II years has transformed our lives. The needs and tastes of the public are shaped by the huge advertisement budgets of a few large corporations. The development of new industries has transformed work, as well as social life. This paper focuses on the technology and the labour process in the fast-food sector of the restaurant industry....

Since the late 1960s, fast-food restaurants have been growing at a much higher rate than independent restaurants, virtually colonizing the suburbs.[1] Local differences in taste and style are obliterated as each town offers the familiar array of trademarked foods: neat, clean, and orderly,

Source: Ester Reiter, "Life in a Fast-Food Factory," from Craig Heron and Robert Storey (eds.), *On the Job: Confronting the Labour Process in Canada*, Kingston and Montreal: McGill Queens University Press, 1986, 309–10, 315–318, 320–322, 324. Reprinted with permission.

the chains serve up the same goods from Nova Scotia to Vancouver Island. The casualties of the phenomenal growth are the small "mom and pop" establishments, rather than the higher-priced, full-service restaurants. Fast-food outlets all conform to a general pattern. Each has a limited menu, usually featuring hamburger, chicken, or fried fish. Most are part of a chain, and most require customers to pick up their own food at a counter.[2] The common elements are minimum delay in getting the food to the customer (hence "fast food") and prices that are relatively low compared to those at full-service restaurants . . . .

[Burger King] store operations are designed from head office in Miami. In 1980, this office commissioned a study to find ways of lowering labour costs, increasing worker's productivity, and maintaining the most efficient inventories. The various components of a restaurant's operations were defined: customer-arrival patterns, manning or positioning strategies, customer/cashier interactions, order characteristics, production-time standards, stocking rules and inventory. Time-motion reports for making the various menu items, as well as corporate standards for service were also included in the calculation, and the data were all entered into a computer. By late 1981, it was possible to provide store managers not only with a staffing chart for hourly sales—indicating how many people should be on the floor given the predicted volume of business for that hour—but also where they should be positioned, based on the type of kitchen design. Thus, although staffing had been regulated since the late 1970s, what discretion managers formerly had in assigning and utilizing workers had been eliminated. The use of labour is now calculated precisely, as is any other objectively defined component of the system, such as store design, packaging, and inventory.[3]

Having determined precisely what workers are supposed to be doing and how quickly they should be doing it, the only remaining issue is that of getting them to perform to specification. "Burger King University," located at headquarters in Miami was set up to achieve this goal. Housed in a remodelled art gallery, the multimillion-dollar facility is staffed by a group of "professionals" who have worked their way up in the Burger King system to the rank of district manager. Burger King trains its staff to do things "not well, but right," the Burger King way.[4] Tight control over Burger King restaurants throughout the world rests on standardizing operations—doing things the "right" way—so that outcomes are predictable. The manager of a Burger King outlet does not necessarily need any knowledge of restaurant operation because the company provides it. What Burger King calls "people skills" are required; thus a job description for a manager indicated that he/she

- Must have good verbal communication skills
- Must have patience, tact, fairness, and social sensitivity in dealing with customers and hourly employees

- Must be able to supervise and motivate team of youthful employees and conduct himself/herself in a professional manner
- Must present a neat, well-groomed image
- Must be willing to work nights, weekends and holidays.[5]

In 1981, a new crew-training program, designed as an outcome of the computer-simulation study was developed. The training program is called "The Basics of Our Business" and is meant to "thoroughly train crew members in all areas of operations and to educate them on how Burger King and the restaurant where they work ... fit into the American free-enterprise system." In addition, the training program involves supervised work at each station, and a new feature that requires every employee to pass a standardized test on appropriate procedures for each station in the store.

Burger King thus operates with a combination of control techniques: technology is used to simplify the work and facilitate centralization, while direct control or coercion is exercised on the floor to make sure the pace of the work remains swift. "If there's time to lean, there's time to clean," is a favourite saying among managers. In fact, workers are expected to be very busy *all* the time they are on shifts, whether or not there are customers in the store. Sitting down is never permissible; in fact, the only chair in the entire kitchen is in the manager's office in a glassed-in cubicle at the rear of the kitchen. From there, the manager can observe the workers at their jobs. The application of these techniques is supported by a legitimizing ideology that calls for "patience, fairness, and social sensitivity" in dealing with customers in order to increase sales and profits "for the betterment of Burger King corporation and its employees."[6]

WORKING AT BURGER KING

I did field work in the fast-food industry by working at a Burger King outlet in suburban Toronto in 1980/1. The Burger King at which I worked was opened in 1979, and by 1981 was the highest volume store in Canada with annual sales of over one million dollars. Everything in the customer's part of the store was new, shiny, and spotlessly clean. Live plants lent a touch of class to the seating area. Muzak wafted through the air, but customers sat on chairs designed to be sufficiently uncomfortable to achieve the desired customer turnover rate of one every 20 minutes. Outside the store, customers could eat at concrete picnic tables and benches in a professionally landscaped setting, weather permitting. Lunches, particularly Thursdays, Fridays, and Saturdays, were the busiest times, and during those periods, customers were lined up at the registers waiting to be served. During the evenings, particularly on Friday nights, families with young children were very much in evidence. Young children, kept amused by the plastic giveaway toys provided by the restaurant and sporting Burger King crowns, sat contentedly munching their fries and sipping their carbonated drinks.

Workers use the back entrance at Burger King when reporting for work. Once inside, they go to a small room (about seven by twelve feet), which is almost completely occupied by an oblong table where crew members have their meals. Built-in benches stretch along both sides of the wall, with hooks above for coats. Homemade signs, put up by management, decorate the walls. One printed, framed, sign read:

WHY CUSTOMERS QUIT
1% die
2% move away
5% develop other friendships
9% competitive reasons
14% product dissatisfactions
68% quit because of ATTITUDE OF INDIFFERENCE TOWARDS
   CUSTOMER BY RESTAURANT MANAGER OR SERVICE PERSONNEL

Another sign reminded employees that only 1/3 ounce of ketchup and 1/9 ounce of mustard is supposed to go on the hamburgers; a crew member using more is cheating the store, while one using less is not giving customers "value" for their dollar.

The crew room is usually a lively place. An AM/FM radio is tuned to a rock station while the teenage workers coming off or on shift talk about school and weekend activities or flirt with each other. Children and weddings are favourite topics of conversation for the older workers. In the evenings, the talk and horsing around among the younger workers gets quite spirited, and now and then a manager appears to quieten things down. Management initiatives are not all geared to control through discipline; social activities such as skating parties, baseball games, and dances are organized by "production leaders" with the encouragement of the managers—an indication that the potentially beneficial effects for management of channelling the informal social relationships at the workplace are understood. Each worker must punch a time card at the start of a shift .... A positioning chart, posted near the time clock, lists the crew members who are to work each meal, and indicates where in the kitchen they are to be stationed.

There are no pots and pans in the Burger King kitchen. As almost all foods enter the store ready for the final cooking process, pots and pans are not necessary .... The major kitchen equipment consists of the broiler/toaster, the fry vats, the milkshake and coke machines, and the microwave ovens .... Even when made from scratch, hamburgers do not require particularly elaborate preparation, and whatever minimal decision making might once have been necessary is now completely eliminated by machines. At Burger King, hamburgers are cooked as they pass through the broiler on a conveyor belt at a rate of 835 patties per hour. Furnished with a pair of tongs, the worker picks up the burgers as they drop off the conveyor belt, puts each on a toasted bun, and places the hamburgers and buns in a steamer. The jobs may be hot and boring, but they can be learned in a matter of minutes.

The more interesting part of the procedure lies in applying condiments and microwaving the hamburgers. The popularity of this tasks among Burger King employees rests on the fact that it is unmechanized and allows some discretion of the worker. As the instructions for preparing a "whopper" (the Burger King name for a large hamburger) indicate, however, management is aware of this area of worker freedom and makes strenuous efforts to eliminate it by outlining exactly how this job is to be performed . . . . Despite such directives, the "Burger and Whopper Board" positions continue to hold their attraction for the workers, for this station requires two people to work side by side, and thus allows the opportunity for conversation. During busy times, as well, employees at this station also derive some work satisfaction from their ability to "keep up." At peak times, a supply of ready-made sandwiches is placed in chutes ready for the cashiers to pick up; the manager decides how many sandwiches should be in the chutes according to a formula involving sales predictions for that time period. At such times, the challenge is to keep pace with the demand and not leave the cashiers waiting for their orders. The managers will sometimes spur the "Whopper-makers" on with cries of "Come on guys, let's get with it," or "Let's go, team."

. . . At Burger King, the goal is to reduce all skills to a common, easily learned level and to provide for cross-training. At the completion of the ten-hour training program, each worker is able to work at a few stations. Skills for any of the stations can be learned in a matter of hours; the simplest jobs, such as filling cups with drinks, or placing the hamburgers and buns on the conveyor belt, can be learned in minutes. As a result, although labour turnover cuts into the pace of making hamburgers, adequate functioning of the restaurant is never threatened by people leaving. However, if workers are to be as replaceable as possible, they must be taught not only to perform their jobs in the same way, but also to resemble each other in attitudes, disposition, and appearance. Thus, workers are taught not only to perform according to company rules, but also are drilled on personal hygiene, dress (shoes should be brown leather or vinyl, not suede), coiffure (hair tied up for girls and not too long for boys), and personality. Rule 17 of the handout to employees underlines the importance of smiling: "Smile at all times, your smile is the key to our success."

While management seeks to make workers into interchangeable tools, workers themselves are expected to make a strong commitment to the store . . . . Workers, especially teenagers, are, then, expected to adjust their activities to the requirements of Burger King. For example, workers must apply to their manager two weeks in advance to get time off to study for exams or attend family functions. Parents are seen by management as creating problems for the store, as they do not always appreciate Burger King's demand for priority in their children's schedules. Thus, the manager warns new trainees to "remember, your parents don't work here and don't understand the situation. If you're old enough to ask for a job, you're old enough to be responsible for coming."[7]

... Making up about 75 percent of the Burger King work force, the youngsters who worked after school, on weekends, and on holidays were called "part-timers." The teenager workers (about half of them boys, half girls) seemed to vary considerably in background....

The daytime workers—the remaining 25 percent of the workforce—were primarily married women of mixed economic backgrounds.... Although they were all working primarily because their families needed the money, a few women expressed their relief at getting out of the house, even to come to Burger King. One woman said: "At least when I come here, I'm appreciated. If I do a good job, a manager will say something to me. Here, I feel like a person. I'm sociable and I like being amongst people. At home, I'm always cleaning up after everybody and nobody ever notices...."[8]

Common to both the teenagers and the housewives was the view that working at Burger King was peripheral to their major commitments and responsibilities; the part-time nature of the work contributed to this attitude. Workers saw the alternatives available to them as putting up with the demands of Burger King or leaving; in fact, leaving seemed to be the dominant form of protest. During my period in the store, on average, eleven people out of ninety-four hourly employees quit at each two-week pay period. While a few workers had stayed at Burger King for periods as long as a few years, many did not last through the first two weeks. The need for workers is constant; occasionally even the paper place-mats on the customer's trays invited people to work in the "Burger King family." "If you're enthusiastic and like to learn, this is the opportunity for you. Just complete the application and return it to the counter." At other times, bounties were offered for live workers. A sign that hung in the crew room for a few weeks read:

Wanna make $10?
   It's easy! All you have to do is refer a friend to me for employment. Your friend must be able to work over lunch (Monday-Friday). If your friend works here for at least one month, you get $20. (And I'm not talking Burger Bucks either.)

Burger King's ability to cope with high staff turnover means that virtually no concessions in pay or working conditions are offered to workers to entice them to remain at Burger King. In fact, more attention is paid to the maintenance of the machinery than to "maintaining" the workers; time is regularly scheduled for cleaning and servicing the equipment, but workers may not leave the kitchen to take a drink or use the bathroom during the lunch and dinner rushes.

The dominant form—in the circumstances, the only easily accessible form—of opposition to the Burger King labour process is, then, the act of quitting. Management attempts to head off any other form of protest by insisting on an appropriate "attitude" on the part of the workers. Crew members must constantly demonstrate their satisfaction with working at Burger King by smiling at all times. However, as one worker remarked, "Why should I smile? There's nothing funny around here. I do my job and that should be good enough for them." It was not, however, and this worker soon quit....

My findings in the fast-food industry are not very encouraging. In contrast to Michael Burawoy,[9] for example, who found that male workers in a unionized machine shop were able to set quotas and thereby establish some control over the labour process, I found that women and teenagers at Burger King are under the sway of a labour process that eliminates almost completely the possibility of forming a workplace culture independent of, and in opposition to, management.

... Unfortunately, there are indications that the teenagers and women who work in this type of job represent not an anomalous but an increasingly typical kind of worker, in the one area of the economy that continues to grow—the service sector. The fast-food industry represents a model for other industries in which the introduction of technology will permit the employment of low-skilled, cheap, and plentiful workers. In this sense, it is easy to be pessimistic and find agreement with Andre Gorz's depressing formulation of the idea of work:

> The terms "work" and "job" have become interchangeable: work is no longer something that one *does* but something that one *has*.
>
> Workers no longer "produce" society through the mediation of the relations of production; instead the machinery of social production as a whole produces "work" and imposes it in a random way upon random, interchangeable individuals.[10]

The Burger King system represents a major triumph for capital: it has established a production unit with constant and variable components that are almost immediately replaceable. However, the reduction of the worker to a single component of capital requires more than the introduction of a technology; workers' autonomous culture must be eliminated as well, including the relationships among workers, their skills, and their loyalties to one another. The smiling, willing, homogenous worker must be produced and placed on the Burger King assembly line....

NOTES

1. Foodservice and Hospitality Magazine, *Fact File—Canada's Hospitality Business*, 4th ed. (Toronto n.d.).

2. This definition comes from the National Restaurant Association and is reprinted in Marc Leepson, "Fast Food, U.S. Growth Industry," *Editorial Research Reports* 7 (1978): 907.

3. "Kitchen design—the drive for efficiency," insert in *Nation's Restaurant News*, 31 August 1981.

4. Personal communications, Burger King "professor," 4 January 1982.

5. Job description handout for Burger King managers, 1981.

6. Handouts to Burger King crew members, 1981.

7. Burger King training session in local outlet, July 1981.

8. Personal communication, Burger King worker, 8 August 1981.

9. Michael Burawoy, *Manufacturing Consent* (Chicago 1979).

10. Andre Gorz, *Farewell to the Working Class* (Boston1982), 71.

Are We Having Sex Now or What?

Greta Christina

In this playful reflection on her own sexual experiences, Christina poses the question of what counts as having sex. Christina invites us to think about what may seem obvious at first glance. But her reflections also point to a fundamental issue in social science research, namely, how one concretely defines what one is studying. Empirical researchers, particularly in surveys, have to define their variables in ways that are observable and measurable in order to gather valid, reliable data. That is, they need to develop *operational definitions* in order to precisely measure what is being studied. For example, in order to find out if people's sexual activity varies according to social class, researchers need to operationally define both having sex and a person's social class. Now surely having sex is a rather obvious matter. Or is it? And if it isn't, then what's the difficulty?

When I first started having sex with other people, I used to like to count them. I wanted to keep track of how many there had been. It was a source of some kind of pride, or identity anyway, to know how many people I'd had sex with in my lifetime. So, in my mind, Len was number one, Chris was number two, that slimy awful little heavy metal barbiturate addict whose name I can't remember was number three, Alan was number four, and so on. It got to the point where, when I'd start having sex with a new person for the first time, when he first entered my body (I was only having sex with men at the time), what would flash through my head wouldn't be, "Oh, baby, baby you feel so good inside me," or "What the hell am I doing with this creep," or "This is boring, I wonder what's on TV." What flashed through my head was "Seven!"

Doing this had some interesting results. I'd look for patterns in the numbers. I had a theory for a while that every fourth lover turned out to be really great in bed, and would ponder what the cosmic significance of the phenomenon might be. Sometimes I'd try to determine what kind of person I was by how many people I'd had sex with. At eighteen, I'd had sex with ten different people. Did that make me normal, repressed, a total slut, a free-spirited bohemian, or what? Not that I compared my numbers with anyone else's—I didn't. It was my own exclusive structure, a game I played in the privacy of my own head.

Then the numbers started getting a little larger, as numbers tend to do, and keeping track became more difficult. I'd remember that the last one was *seventeen* and so this one must be *eighteen*, and then I'd start having doubts about whether I'd been keeping score accurately or not. I'd lie awake at night thinking to myself, well, there was Brad, and there was that guy on my birthday, and there was David and ... no, wait, I forgot that guy I got drunk with at the social my first week at college ... so that's seven, eight, nine ... and by two in the morning I'd finally have it figured out. But there was always a nagging suspicion that maybe I'd missed someone, some dreadful tacky little scumball that I was trying to forget about having invited inside my body. And as much as I maybe wanted to forget about the sleazy little scumball, I wanted more to get that number right.

It kept getting harder, though. I began to question what counted as sex and what didn't. There was that time with Gene, for instance. I was pissed off at my boyfriend, David, for cheating on me. It was a major crisis, and Gene and I were friends and he'd been trying to get at me for weeks and I hadn't exactly been discouraging him. I went to see him that night to gripe about David. He was very sympathetic of course, and he gave me a backrub and we talked and touched and confided and hugged, and then we started kissing, and then we snuggled up a little closer, and then we started fondling each other, you know, and then all heck broke loose, and we rolled around on the bed groping and rubbing and grabbing and smooching and pushing and pressing and squeezing. He never did actually get it in. He wanted to, and I wanted to too, but I had this thing about being faithful to my boyfriend, so I kept saying, "No, you can't do that, Yes, that feels so good, No, wait that's too much, Yes, yes, don't stop, No, stop that's enough." We never even got our clothes off. Jesus Christ, though, it was some night. One of the best, really. But for a long time I didn't count it as one of the times I'd had sex. He never got inside, so it didn't count.

Later, months and years later, when I lay awake putting my list together, I'd start to wonder: Why doesn't Gene count? Does he not count because he never got inside? Or does he not count because I had to preserve my moral edge over David, my status as the patient, ever-faithful, cheated-on, martyred girlfriend, and if what I did with Gene counts then I don't get to feel wounded and superior?

Years later, I did end up fucking Gene and I felt a profound relief because, at last, he definitely had a number, and I knew for sure that he did in fact count.

Then I started having sex with women, and, boy, howdy, did *that* ever shoot holes in the system. I'd always made my list of sex partners by defining sex as penile-vaginal intercourse—you know, screwing. It's a pretty simple distinction, a straightforward binary system. Did it go in or didn't it? Yes or no? One or zero? On or off? Granted, it's a pretty arbitrary definition, but it's the customary one, with an ancient and respected tradition behind it, and when I was just screwing men, there was no compelling reason to question it.

But with women, well, first of all there's no penis, so right from the start the tracking system is defective. And then, there are so many ways women can have sex with each other, touching and licking and grinding and fingering and fisting—with dildoes or vibrators or vegetables or whatever happens to be lying around the house, or with nothing at all except human bodies. Of course, that's true for sex between women and men as well. But between women, no one method has a centuries-old tradition of being the one that counts. Even when we do fuck each other there's no dick, so you don't get that feeling of This Is What's Important, We Are Now Having Sex, objectively speaking, and all that other stuff is just foreplay or afterplay. So when I started having sex with women the binary system had to go, in favor of a more inclusive definition.

Which meant, of course, that my list of how many people I'd had sex with was completely trashed. In order to maintain it I would have had to go back and reconstruct the whole thing and include all those people I'd necked with and gone down on and dry-humped and played touchy-feely games with. Even the question of who filled the all-important Number One slot, something I'd never had any doubts about before, would have to be re-evaluated.

By this time I'd kind of lost interest in the list anyway. Reconstructing it would be more trouble than it was worth. But the crucial question remained: What counts as having sex with someone?

It was important for me to know. You have to know what qualifies as sex because when you have sex with someone your relationship changes. Right? *Right?* It's not that sex itself has to change things all that much. But knowing you've had sex, being conscious of a sexual connection, standing around making polite conversation with someone while thinking to yourself, "I've had sex with this person," that's what changes things. Or so I believed. And if having sex with a friend can confuse or change the friendship, think how bizarre things can get when you're not sure whether you've had sex with them or not.

The problem was, as I kept doing more kinds of sexual things, the line between *sex* and *not-sex* kept getting more hazy and indistinct. As I brought more into my sexual experience, things were showing up on the dividing line demanding my attention. It wasn't just that the territory I labeled *sex* was expanding. The line itself had swollen, dilated, been transformed into a vast gray region. It had become less like a border and more like a demilitarized zone.

Which is a strange place to live. Not a bad place, just strange. It's like juggling, or watchmaking, or playing the piano—anything that demands complete concentrated awareness and attention. It feels like cognitive dissonance, only pleasant. It feels like waking up from a compelling and realistic bad dream. It feels like the way you feel when you realize that everything you know is wrong, and a bloody good thing too, because it was painful and stupid and it really screwed you up.

But, for me, living in a question naturally leads to searching for an answer. I can't simply shrug, throw up my hands, and say, "Damned if I know." I have to explore the unknown frontiers, even if I don't bring back any secret treasure. So even if it's incomplete or provisional, I do want to find some sort of definition of what is and isn't sex.

I know when I'm *feeling* sexual. I'm feeling sexual if my pussy's wet, my nipples are hard, my palms are clammy, my brain is fogged, my skin is tingly and super-sensitive, my butt muscles clench, my heartbeat speeds up, I have an orgasm (that's the real giveaway), and so on. But feeling sexual with someone isn't the same as having sex with them. Good Lord, if I called it sex every time I was attracted to someone who returned the favor I'd be even more bewildered than I am now. Even *being* sexual with someone isn't the same as *having* sex with them. I've danced and flirted with too many people, given and received too many sexy, would-be-seductive backrubs, to believe otherwise.

I have friends who say, if you thought of it as sex when you were doing it, then it was. That's an interesting idea. It's certainly helped me construct a coherent sexual history without being a revisionist swine: redefining my past according to current definitions. But it really just begs the question. It's fine to say that sex is whatever I think it is; but then what do I think it *is*? What if, when I was doing it, I was *wondering* whether it counted?

Perhaps having sex with someone is the conscious, consenting, mutually acknowledged pursuit of shared sexual pleasure. Not a bad definition. If you are turning each other on and you say so and you keep doing it, then it's sex. It's broad enough to encompass a lot of sexual behavior beyond genital contact/orgasm; it's distinct enough *not* to include every instance of sexual awareness or arousal; and it contains the elements I feel are vital—acknowledgment, consent, reciprocity, and the pursuit of pleasure. But what about the situation where one person consents to sex without really enjoying it? Lots of people (myself included) have had sexual interactions that we didn't find satisfying or didn't really want and, unless they were actually forced on us against our will, I think most of us would still classify them as sex.

Maybe if *both* of you (or all of you) think of it as sex, then it's sex whether you're having fun or not. That clears up the problem of sex that's consented to but not wished-for or enjoyed. Unfortunately, it begs the question again, only worse: now you have to mesh different people's vague and inarticulate notions of what is and isn't sex and find the place where they overlap. Too messy.

How about sex as the conscious, consenting, mutually acknowledged pursuit of sexual pleasure of *at least one* of the people involved. That's better. It has all the key components, and it includes the situation where one person is doing it for a reason other than sexual pleasure—status, reassurance, money, the satisfaction and pleasure of someone they love, etc. But what if *neither* of you is enjoying it, if you're both doing it because you think the other one wants to? Ugh.

I'm having trouble here. Even the conventional standby—sex equals intercourse—has a serious flaw: it includes rape, which is something I emphatically refuse to accept. As far as I'm concerned, if there's no consent, it ain't sex. But I feel that's about the only place in this whole quagmire where I have a grip. The longer I think about the subject, the more questions I come up with. At what point in an encounter does it *become* sexual? If an interaction that begins nonsexually turns into sex, was it sex all along? What about sex with someone who's asleep? Can you have a situation where one person is having sex and the other isn't? It seems that no matter what definition I come up with, I can think of some real-life experience that calls it into question.

For instance, a couple of years ago I attended (well, hosted) an all-girl sex party. Out of the twelve other women there, there were only a few with whom I got seriously physically nasty. The rest I kissed or hugged or talked dirty with or just smiled at, or watched while they did seriously physically nasty things with each other. If we'd been alone, I'd probably say that what I'd done with most of the women there didn't count as having sex. But the experience, which was hot and sweet and silly and very, very special, had been created by all of us, and although I only really got down with a few, I felt that I'd been sexual with all of the women there. Now, when I meet one of the women from that party, I always ask myself: Have we had sex?

For instance, when I was first experimenting with sadomasochism, I got together with a really hot woman. We were negotiating about what we were going to do, what would and wouldn't be ok, and she said she wasn't sure she wanted to have sex. Now we'd been explicitly planning all kinds of fun and games—spanking, bondage, obedience—which I strongly identified as sexual activity. In her mind, though, *sex* meant direct genital contact, and she didn't necessarily want to do that with me. Playing with her turned out to be a tremendously erotic experience, arousing and stimulating and almost unbearably satisfying. But we spent the whole evening without even touching each other's genitals. And the fact that our definitions were so different made me wonder: Was it sex?

For instance, I worked for a few months as a nude dancer at a peep show. In case you've never been to a peep show, it works like this: the customer goes into a tiny, dingy black box, kind of like a phone booth, puts in quarters, and a metal plate goes up; the customer looks through a window at a little room/stage where naked women are dancing. One time a guy came into one of the booths and started watching me and masturbating. I came over and squatted in front of him and started masturbating too, and we grinned at each other and watched each other and masturbated, and we both had a fabulous time. (I couldn't believe I was being paid to masturbate—tough job, but somebody has to do it . . . .) After he left I thought to myself: Did we just have sex? I mean, if it had been someone I knew, and if there had been no glass and no quarters, there'd be no question in my mind. Sitting two feet apart from someone, watching each other masturbate? Yup, I'd call that sex all right. But this was different, because it was a stranger, and because of the glass and the quarters. Was it sex?

I still don't have an answer.

The Educational Implications of Our "Technological Society"

James Turk

In this article Turk asks directly: Why are you studying what you are studying? Why do we assume that as workplace technology becomes more sophisticated the worker must become more technologically proficient as well? Microelectronics and the shift from a manufacturing to a service-based economy are two of the major developments that are altering the nature and organization of work today. But Turk reminds us that we can and must learn from history. In the past, each time new technology has been introduced, the worker has become less skilled. Who is designing curricula in colleges and universities? What functions are these curricula designed to perform?

In this talk, I want to do three things. One is to address a prevalent myth about new technologies and their implications for education. The second is to attempt to clarify some terms that are essential for any meaningful discussion of the issues before us. The third is to focus on the educational implications of new technologies for workers—both production workers and salaried workers.

As you may have already surmised, one difference of a labour perspective is that we do not assume a session on "Career Implications in Technology" need focus primarily or solely on management. Workers have "careers" too—jobs which they hope to pursue and do well. And post-secondary institutions like Ryerson have had, and should increasingly have, a role to play in the education of these workers for their "careers." More about that later.

MYTHS ABOUT THE NEW "TECHNOLOGY SOCIETY"

The context for this conference is the oft-repeated and commonly held notion that the dramatic outpouring of new and sophisticated microelectronic

Source: James Turk, "The Educational Implications of Our Technological Society," from *Catalyst: Newsletter of Local 556,* George Brown College, February, 1991. Reprinted with permission from the author.

technologies means that our educational system needs to be reshaped. Workers will need more sophisticated job skills, and schools, from the primary to the post-secondary levels, must prepare workers for the new high-tech age by giving greater emphasis to science, by making "computer literacy" a priority, and so forth.

The underlying view is that the employment future lies with those able to perform professional and technically sophisticated work.

I want to call into question much of this conventional wisdom. Let me begin with a myth which does not serve us well in our discussions of education and technology, namely, that the new microelectronic technologies will require a more highly skilled, better trained workforce.

Generally, the opposite is the case. The history of the development of the microelectronic technologies, and of their subsequent use, is a history of designing and using machines which deskill work and diminish the role of workers. Insofar as possible, decision making, which formerly was undertaken on the shop or office floor, is removed to the confines of management.

The deskilling is not inherent in new technologies. There is nothing natural or inevitable about deskilling. The new technologies have been consciously designed to deskill work—to allow employers to draw from a larger (and therefore less highly paid) labour pool. Technologies could be designed which enhance and make use of workers' skills, but designers and purchasers of new technologies have little interest in such approaches.[1]

The result is that the design and use of the new technologies is creating a pear-shaped distribution of skills. On the one hand, jobs are being created for a relatively small number of highly skilled people to design, program, and maintain the equipment. On the other hand, the present skills of the great majority of workers are being diminished, and many of their jobs eliminated.

This pattern applies across the board. Let me give you three examples.

In manufacturing operations, machinists have been one of the more highly skilled trades. Roger Tulin, a skilled machinist who has spent his evenings getting a Ph.D. in social sciences, has written on the changing machinist's work with computer-controlled machine tools:

> For many jobs, the new machines are better and more reliable than conventional methods of machining... computer-numerical-controls could allow skilled machinists who can program, set up, and operate these machines to reach new levels of craftsmanship. The most highly skilled metal workers like to make perfect parts. That's the source of their satisfaction. The new technology could allow them to conceive and execute work that was previously beyond anyone's reach.

However, this hasn't been what shop managers have wanted ... their interest is to get the work out with the least amount of labour time possible. So the programming and setting up is usually done by a small group of specialists. "Operators," at lower wages and skill levels, run the production cycles. They are given only the bits of information necessary to keep the cycle running. It's the unused capability, the frustration of the human potential for creative work, that makes the reality of work life so dismal for large numbers of NC and CNC operators.

As machine tools have been made more and more fully automatic, the areas of production which require a full set of conventional skills have been cut back further and further. The "monkey" in the machine shop, who pushes buttons on a task that's broken down to fit so-called monkey intelligence, [this comment is based on a popular ad for CNC equipment which shows a monkey producing "skilled" work] is but a symbol of how management sees the future.[2]

David Noble, in his exceptional work on the history of technology,[3] shows in painstaking detail the history of the development of computer-controlled machine tools and how, at each step in their development and use, the priority was to take skill away from the operator and subject the operator to more direct management control.

One can see the same deskilling in the development and use of office technologies. Evelyn Glenn and Roslyn Feldberg of Boston University have undertaken extensive examinations of the changing character of clerical work. Their conclusions are clear-cut:

... narrow, largely manual skills displace complex skills and mental activity ... close external control narrows the range of worker discretion ... impersonal relationships replace social give and take.[4]

Their study of a number of different organizations adds that "the larger organizations are leading the changes by developing technologies and organizational techniques [for achieving these ends]."[5]

The same pattern of deskilling has also been identified within technical professions. Phillip Kraft, of the State University of New York at Binghamton, has carefully examined the changing nature of programming or software production. His conclusions are remarkably similar to Tulin's, Noble's and Glenn and Feldberg's:

What is most remarkable about the work programmers do is how quickly it has been transformed. Barely a generation after its inception, programming is no longer the complex work of creative and perhaps even eccentric people. Instead, divided and routinized, it has become mass-production work parcelled out to interchangeable detail workers. Some software specialists still engage in intellectually demanding and rewarding tasks ... but they make up a relatively small and diminishing proportion of the total programming workforce. The great and growing mass of people called programmers ... do work which is less and less distinguishable from that of clerks or, for that matter, assembly line workers.[6]

The point of these comments is to argue that contrary to the widely held (and widely perpetrated) view that the new technologies are increasing the demand for a more highly skilled workforce, the opposite is the case.

Evidence for this claim comes not only from scholars studying the workplace, but also from organizations like the U.S. Bureau of Labor Statistics which projects job growth over the next decade or so.

Its projections, the most sophisticated in North America, are quite startling for proponents of the high-tech future. Not one technologically sophisticated job appears among their top 15 occupations which are expected to experience the largest job growth.

The category which will contribute the most new jobs through 1995 is janitors—alone accounting for 775,000 new jobs or 3% of all new jobs created in the United States. Following janitors, in order, are cashiers, secretaries, office clerks, sales clerks, nurses, waiters and waitresses, primary school teachers, truck drivers, nursing aides and orderlies.

If you want to go down the list further, the eleventh occupation with the most substantial growth is salespeople, followed by accountants, auto mechanics, supervisors of blue-collar workers, kitchen helpers, guards and doorkeepers, fast food restaurant workers.[7]

In a separate examination of high technology sectors, the Bureau concludes,

> It should be reiterated that even when high tech is very broadly defined ... it has provided and is expected to provide a relatively small proportion of employment. Thus, for the foreseeable future the bulk of employment expansion will take place in non-high tech fields.[8]

In short, the persistent deskilling of the majority of existing jobs, and the best forecasts for the nature of future jobs, lead to the same conclusion: a pear, rather than an inverted pyramid, describes the emerging skills distribution in our "technological society."

THE MEANING OF "SKILLS"

Before, talking about educational implications, I mentioned that I wanted to say a word about definitions. The key term in much of this discussion is "skills."

Many who would dissent from my argument would point to the fact that workers are (and presumably therefore need to be) better educated now than twenty years or forty years ago. Certainly workers today—from the shop floor to the manager's office—on average, have far more schooling than in the past. But that is no evidence that they are, or need be, more skilled. The lengthening of the average period of schooling has relatively little to do with changing occupational requirements for most workers. Rather the lengthening of years in school has resulted from attempts to decrease unemployment levels (beginning in the 1930s), to use the educational system to absorb some of the returning service personnel after World War II, to changing social expectations about the right to more education, and so forth.

In response to the higher level of average grade attained, employers have introduced higher minimum levels of education as requirements for hiring—whether it be a retail clerk at Eaton's, a machine operator at Canadian General Electric or an entry-level management trainee at General Motors.

But there has been no study which has demonstrated that the higher levels were a result of the changing nature of the jobs rather than an increased supply of people who had spent longer in school.[9]

Moreover, one must recognize that traditional designations of "skill" have only an inexact relation to what we would commonly mean by "skill." To put it differently, the definition of "skill" must be understood politically as well as descriptively. For example, things that are required in jobs done primarily by women tend to be defined less as skill than things required in jobs done traditionally by men.[10]

Similarly, there are often necessary "skills" required in the most "unskilled" work—a point employers often discover when they open a new plant in a low-wage area and find that they cannot get the production they expected initially because the inexperienced workforce does not have the "skills" required by the "unskilled" work.

I mention this only to highlight for you the fact that the definition of "skill" is more problematic than we conventionally take it to be. When I have argued that work is being deskilled, I am not referring to job classifications of skill, nor to educational requirements imposed by employers, but to the mastery of craft, that is the knowledge of processes and materials; the ability to conceptualize the product of one's labour and the technical ability to produce it.

As Braverman notes, most discussions of skill use the term as "a specific dexterity, a limited and repetitive operation, 'speed as skill', etc."[11] He goes on to say that the concept of skill has been degraded to the point that:

> ... today the worker is considered to possess a 'skill' if his or her job requires a few days' or weeks' training, several months of training is

regarded as unusually demanding, and the job that calls for a learning period of six months or a year—such as computer programming—inspires a paroxysm of awe. (We may compare this with the traditional craft apprenticeship, which rarely lasted less than four years and which was not uncommonly seven years long.)[12]

To this point I have attempted to argue that new technologies in workplaces from a manufacturing plant floor to software production houses to offices are designed and used to deskill the work of the vast majority of workers, and, concomitantly, the definition of skill is also being degraded, giving the impression that the real degradation of skill is not as stark as it is.

What has come to be defined as skills training is a distorted and narrow kind of job training of the sort described many years ago by the Gilbreths in the *Primer* on scientific management:

> Training a worker means merely enabling him to carry out the directions of his work schedule. Once he can do this, his training is over, whatever his age.[13]

Even today, with all the mystifying hype about job enrichment and new forms of work organization, Frank Gilbreth's characterization of training is a perfect description of most so-called "skills training."

EDUCATIONAL IMPLICATIONS

The implications of all this are what concern us today.

The most obvious and important implication is that there is little foundation to the view that rising skill levels for the labour force as a whole demand the reshaping of school, college and university curricula to provide more emphasis on mathematics, computer science, and technical training.

While some jobs will require a significant amount of this type of education, the great majority (and a growing percentage) will require little of this knowledge in order to fulfil the requirements of the work. If anything, on average, there will be a diminution of the need for this kind of technical education as essential job prerequisite.

The dangers of a misplaced emphasis on more technical knowledge at all levels of the educational system are several.

First, false expectations are being created. Students will be primed with the myth about the skills their future jobs will require, and then, when they get jobs (if they get jobs), they will discover the cruel joke of their skilled training for what they find to be deskilled jobs.

Second, the rush to emphasize computer literacy and a more technical curriculum can force a de-emphasis of more important educational priorities that today's and tomorrow's students will require, not only for their jobs but for greater fulfilment in their lives.

The deskilling of work means that people will have increasingly to find meaning outside their work. The rapidity of technological change means that people will likely shift jobs (regardless of whether they shift employers) more frequently in their working lives. The greater availability of information and the burgeoning quantity of that information will put greater pressures on people who want to be informed and active participants in their society.

All of these facts mean that the priorities for education from kindergarten through university, including technical and vocational programs, must be to provide people with the capabilities to think critically, and to develop their cognitive, expressive and analytical skills to the fullest. It must, as well, provide people with extensive knowledge of their social, cultural, political and economic institutions, and prepare and encourage them to participate actively in the shaping of decisions that affect their lives.

Far from de-emphasizing a solid general education in the humanities, social and natural sciences, the implications of the emerging "technological society" are that we should be stressing this type of education more than ever.

Certainly there is a necessary place for people specializing in technical matters, but that may be no greater a need in the future than it has been in the past. More likely, there will be a lesser need for such specialized education. Given the power of what can be done with the new technologies, even our scientists will need a sound, general education more than ever. It will be essential for them to have a humanistic perspective from which they pursue their scientific achievements. The quality of our everyday lives, even the future of humankind, is dependent on scientists realizing the broader implications of what they are doing.

Our production and office workers will need narrow job training, which should be provided by the employer. Our skilled craftspeople that survive the deskilling mania of technology designers will continue to need proper apprenticeships (which have increasingly disappeared over the past forty years).

But all will need, as well, a tough, critical, informative general education—beginning at the primary level through to the highest levels—if we are to achieve our fullest potential as individuals and as a society.

NOTES

1. See Noble, David. 1984. *Forces of Production: A Social History of Industrial Automation.* New York: Knopf; and Zimbalist Andrew, (ed.). 1979. *Case Studies on the Labor Process.* New York: Monthly Review Press.

2. Tulin, Roger. 1984. A Machinist's Semi-Automated Life. San Pedro, California: Singlejack Books, p. 14.

3. Noble, *op. cit.*

4. Glenn, Evelyn and Roslyn Feldberg. 1977. "Degraded and Deskilled: The Proletarianization of Clerical Work." *Social Problems 24*: 42.

5. *Ibid.*, p. 52.

6. Kraft, Phillip. 1977. *Programmers and Managers: The Routinization of Computer Programming in the United States.* New York: Springer-Verlag, p. 97. See also Greenbaum, Joan. 1979. *In the Name of Efficiency: Management Theory and Shopfloor Practices in Data Processing Work.* Philadelphia: Temple University Press.

7. U.S. Bureau of Labor Statistics, "Occupational Employment Projections Through 1995." *Monthly Labor Review* (Nov. 1983): 37–49.

8. U.S. Bureau of Labor Statistics. "High Technology Today and Tomorrow: A Small Slice of the Employment Pie." *Monthly Labor Review* (Nov. 1983): 58.

9. See Berg, Ivar. 1971. *The Great Training Robbery.* Boston. See also Braverman, Harry. 1974. *Labor and Monopoly Capital.* New York: Monthly Review Press, pp. 424–449.

10. Gaskell, Jane. 1983. "Conceptions of Skill and the Work of Women: Some Historical and Political Issues." *Atlantis 8*(2): 11–25.

11. Braverman, p. 443–444.

12. Braverman, p. 444.

13. Quoted in Braverman, p. 447.

A Child Is Born

Germaine Greer

In this selection, we are confronted with an uncomfortable charge: We, in the developed Western world, don't like children. Despite all of the rhetoric in the popular media surrounding family values, and our purported concern for children's wellbeing, Germaine Greer says that we just do not like kids. To appreciate Greer's argument we need to recognize that there may be a difference between an individual's personal feelings about a topic, such as children and the environment, and the implications of the cultural practices they are involved in enacting as they actually go about their everyday life. In this sense, there may be a gap, or even a conflict, between a group's "ideal culture" (what members claim to value, believe in, and live by) and "real culture" (what members' actual activities, perceptions, and attitudes indicate they do value, believe in, and live by).

   Greer's polemical argument also provokes us to examine the decadence of consumer society and the effect our cultural views and economic practices have, not only on ourselves, but also on those upon whom we foist our culture and economic development. How does she challenge us to think differently about the poor, (over)population, the meaning of childbirth, and the place of children in our society as well as in others'?

In World Population Year, the location chosen for the international conference was Bucharest. The choice was perhaps unfortunate, for the *raison d'être* behind the whole jamboree was fear of the population explosion and the promotion of birth-control programmes. Just the year before the Rumanian government had outlawed abortion and banned the sale of contraceptives, because the decline in population growth and the increasing senility of the population have been construed as a threat to the country's economic future....

   The Rumanian example is just one of many which could be cited to show that even the government of totalitarian countries cannot counteract the profound lack of desire for children which prevails in Western society, especially among upwardly mobile social groups.... Historically, babies

Source: Germaine Greer, "A Child is Born," from *Sex and Destiny: The Politics of Human Fertility* (London: Picador/Martin Secker and Warburg, 1984). © Germaine Greer, 1984. Reprinted with permission from Gillon Aitken Associates. This material has been edited for length. For a complete reading of Prof. Greer's argument with its supporting evidence, please see the original publication.

have been welcome additions to society; their parents derived prestige and joy and satisfaction from their proximity and suffered little or no deterioration in the quality of their lives, which could even have been positively enhanced by the arrival of children. Parents, themselves still relatively junior in the social hierarchy, had no need to cudgel their brains to decide if they were ready for the experience, for they were surrounded by people who watched their reproductive career with passionate interest, who would guide them through the fears and anguish of childbirth and take on a measure of responsibility for child-rearing. Historically, human societies have been pro-child; modern society is unique in that it is profoundly hostile to children. We in the West do not refrain from childbirth because we are concerned about the population explosion or because we feel we cannot afford children, but because we do not like children.

Conventional piety is still such that to say such a thing is shocking. Parents will point angrily to the fact that they do not beat or starve or terrorise their children but struggle to feed, house, clothe and educate them to the best of their ability. Our wish that people who cannot feed, house, clothe and educate their children adequately should not have them is born of our concern for the children themselves, or so most of us would claim. At the heart of our insistence upon the child's parasitic role in the family lurks the conviction that children must be banished from adult society. Babies ought not to be born before they have rooms of their own: when they are born they must adhere to an anti-social time-table. Access to the adult world is severely rationed in terms of time, and in any case what the child enters is not the adult's reality but a sort of no-man's land of phatic communication. Mothers who are deeply involved in exploring and developing infant intelligence and personality are entitled to feel that such a generalisation is unjust, but even they must reflect that they share the infant's ostracised status. No one wants to hear the fascinating things that baby said or did today, especially at a party. Mother realises she is becoming as big a bore as her child, and can be shaken by the realisation. The heinousness of taking an infant or a toddler to an adult social gathering is practically unimaginable; as usual the discomfort and uneasiness are manifested as concern. A baby may be produced and brandished momentarily but then it must disappear, otherwise well-meaners begin cooing about it being time for bed: the more baby chirps and chatters and reaches for necklaces and earrings, the more likely it is to be told that it is a poor little thing.…

Adults cannot have fun while kids are around; kids must be "put down" first. Drinking and flirting, the principal expressions of adult festivity, are both inhibited by the presence of children. Eventually our raucousness wakes them and they watch our activities through the stair-rails and learn to despise us. In lieu of our real world we offer them a fake one, the toy world. Parents shocked by some family crisis try to mend the cracks in the nuclear family by dating their children, abjectly courting them. The scale and speed of our world is all anti-child; children cannot be allowed to roam the streets, but must run a terrifying gauntlet to get to the prime locus of their segregation, school. They cannot open doors or windows, cannot see on top of counters, are stifled and trampled in crowds, hushed when they speak or cry before strangers, apologised for by harassed mothers condemned to share their ostracised condition....

If the truth is, we of the industrialised West do not like children, the corollary is equally true, our children do not like us. It is blasphemy to deny that parents love their children (whatever that may mean) but it is nevertheless true that adults do not like children. People of different generations do not consort together as a matter of preference: where a child and an old person develop any closeness, we are apt to suspect the motives of the older person. Most social groupings tend to be formed of individuals in the same age set and social circumstances, and even within the family, parents and children spend very little time in each other's company....

The gulf that yawns between adult society and the world of children in the "Anglo-Saxon" West is by no means universal. There are societies where adults and children laugh at the same jokes, where adults would not dream of eating their evening meal without their children about them and would not inhibit discussion of serious matters because children were present. In fact such societies are still more populous than our own. There are huge cities which are practically run by children, children who support their parents and their brethren by their skills and initiative, where children and adults inhabit the same cruel world and survive by clinging to each other. But these are the societies whose children, we think, should not be born....

Sophisticated Caucasians are a shrinking proportion of all human beings, for reasons which should be becoming obvious. People for whom pregnancy is not a strange and disorienting condition already outnumber us and threaten to do so by an ever-increasing margin. In their societies a woman's body is not the more admired the less of it there is. Even in comparatively sophisticated Tuscany, a woman with a gap between her thighs is called *secca*, dry; the word is the same as is used for dead plants.[1] Only in consumer society is the famished female type admired, partly because it is so rare....

The ways of managing childbirth in traditional societies are many and varied; their usefulness stems directly from the fact that they are accepted culturally and collectively so that the mother does not have the psychic burden of re-inventing the procedure.[2] Even though the potential catastrophes are alive in the memory of her community and the index of anxiety high, a ritual approach to pregnancy which hems the pregnant woman about with taboos and prohibitions helps to make the anxiety manageable. A woman who observes all the prohibitions and carries out all the rites will be actively involved in holding the unknown at bay. She will have other reinforcements, for many of the ritual observances of pregnancy involve the participation of others who should support her, primarily her husband, then her kinfolk and then the other members of her community....

It would be clearly absurd to maintain that traditional childbirth is more efficient than modern obstetric techniques in keeping perinatal mortality to a minimum. Some of the procedures appear instead to be calculated to cull the newborn, such as cutting the umbilical cord with a dirty sickle or branding the infant with a hot iron. People living at subsistence level or below know that there are worse fates than death; slow starvation is a more painful end than extinction on the threshold of life. Traditional societies are aware of a wider range of anguish than we of the cushioned West. At the same time that the child confronts artificial hazards at birth, care is taken to see that nothing threatens the establishment of suckling and lactation. Thus the newborn is not separated from its mother, not "monitored" for twenty-four hours to see that all its systems are functioning and no intervention is required. If such intervention is in fact required, the infant will die....

Childbirth has been transformed from an awesome personal and social event into a medical phenomenon, from a heroic ordeal into a meaningless and chaotic one; physical pain which we can bear has been transformed into mental stress, which we are less well geared for. The management of pregnancy, childbed and child raising was the principal expression of the familial and societal network of women, itself one of the essential cohesive elements in any society and a necessary leaven to the competitive hierarchies of men. The institutionalisation of child production has destroyed this alternative structure. It is largely as an unconscious reaction to this diminution of women's role that women are now exerting such pressure to be allowed into the competitive male hierarchy. It is probably inevitable that such women, who have the considerable advantages of literacy and articulacy, should see in the lives of women still living in the web of fertility and continuity the meaninglessness and bewilderment that they themselves would experience in such roles in a society that sets no store by them. The women who are active in international organisations are likely to assume that no one who had a choice would go through childbirth more than once or twice and at a relatively advanced age; the women who compile the Knowledge-Attitude-Practice studies, beloved of family planning organisations, noting the correlation between education level and decline in fertility, may be convinced that their "own-society" values are valid for all times and places. The women they are seeking to help might feel sorrier for them than they can well imagine. The majority of the world's women have not simply been entrapped into motherhood: in societies which have not undergone demographic transition, where children are a priceless resource, the role of mother is not a marginal one but central to social life and organisation....

This discussion of the difference in the role of mother in highly industrialised bureaucratic communities and in traditional agricultural communities is not meant as a panegyric for the disappearing world, but simply to indicate something of the context in which the birthrate in the developed world has fallen. There is little point in feeling sorry for Western mothers who are most often as anxious to be freed of their children as their children are to be freed of them. The inhabitants of old people's homes and retirement villages do not sit sobbing and railing against destiny, although they do compete with each other in displaying the rather exiguous proofs of their children's affection. As the recession bites deeper and unemployment rises, more and more adult children are having to remain dependent on their parents, who are lamenting loudly and wondering more vociferously than usual why they let themselves in for such a thankless task as parenting. The point of the contrast is simply to caution the people of the highly industrialised countries which wield such massive economic and cultural sway over the developing world against assuming that one of the things they must rescue the rest of the world from is parenting. That motherhood is virtually meaningless in our society is no ground for supposing that the fact that women are still defined by their mothering function in other societies is simply an index of their oppression. We have at least to consider the possibility that a successful matriarch might well pity Western feminists for having been duped into futile competition with men in exchange for the companionship and love of children and other women.

There is no possibility of return to the family-centred world. Groups of individuals may attempt to live in the electronic age by the values of an earlier time: they may go back to the land, live in artificial extended families, and give birth at home according to rituals they have learned from anthropology books or from Lamaze or Leboyer, but their freedom to do so is itself dependent upon the wealth created by the workers who live in the mobile nuclear families which the communards despise. If they fail in their chosen lifestyle, the safety net of consumer society will catch them: if their home birth goes wrong, they can load the suffering woman into a car and speed down the tarmac road to the nearest hospital. The women giving birth in African or Asian villages cannot telephone for emergency squads to speed to their aid, supposing there existed a road suitable for travelling on at speed. Yet the chances are that women in traditional societies will cling to their own methods, for such methods are among the ways that peoples define themselves. Humane and intelligent planning could devise ways of decreasing perinatal morbidity without destroying the character and significance of the experience, but international aid does not come in such forms. There are too many indications that the impact of our medical technology on traditional mothering has been disastrous, to the point of raising the question of whether it has not been our subconscious intention to discourage parenting among foreigners even more effectively than we have discouraged it among ourselves ....

If we turn birth from a climatic personal experience into a personal disaster, it matters little that the result is more likely to be a live child. Women will not long continue to offer up their bodies and minds to such brutality, especially if there is no one at home to welcome the child, to praise the mother for her courage and to help her raise it. In fact peasant communities are more level-headed and sceptical of us and our methods than we realise and they have resisted the intrusion of our chromium-plated technology more successfully than we like to think. They know that death attends too frequently in the traditional birthplace, but they also know that there are worse fates than death. Nevertheless, all that stops our technology from reaching into every hut and hovel is poverty: the cultural hegemony of Western technology is total.

The voices of a few women raised in warning cannot be heard over the humming and throbbing of our machines, which is probably just as well, for if we succeed in crushing all pride and dignity out of child bearing, the population explosion will take care of itself.

THE MYTH OF OVERPOPULATION

... Is the world overpopulated? If I must adopt some position on this point, it will be a highly compromised one. I have been to old Delhi, as Paul Ehrlich has, but somehow, as usual, I saw the wrong thing. Here follows the apocalyptic vision of Paul Ehrlich, set out on page one of his bestseller, under the heading "The Problem."

> I have understood the population problem intellectually for a long time. I came to understand it emotionally one stinking hot night in Delhi a few years ago. My wife and daughter and I were returning to our hotel in an ancient taxi. The seats were hopping with fleas. The only functional gear was third. As we crawled through the city, we entered a crowded slum area. The temperature was well over 100°F; the air was a haze of dust and smoke. The streets seemed alive with people. People eating, people washing, people sleeping. People visiting, arguing and screaming. People thrusting their hands through the taxi window begging. People defecating and urinating. People clinging to buses. People herding animals. People, people, people, people. As we moved slowly through the mob, hand horn squawking, the dust, noise, heat

and cooking fires gave the scene a hellish aspect. Would we ever get to our hotel? All three of us were, frankly, frightened. It seemed that anything could happen—but, of course, nothing did. Old India hands will laugh at our reaction. We were just some over-privileged tourists, unaccustomed to the sights and sounds of India. Perhaps, but since that night I've known the feel of over-population.[3]

This then is the problem that Ehrlich and his paymasters and the gullible public are setting out to solve....

If we exclude the carping about the taxi and the temperature, we are left with people, people, people, people. This slum is rather odd in that there are buses and people herding animals in it, so we might guess that Ehrlich did not get himself inside a real slum, where there is no room for buses, taxis or herds, just huts. What he seems to have seen were settlements. The area was certainly not more crowded that Manhattan at three o'clock on a weekday afternoon: the difference is that in Delhi the people were all at pavement level. If they had been nicely shut up in high-rises Ehrlich would not have troubled his head about them, even if he had heard that most of them were drugging themselves with heroin or alcohol or doctors' substitutes. He saw no drunks, no crazy people, no obese people, either, I'll be bound. If he did, he does not say. He does not say that he saw anyone laughing, or men and women playing with their babies. Some of the smells around those tiny fires, made by burning the tips of scrap wood, were spicy and good. Those people had not come there as a couple, years ago, and multiplied until they filled up the area. If he had been less of a ninny and got out of the taxi to talk to the people (who are better at speaking his language than he is at speaking theirs) he might have found out how they got there.

But no. Intellectual understanding precludes investigation. If he had gone to the shanties, he would have been surprised to find that the earthen floors were swept smooth, that the family possessions, most of the wife's dowry, were neatly hung on poles or standing on a narrow shelf, the brass bowls polished till they shine in the darkness, reflecting every sliver of available light, polished with earth. He did not notice that the people doing their washing-up were not using detergents. Cow dung smoke and dust are hard on the sinuses but they are less deadly than industrial effluent and exhaust fumes, which Dr. Ehrlich seems to prefer. He did not notice the complete absence of the United States' principal product, trash. The greatest insult about all these people living shoulder to shoulder, the visible counterpart of the population planner's nightmare, is that they make do with so little. Such low purchasing power is anathema to men of Ehrlich's kidney. He does not tell us how he treated the beggars, whose presence seems to indicate that he did not get very far off the beaten track. If he had gone to some parts of Delhi in 1970 he would have found American teenagers begging to support their drug habit, much as they do in America. Still, one can sympathise, for the good doctor has had himself sterilised after the birth of one daughter; this matter has been made public presumably in order to show the utter sincerity of his preaching of Zero Population Growth, only to have to contemplate the ghastly vision of the world being taken over by thin brown people who eat, wash, defecate and cling to buses in defiance of him.

In order to feel in his bowels the reality of overpopulation Ehrlich had to go to India; others feel it more strongly when they come across discarded aluminium beer cans in the wilderness. If we agree that the world is over-populated we have still to decide what we mean by the term and what the phenomenon consists in. An Indian Paul Ehrlich might see the sudden exponential increase in the global human population as the result of an ecological disaster which happened about five hundred years ago, namely the explosion of Europe. This was not caused by population pressure, although population pressure there was and always will be, but by the demands of the European trade economy . . . .

Whether we believe that the world is over-populated or not depends to some extent on how we think people should live. If we in the West think that only our kind of life is worth living, then clearly the numbers that the earth supports will have to be substantially reduced. The world could become a vast luxury hotel, complete with recreational space for us to hunt and ski and mountaineer in, but it must not be forgotten that our luxurious lifestyle demands the services of a huge number of helots, who cannot be paid so much that they can afford rooms in the hotel for themselves. Like Ricardo, we would like to see the supply of helots kept constant, neither falling so low that we have to take out the trash ourselves or becoming so high that we shake in our shoes fearing insurrection in the compound. The official ideology is that the guests in the hotel create all the wealth; only by the extraordinary efficiency of their wealth-creation are all the rest able to survive by merely drudging in the kitchens and the lavatories and the market gardens. At very little cost to himself the guest creates the wealth which is apportioned to them for these worthless but indispensable and time-consuming activities. If this is so, if the capitalist system is actually the best system for creating wealth ever devised, perhaps it could be made less spectacularly unjust; for example, perhaps the cultivation of inessentials which are regarded as essentials, but for which we will not pay a price commensurate with the human labour that they absorb, should be made illegal. Perhaps we should impose the same penalties on the consumption of sugar, tobacco and tea as we do on heroin, so that people brutalised by this kind of cultivation could go back to farming food crops. Perhaps we should outlaw speculation on commodities, which, if it is a way to maintain a "fair" market price, seems to have got a very odd idea of fairness. When primary producers try to do a little speculation on their own behalf, we quickly decide that speculative buying of commodities is not at all fair, as when the London Metal Exchange closed trading in tin, when it was suspected that the government of Malaysia was bidding to force up the price of a ship-load of tin already on the water.

I don't know how many people the earth can support, and I don't believe that anybody else does either; it can certainly support more people on a low calorie intake than it can on a high calorie intake, but as the world is not a huge soup-kitchen the fact is irrelevant. It is quite probable that the world is overpopulated and has been so for some time but getting into a tizzy about it will not prove helpful. Nothing good can come of fear eating the soul. We cannot take right decisions if we are in a funk. We do not have access to our imagination, if we are convinced that catastrophe lurks just around the corner. We may be living in catastrophe now; perhaps we shall have to adapt to it, or go under. Perhaps catastrophe is the natural human environment, and even though we spend a good deal of energy trying to get away from it, we are programmed for survival amid catastrophe. It is an odd thing that people living precariously have more commitment to the continuity of their line than people safely ensconced in plenty. If this is the case, there is not much we can do about it, for we cannot design a political system which will supply the right proportions of potential catastrophe. If we are to deal with the problem of people at all gracefully, we will have to stop rushing into situations we do not understand encumbered with all kinds of non-solutions. In the past we have tried to avoid this by undertaking all kinds of research, which cost many times what practical help would have cost, and came up with conclusions that were no use to anyone. What does it help to know that when all the statistical correctives are applied to the reproductive histories of a small number of Egyptian women that the death of a child made no difference to overall reproductive performance, did not lead to shorter birth intervals, etc. etc? As we have no plan to kill Egyptian babies, or keep them alive, there is no way we can use this information . . . .

Let us therefore abandon the rhetoric of crisis, for we *are* the crisis. Let us stop wasting energy in worrying about a world crammed with people standing shoulder to shoulder and counting the babies born every minute (one in every five of them a Chinese and just about all of them foreign) and begin to use our imagination to understand how it is that poverty is created and maintained. Let us get to know Lady Poverty up close, so that we lose our phobia about the poor. If we must be afraid, let us rather be afraid that man, the ecological disaster, now has no enemy but his own kind. Rather than being afraid of the powerless, let us be afraid of the powerful, the rich sterile nations, who, whether they be of the Eastern or the Western variety, have no stake in the future. The birth of every unwanted child is a tragedy, for itself and for the unwilling parents, but in spite of all the attention we have given to the matter, more unwanted children are born to us, the rich, than to them, the poor. This may seem a paradox, but the times gives it proof.

NOTES

1. Among the Sande women of Sierra Leone, "The opposite of fat is not thin, but dry, connoting among other things, a dry and barren uterus." C. P. McCormick, "Health, Fertility and Birth in Moyamba District, Sierra Leone," *Ethnography of Fertility and Birth*, ed. C. P. McCormick (London, 1982), p. 122.

2. Although there is very much less literature on pregnancy and childbirth management in other cultures than there would have been if more anthropologists had been women, and women interested in female society, some context for these remarks can be got from H. H. Ploss and M. and P. Bartels, *Woman* trans, E. Dingwall (London, 1935) Vol. 2, *passim;* K. E. Mershon, *Seven plus Seven: Mysterious Life-Rituals in Bali* (New York, 1971); Richard Hessney, "Birth Rites—a Comparative Study", *Eastern Archaeologist*, XXIV, 2, May-August, 1971; Jean Lois Davitz, "Childbirth Nigerian Style", *R.N. Magazine,* IV, March 1972; F. Landa Jocano, "Maternal and Child Care Among the Tagalos in Bay, Laguna, Philippines", *Asian Studies,* VIII, 3 December 1970; Joel Simmons Kahn, "Some Aspects of Vietnamese Domestic and Communal Ritual", unpublished M.Phil. thesis, London School of Economics, 1969; Joyce S. Mitchell, "Life and Birth in New Guinea" *Ms*, 1, May, 1973, pp. 21–3; "Rites de la Naissance", *France-Asie* XII, March-May 1956; Helen Gideon, "A Baby is Born in the Punjab", *American Anthropology*, LXIV, pp. 1220–1234.

3. Paul Ehrlich, *The Population Bomb* (London, 1971).

Portions of this article have been removed; at the request of Germaine Greer, readers should refer to the original text in order to read the entire argument with its supporting evidence.

PART TWO

The Sociological Tradition: Understanding the Social World

What does it mean to understand something "sociologically"? From whom is the sociological tradition derived? This section of readings continues to explore sociology's particular way of examining objects, events, and activities as *social phenomena,* and to highlight the historical development of the sociological perspective.

Generally, there are two broad orientations used to introduce students to the sociological perspective. The first draws on Mills' famous essay, "The Promise," and presents sociology as an imaginative, critical inquiry focusing on the individual's experience of personal troubles and the formulation of them as public issues. This approach inquires into how these troubles and issues are connected to historical transformations and the institutional arrangements or systems (such as the economic, political, family, or technological systems) within which they are embedded. The second approach draws on the work of Peter Berger and presents sociology as a skeptical form of consciousness that focuses on the world of social phenomena.

Events, interactions, objects, problems, etc., can be regarded in myriad ways. A nail or a cell phone can be viewed and understood as a physical, technological, or economic phenomenon. The event of dying can be studied as a biological or a psychological phenomenon. But to study any phenomenon—nails, cell phones, dying, families, work—sociologically is to examine how it is constructed and organized in relation to culture and social structure. At the same time, to understand any phenomenon as a social phenomenon is to imagine how the phenomenon itself—be it marriage or the microwave oven—is a response to a deep social need or problem which occasioned it.

The aim is not simply to describe these problems, but to see them as sociological problems. This involves seeing how the problems reflect strains and contradictions within the cultural meaning systems, institutional arrangements, or historical changes in which they are embedded. This is the first step to understanding what has made them possible. As Peter Berger says,

> The problems that will interest the sociologist are not necessarily what other people may call "problems.". . . People commonly speak of a "social problem" when something in society does not work the way it is supposed to according to official interpretations . . . . It is important . . . to understand that a sociological problem is some-thing quite different from a "social problem" in this sense . . . . The sociological problem is always the understanding of what goes on here in terms of social interaction. Thus, the sociological problem is not so much why some things "go wrong" from the viewpoint of the authorities and the management of the social scene, but how the whole system works in the first place, what are its presuppositions, and by what means it is held together.[1]

But how do we explain and understand the many troubles, issues, events, and other phenomena of everyday life in relation to culture, social structure, and history? The following section begins with two articles examining sociology as a social science (the sociological perspective) and as a distinctive imaginative activity. The remainder of the articles represent selections from the works of key thinkers who have made significant theoretical contributions to understanding the world sociologically. The selections generally highlight classical theorists whose work has given rise to some of the more popular theoretical approaches in sociology—structural-functionalism, conflict theory, symbolic interactionism, feminism, and structuralism—which are central to the sociological tradition. It concludes with an article which addresses the question of whether we are witnessing the rise of a postmodern society, and whether new forms of researching and theorizing are required to understand what is occurring in the 21st century.

Common to all of the theorists is a struggle to use the examination of sociological topics—whether personal troubles and public issues, everyday experiences or social problems, emerging social trends or ongoing social changes—to explore the "sociological problems" which underlie them. Many of these problems extend beyond the particular society in which they arise and are fundamental to the very nature of society itself. How is social order possible? What is the nature of human sociality? Is social inequality necessary? As Dennis Wrong put it,

> Unlike particular "social problems" which perhaps can be studied, dealt with and set aside, sociological theory concerns itself with questions arising out of problems inherent in the very existence of human societies and that cannot be finally solved. They are problems for human societies.[2]

NOTES

1. Peter Berger, *Invitation to Sociology*, Anchor, 1963, pp. 36–37.
2. Dennis Wrong, "The Oversocialized Conception of Man in Modern Sociology," in Lewis Coser (ed.), *The Pleasures of Sociology*, Mentor/New American, 1980, p. 9.

The Sociological Perspective

William Levin

Sociology represents the modern interest in understanding human society and its historical changes, the nature of social interactions, and us as human beings. And it is usually described not as a science, but as a "social science." What does this mean? In this article Levin outlines the nature and subject matter of the sociological perspective and its orientation to scientific standards in developing both quantitative and qualitative approaches to doing research. Insofar as the subject matter of sociology is social action and social relationships, what particular challenges does the sociologist face? How does the sociologist not just document patterns of interaction but arrive at an *understanding* of social relationships and social organization?

. . . "What am I doing here?" I ask it countless times while I am typing away indoors, knowing that outdoors there is hiking to be done or tennis to be played. You may be asking it of yourself at this very moment. How is it that you are enrolled in a sociology course? (I must assume that you are, or you would not be reading this book.) What forces have combined to place you in this particular spot at this scheduled time with all these other people? By developing an understanding of the sociological perspective you can add to your ability to identify some important forces that shape your behavior, forces of which you have probably been quite unaware.

. . .

The sociological perspective focuses on the patterns of relationships among individuals rather than solely on the individuals themselves. It gives us a unique way of seeing human behavior. It alerts us to the fact that a great deal of our behavior is shaped by our membership in groups with other people. Social forces, like physical forces, strongly influence our behavior. The analogy with physical forces is helpful, because in both categories the forces may normally be invisible to us but their existence can be made obvious if we know how to look at behavior or know what

Source: From *Sociological Ideas: Concepts and Applications* 2nd edition by Levin. 1988. Reprinted with permission of Wadsworth, a division of Thomson Learning. www.thomsonrights.com. Fax 800-730-2215.

kinds of questions to ask. Gravity is invisible until its effect on a dropped object is observed and measured. Air becomes "visible" when we see its influence on leaves or we turn an empty glass over and push it into a bowl of water—the water can't get in because there is something already there. In just this way social forces can be made visible by asking sociological questions, observing and measuring human social interactions, and then guessing about the nature of the underlying social forces that caused the observed behavior.

Just as there are physical facts that we learn when studying the workings of the physical world, so there are social facts that sociologists have discovered by studying human social behavior. Émile Durkheim, a French sociologist, coined the term *social fact* (Durkheim, [1893] 1958) to describe the forces that constrain (or control) human behavior and that result from our membership in groups rather than from what we are like as individuals. It is important to keep in mind that, when sociologists try to explain human behavior, the individual characteristics of the people involved are *not* of primary importance.

. . .

SOCIOLOGY AND SOCIAL ORDER

Just as the natural sciences developed to deal with specific questions we have about the natural order, so sociology began when we started to ask questions about the forces underlying human social order. Although there have been social thinkers at various times throughout history, sociology is a relatively new discipline. Auguste Comte (1798–1857) did not coin the term *sociology* until the middle of the nineteenth century (Comte, [1848] 1957), and the thinkers who gave the new field its main theoretical direction, Karl Marx (1818–1883), Max Weber (1864–1920), and Émile Durkheim (1858–1917), produced their most influential work between the mid-nineteenth and early twentieth centuries. Why did we suddenly become aware of social forces, even though they had certainly existed for as long as thinking humans have lived in organized groups?

The answer to this question illustrates something about the way social science proceeds. We generally do not notice orderliness. It is taken to be normal, everyday, the way things "ought to be." But occasional disruption of the normal order of events draws our attention and, strange as it may seem, points to the underlying rules of orderliness. The greater the disruption, the more attention it draws. While the exception may not actually prove the rule, it certainly starts us thinking about what the rules might be. Attempts at proof come later.

Marx, Weber, Durkheim, and others discussed in this book were reacting to the greatest disruption of all, world revolution. For more than a century the Western world had been experiencing severe and rapid changes in its basic structures. The feudal, aristocratic structure that had dominated for centuries in France, England, and the West in general was overthrown, replaced by national governments that were to be run, to varying degrees, according to the principles of representative republics. The populations became increasingly urbanized and industrialized. The Industrial Revolution was not just a matter of replacing hand work with machines or animal power with steam power; it was also a revolution in the way people evaluated their own worth. To the extent that their living came to depend upon their labor, they could now be judged on the basis of what they did, rather than on the basis of the village, religion, or family into which they had been born. The feudal order of things had been so stable for so long that questions about the forces that had created that order were unlikely to arise. But in the turbulence of social and economic revolution, and in its aftermath, such questions had to be asked. What was causing the changes? Would a new stable form of social organization develop, or would the upheaval continue forever? If a new order did arise, what would it be like? How long could it be expected to last? Would there be more monarchs and serfs in the new order? And the most important question of all: How can we discover basic laws of social order that will allow us not only to predict social change but also to influence or even control it in the future?

So sociology really developed in response to a problem in the social order, and our understanding of some rules of social order was a consequence (Nisbet, 1966). Sociology ever since has been a problem-oriented discipline. It is not the stable and orderly segments of society that attract our attention, but the areas we define as problems. . . .

As a consequence of this problem orientation, sociology has developed a reputation as a critical discipline. Sociologists tend to look beneath the official explanations of how things work to discover where problems may be hiding (Berger, 1971). Why are there so many poor people, and why do they tend to share so many characteristics—age, sex, race, ethnicity, educational level, geographical distribution, and so on? Such questions focus on problems. But they have the benefit of helping us try to make our social order live up to its highest goals.

Sociologists are not just critics of society's problems. A major goal of the sociological perspective is to understand the basic principles of social order. Our constant attention to disruptions of order gives us the most direct route to understanding order itself. We want to test our guesses about how social order works. We want to build general social theories that can explain in the simplest possible terms a wide variety of behaviors.

To do this we try to be as insightful and as objective as possible. Just as other disciplines—medicine, psychology, and political science—developed during and after the Industrial Revolution, sociology has tried, when appropriate, to emulate the methods of the natural sciences. . . . The objectivity of sociology depends on applying this method of inquiry. Basically, sociological inquiry becomes more objective when it is taken from the personal control of subjective individuals and conducted according to publicly agreed upon methods for the objective testing of ideas. By putting our guesses about the social world in the form of hypotheses that state clearly how everything is to be measured, we make it possible for anyone to verify (or contradict) our results by trying the same study (a process called *replication*).

Sociology as a scientific enterprise differs from other sciences in the types of forces on which it focuses and in the theories it attempts to build to explain them. Social behavior is very complex. Just trying to explain your presence in a sociology course clearly involves psychological, economic, and historical forces, to name just a few. So sociology often requires an awareness of a number of other perspectives. Sociology is *multicausal* in its approach, that is, it alerts us to the fact that virtually any social behavior has many contributing causes, all of which must be included in our explanation.

The nature of social order may be quite different from the nature of other systems of order. Humans, after all, are conscious actors who are capable of evaluating the meaning of their experiences and reacting in a variety of ways. In some circumstances, therefore, it is misleading to apply the methods of the natural sciences to human social behavior. . . .

SCIENCE AND VERIFICATION

Until the development of the scientific perspective and quantitative research methods, belief in the truth of some explanation of how things worked depended solely on either the power of someone to enforce that belief (it was dangerous to contradict the king or priest) or the beauty or persuasiveness with which the idea was expressed. Of course we are still subject to such influences today. But the scientific perspective is our main way of removing the process of discovery from the control of powerful or persuasive individuals and making it more of an objective, communal search.

The first step in this process is objective observation. Objectivity requires that we focus on only those characteristics in the world that can be sensed, rather than on our beliefs or feelings about them. The best check on whether we have observed objectivity is whether another individual, observing the same thing and using the same measurement instruments, will see the same thing. Can an observation be verified (or contradicted)? . . .

When we speak in normal conversation we use concepts, abstractions drawn from observed events that make up our everyday vocabulary. . . . The dictionary is loaded with such definitions. To increase the accuracy of observation and make it possible for others to observe in exactly the same way, scientists use what are called operational definitions, which specify exactly how a defined concept is to be measured. . . .

Operational definitions help make observations comparable when we are studying something concrete. . . . They are even more important when we are studying more abstract concepts, such as ambition or social class membership. It is much easier for us to be mistakenly talking in differing terms when we are discussing more abstract concepts. Do we always mean the same thing when we are talking about social class? You may think of it only in terms of yearly income, while I may want to include level of education and quality of home. Any operational definition of social class must specify exactly what elements are included and exactly how they will be calculated.

Observations made using clear operational definitions not only make it possible to check the studies that others have done but also help organize the way we see the world. Once we begin to observe things this way, patterns quickly emerge. The patterns of nature, whether in form or in behavior, surround us in mind-boggling abundance. The types of questions facing naturalists, psychologists, or sociologists can also be clearly differentiated. For the naturalist: "Why do certain birds migrate to specific places on specific dates, and how do they do it?" For the psychologist: "Why do some individuals develop such inability to confront authority that they greatly exaggerate its importance and identify strongly with it?" For the sociologist: "Why do Protestants commit suicide at a greater rate than Catholics, single and divorced people at a greater rate than married people, and army officers at a greater rate than enlisted personnel?" Each question resulted from the observations of naturalists, psychologists, or sociologists, and each forms a hypothesis worth testing in the real world. The process of testing such hypotheses against reality is called empiricism, and it is a primary characteristic of the scientific perspective . . . .

A number of research methods have been developed in sociology, each appropriate for one type of question being investigated.

Descriptive hypotheses are usually tested by conducting survey research. It is the kind of social research familiar to most people and emphasizes accuracy and reliability of measurement. Election-night polling and marketing research ("Which kind of soap do you prefer?") are two common examples. . . .

By contrast, causal hypotheses are usually tested by experimental research. . . . An experimental study of the relationship between hours of study and test grades might work as follows: Divide a class of students into two equal groups, by random assignment, to ensure that the groups do not differ in important characteristics, such as intelligence or motivation. Control the moderating variables by giving all students the same quality of notes from which to study, making sure that everyone gets the same amount of rest, and having everyone study under the same conditions (light, noise, etc.). Then manipulate the independent variable by allowing one group to study for only one hour and having the other group study for three hours. Give all the students the exact same test, making sure that both groups take it under the same conditions (control of the experimental environment). If hours of study (the independent variable) is causally related to test grade (the dependent variable), then the group that was made to study more hours should have higher grades than the group that studied less.

In this discussion on the scientific perspective, I have focused on the quantitative methods of sociology and other sciences. Quantitative methods are aimed at trying to evaluate numerically the qualities being investigated. It usually seems reasonable to attach numbers to variables such as hours of study or test grades, but it is somewhat more difficult to evaluate concepts such as group cohesion or social class membership numerically. This is the major problem sociology has in using the methods of the natural sciences. All quantitative scientific research proceeds under the assumption (called positivism) that what is being studied has a stable reality that can be measured from the outside by an objective observer. But there are a considerable number of questions in sociology that do not lend themselves to this kind of approach.

ANOTHER APPROACH TO SOCIAL RESEARCH

One of the most interesting ideas in sociology is symbolic interaction.... Basically it suggests that the behavior of humans in social situations can be understood only from the point of view of the actor, since that individual is constructing the meaning of the behavior for each specific situation in which it occurs. Symbolic interactionists generally argue that the quantitative methods on which I have focused here cannot fully represent the meaning of social behavior. The problem is that attaching the same numerical evaluation to two different people who behave the same way can misrepresent the meanings of those actions to them....

One of my favourite films of all time is *Rashomon* (1951), by the great Japanese filmmaker Akira Kurosawa. It tells the story of four people who are involved in a rape and murder. We see the events unfold as each survivor tells the story of what happened, but each tells a different tale. At first it seems to be a mystery in which we are being asked to discover the liars. But the genius of the film is that Kurosawa compels us to realize that each of the conflicting stories is actually true. It is not just that each participant remembers events differently but also that, from the point of view of each person, the events *actually were* experienced differently. (An American version of *Rashomon*, called *The Outrage*, is almost as good as the original and easier to find on late night television.) These films aren't just examples of how filmmakers sometimes play fast and loose with the truth. There is a very good sociological explanation for what these films say about how humans construct reality.

SYMBOLIC INTERACTION AND THE CONSTRUCTION OF SOCIAL REALITY

Human beings communicate and interact with one another at the symbolic level. That is, we have the ability to abstract the qualities of objects and experiences and to assign symbols such as words or gestures to them. ...

G. H. Mead, an extremely important figure in the development of symbolic interactionism, proposed in *Mind, Self and Society* (1934) that the unique ability of humans to symbolize led to the construction of a special kind of social order. According to Mead, our ability to communicate by symbols means that we can contemplate possible actions, guess about their outcomes, compare the qualities of different objects; in short, we can think. Mead referred to this ability as mind.

Just as we are able to evaluate our own potential actions, we can also evaluate the actions of others. (Mead called this "taking the role of a particular other.") Mead concluded that by accumulating such evaluations we develop a self that is, a cumulative idea of who we are that is constructed from the actions of others toward us. Mead spent a good deal of time discussing the way the development of the self operates in the process of socialization.... It is enough to say here that Mead believed that there is no inborn, or unvarying, human character (or self). Rather, since each of us experiences a unique set of interactions with people and the environment, each necessarily develops a unique self. Ultimately humans develop the ability to evaluate and internalize the expectations of a group of people simultaneously. (Mead called this "taking the role of the generalized other.") Accordingly, society is the sum total of the ways in which people agree to act in given situations, which are internalized by its members.

SOCIAL REALITY AND THE DEFINITION OF THE SITUATION

There is nothing in social order that is real unless we participants in society agree that it is real. Or, to use the now famous words of W. I. Thomas, "If [people] define situations as real, they are real in their consequences" (Thomas & Thomas, 1928:572). Shaking hands is a friendly greeting only because we define it as such. In another time or culture it might mean hostility (for example, "What are you hiding in your hand?").

In everyday interaction, having agreed rules for behavior makes things predictable and, therefore, relatively stable. But everyone is taught the rules of everyday behavior under different circumstances, and everyone lives a unique life. Even if we were all taught pretty much the same rules of behavior, we would inevitably learn them differently. Also, we sometimes face contradictory rules or ambiguous situations in which the proper rule is not apparent. In short, there is a great deal of room for individual evaluations of the meaning of a social situation to differ. At a party, a casual comment taken as a compliment by a woman is later recalled as an insult by her husband. They may discuss it, and one may convince the other; but, if they don't, the "real" meaning of the original comment will remain undisturbed in the mind of each person forever. In the film *Rashomon*, what was a rape to one character was a seduction to another, because that was the meaning applied to the situation by each.

Symbolic interactionists emphasize how humans decide on the meaning of every situation in which they interact. Thus, the agreements that are internalized in the process of socialization are only guidelines for social behavior, not unvarying, narrow commands. Herbert Blumer (1969), whose work has been important in shaping modern symbolic interactionism, described social order as a process of continual creation. Every time we interact, we create a new version of social reality. When we find ourselves in a specific situation, we do not act automatically, as if we were born with instincts or had been conditioned to respond a certain way. Rather, we evaluate the stimulus and decide what it means before we act. Two people can obviously interpret the same stimulus differently. When the boss told me "Please sit down," it sounded like a command to me, but my coworker thought the boss was just being courteous. The symbolic interactionist view of social reality has serious consequences for the ways in which we study human social behavior.

PARTICIPANT OBSERVATION AND THE VERSTEHEN METHOD

In the natural sciences, the measurement of objects, including their behavior and relationship to one another, is overwhelmingly quantitative. That is, researchers use numerical evaluations such as weight in grams, velocity in meters per second, length in centimeters, and energy in joules. This is an effective approach, because there seems to be a stable, physical reality in the natural world that yields consistent measurement results (when we know how to measure it), and that does not seem to mind the process at all. Of course, rocks, electrons, and red blood cells have no consciousness with which to respond to the measurement they undergo. But people do.

Weigh a rock and it does not react, but a human might try to weigh less or more (depending on his or her self-image) or might refuse to be weighed at all. The problem becomes even more acute when the quality being measured is more abstract, such as an opinion about welfare or prayer in schools. If the person being interviewed does have an opinion (we cannot assume that the quality being measured is present in everyone), she or he might lie about that opinion, distort it, or even refuse to reveal it. One possible solution is to avoid asking people to respond to questions. That way they cannot distort answers. Instead we might just observe their behavior and evaluate it. But, as the symbolic interactionists point out, the exact same situation can mean entirely different things to different participants. So it can be extremely misleading to apply numerical evaluations to human statements or behaviors. In many situations the objective, quantitative methods of the natural sciences are completely inappropriate for studying human social behavior. How, then, can it be studied?

The answer of the symbolic interactionists is that, to understand the meaning of a particular social situation, we must both observe and take part. The idea is to put ourselves in a position to construct the meaning of the situation just as the participants do. Max Weber ([1925] 1946) referred to this approach by the term verstehen (German for *understanding*). It is the process through which subjective reality can be discovered. The researcher must see a situation from the point of view of the participants, which makes the fact that the researcher is a subjective interpreter of events actually an advantage rather than a detriment to the research process. This is exactly opposite to the quantitative approach of the natural sciences, in which every effort is made to remove the influence of the researcher's evaluations from the measurement process. The difference in the two approaches results from their different views of the nature of social reality. Positivists (who use the methods of the natural sciences) see social order as relatively fixed and measurable, like the order of the rest of the natural world. . . . By contrast, symbolic interactionists see social order as constantly being created, subject to the meanings and evaluations of the participants in each interaction. Thus, social order cannot be measured quantitatively, since it is the *quality* of meaning that lies at its core. The qualitative approach of symbolic interactionists requires close, careful, and systematic observation and thus is, in an important sense, scientific. But the methods employed by symbolic interactionists are quite consciously not those of the positivists.

QUALITATIVE APPROACHES

Many of the tools used by qualitative researchers are the result of common sense. For example, when observing social interaction it is a good idea to describe and record the setting, participants, actions, and verbal exchanges carefully. But, when people know they are being observed they are likely to act differently from the way they normally would. So the researcher may not reveal that he or she is observing. (Of course, this raises important ethical problems of deception that are a continuing issue for participant observation.) But beyond the mechanics of participant observation there are a number of theoretical models within symbolic interactionism that have led to important insights into the way social reality is constructed.

Erving Goffman developed an approach he called dramaturgy, in which social interaction is viewed as a series of small plays, or dramas. Each interaction is, in a sense, scripted by the roles the participants play. The lines are not as specific as in scripts, but, once the roles are chosen for the situation, the participants enact lines appropriate for the characters being portrayed. We have these roles available because we are socialized to know what others expect of our behavior in given situations. For example, if I decide that I am expected to play the role of teacher in a situation, I will say lines appropriate for a teacher. We become quite skilled at playing a variety of roles and at deciding when to play each role.

Goffman's numerous studies (for example, *The Presentation of Self in Everyday Life* [1959], *Asylums* [1961], *Stigma* [1963], and *Relations in Public* [1971]) uncovered how individuals present themselves to others within the framework of social roles. Goffman's work revealed previously unrecognized rules that we use everyday to make sense of our relations with one another. Among these are the rules used by physically handicapped persons and mental patients to cope with their stigma and the rules people use to maintain socially appropriate distance between themselves and those with whom they interact. ... As a methodological key for the participant observer, the idea of dramaturgy can be extremely helpful. Imagine observing the interaction of several people in a bar. Without some organizing principle, the interaction might seem to be a disorganized random mass. But if you ask yourself what role each person is playing in each separate interaction, a number of patterns are likely to unfold. You might identify the role of "old hand," or "protector," or "shark," or "lost lamb," or "no-nonsense drinker," or "good-time Charlie" (or Charlene). Once the various roles are specified, the way each person deals with another becomes clearer. For example, what kind of conversation would you expect between a "shark" and a "lost lamb"? It would most certainly be different from the encounter of the "no-nonsense drinker" and the "shark," who probably would have nothing at all to say to one another.

GARFINKEL'S ETHNOMETHODOLOGY

Another approach to the understanding of social interaction is provided by the work of Harold Garfinkel (1967). Ethnomethodology is an approach to understanding interaction based on the assumption that social reality is the result of our agreement to agree with one another.

That is, we negotiate reality by exchanging accounts of what is going on between us, with the unstated assumption that we will reach an agreement eventually. A methodological key used by ethnomethodologists is intentional disruption of the process of reality negotiation or meaning construction. When a researcher upsets the process by which the meaning of a situation is negotiated, the normally hidden methods of the negotiation can be starkly highlighted. For example, suppose you have just been introduced to a woman and you stand with your face only inches from hers and stare unflinchingly into her eyes. The normal process by which two newly introduced people negotiate the meaning of an interaction would definitely be disrupted by your behavior. There are likely to be some awkward moments during which the "target" person scrambles for a way to construct a new meaning for the situation ("What is this, are you crazy?" or "Are you from another country?" or "Did Ted put you up to this?"). The normal rules that have temporarily been abandoned may become clearer by their very absence. . . .

SUMMARY

One of the major perspectives in sociology is symbolic interaction. It is based on the belief that social reality is constructed by people every time they interact. Through a process of negotiation, people come to agreement about the meaning of the situation in which they are involved. Once this definition of the situation is in place, further interaction can occur according to its specific, agreed rules. Accordingly, human social interaction takes place primarily at the level of symbols, which are ways of expressing the meaning of an experience. . . . Since symbolic interactionists believe that reality can be understood only from the point of view of the actors in social situations (since the actors define the meaning of the situation), special methods are needed for the study of human interaction. Unlike the methods of the natural sciences, which are based on the idea that the nature of a phenomenon can be measured accurately from an objective, external point of view, the methods of the symbolic interactionists are based on the idea that no such objectivity is possible. Instead, they use participant observation, taking part in the interaction so that it can be understood from the point of view of the participants.

References

Berger, Pete L. 1971. "Sociology and Freedom" *American Sociologist* 6 (February): 1–5.
Blumer, Herbert. 1969. *Symbolic Interactionism: Perspective and Method.* Englewood Cliffs, N. J.: Prentice-Hall.
Comte, Auguste. [1848] 1957. *A General View of Positivism.* New York: Speller.
Durkheim, Emile. [1893] 1958. *The Rules of Sociological Method.* Glencoe, Ill.: Free Press.

Garfinkel, Harold. 1967. *Studies in Ethnomethodology*. Englewood Cliffs, N. J.: Prentice-Hall.

Goffman, Erving. 1959. *The Presentation of Self in Everyday Life*. Garden City, N. Y.: Doubleday.

_____ . 1961. *Asylums: Essays on the Social Situation of Mental Patients and Other Inmates*. Chicago: Aldine.

_____ .1963. *Stigma: Notes on the Management of a Spoiled Identity*. Englewood Cliffs, N. J.: Prentice-Hall.

_____ . 1971. *Relations in Public*. New York: Basic Books.

Gouldner, Alvin. 1970. *The Coming Crisis of Western Sociology*. New York: Avon Books.

Mead, G. H., 1934. *Mind, Self and Society*. Chicago: University of Chicago Press.

Nisbet, Robert. 1966. *The Sociological Tradition*. New York: Basic Books.

Thomas, W. I., and D. S. Thomas. 1928. *The Child in America*. New York: Alfred A. Knopf.

Weber, Max. [1925] 1946. From *Max Weber: Essays in Sociology*, trans. And ed. Hans H.Gerth and C. Wright Milss. New York: Oxford University Press.

The Promise

C. Wright Mills

Mills' introduction to the sociological imagination is an essay to be studied rather than skimmed for information. In it, Mills both highlights significant characteristics of the sociological imagination and exercises this imagination in a study of some of the central issues of contemporary North American society. Some guiding questions for the reader are: What is the sociological imagination? What is the difference between a personal trouble and a public issue? What is the key issue of our times according to Mills? And what are the societal and historical roots of this issue?

Nowadays men often feel that their private lives are a series of traps. They sense that within their everyday worlds, they cannot overcome their troubles, and in this feeling, they are often quite correct: What ordinary men are directly aware of and what they try to do are bounded by the private orbits in which they live; their visions and their powers are limited to the close-up scenes of job, family, neighbourhood; in other milieux, they move vicariously and remain spectators. And the more aware they become, however vaguely, of ambitions and of threats which transcend their immediate locales, the more trapped they seem to feel.

Underlying this sense of being trapped are seemingly impersonal changes in the very structure of continent-wide societies. The facts of contemporary history are also facts about the success and the failure of individual men and women. When a society is industrialized, a peasant becomes a worker; a feudal lord is liquidated or becomes a businessman. When classes rise or fall, a man is employed or unemployed; when the rate of investment goes up or down, a man takes new heart or goes broke. When wars happen, an insurance salesman becomes a rocket launcher; a store clerk, a radar man; a wife lives alone; a child grows up without a father. Neither the life of an individual nor the history of a society can be understood without understanding both.

Yet men do not usually define the troubles they endure in terms of historical change and institutional contradiction. The well-being they

Source: C. Wright Mills, "The Promise," from *Sociological Imagination*, pp. 3–13 © 1999. By permission of Oxford University Press, Inc.

enjoy, they do not usually impute to the big ups and downs of the societies in which they live. Seldom aware of the intricate connection between the patterns of their own lives and the course of world history, ordinary men do not usually know what this connection means for the kinds of men they are becoming and for the kinds of history-making in which they might take part. They do not possess the quality of mind essential to grasp the interplay of man and society, of biography and history, of self and world. They cannot cope with their personal troubles in such ways as to control the structural transformations that usually lie behind them.

Surely it is no wonder. In what period have so many men been so totally exposed at so fast a pace to such earthquakes of change? That Americans have not known such catastrophic changes as have the men and women of other societies is due to historical facts that are now quickly becoming "merely history." The history that now affects every man is world history. Within this scene and this period, in the course of a single generation, one sixth of mankind is transformed from all that is feudal and backward into all that is modern, advanced, and fearful. Political colonies are freed; new and less visible forms of imperialism installed. Revolutions occur; men feel the intimate grip of new kinds of authority. Totalitarian societies rise, and are smashed to bits—or succeed fabulously. After two centuries of ascendancy, capitalism is shown up as only one way to make society into an industrial apparatus. After two centuries of hope, even formal democracy is restricted to a quite small portion of mankind. Everywhere in the underdeveloped world, ancient ways of life are broken up and vague expectations become urgent demands. Everywhere in the overdeveloped world, the means of authority and of violence become total in scope and bureaucratic in form. Humanity itself now lies before us, the super-nation at either pole concentrating its most coordinated and massive efforts upon the preparation of World War Three.

The very shaping of history now outpaces the ability of men to orient themselves in accordance with cherished values. And which values? Even when they do not panic, men often sense that older ways of feeling and thinking have collapsed and that newer beginnings are ambiguous to the point of moral stasis. Is it any wonder that ordinary men feel they cannot cope with the larger worlds with which they are so suddenly confronted? That they cannot understand the meaning of their epoch for their own lives? That—in defence of selfhood—they become morally insensible, trying to remain altogether private men? Is it any wonder that they come to be possessed by a sense of the trap?

It is not only information that they need—in this Age of Fact, information often dominates their attention and overwhelms their capacities to assimilate it. It is not only the skills of reason that they need—although their struggles to acquire these often exhaust their limited moral energy.

What they need, and what they feel they need, is a quality of mind that will help them to use information and to develop reason in order to achieve lucid summations of what is going on in the world and of what may be happening within themselves. It is this quality, I am going to contend, that journalists and scholars, artists and publics, scientists and editors are coming to expect of what may be called the sociological imagination.

I

The sociological imagination enables its possessor to understand the larger historical scene in terms of its meaning for the inner life and the external career of a variety of individuals. It enables him to take into account how individuals, in the welter of their daily experience, often become falsely conscious of their social positions. Within that welter, the framework of modern society is sought, and within that framework the psychologies of a variety of men and women are formulated. By such means the personal uneasiness of individuals is focused upon explicit troubles and the indifference of publics is transformed into involvement with public issues.

The first fruit of this imagination—and the first lesson of the social science that embodies it—is the idea that the individual can understand his own experience and gauge his own fate only by locating himself within his period, that he can know his own chances in life only by becoming aware of those of all individuals in his circumstances. In many ways it is a terrible lesson; in many ways a magnificent one. We do not know the limits of man's capacities for supreme effort or willing degradation, for agony or glee, for pleasurable brutality or the sweetness of reason. But in our time we have come to know that the limits of "human nature" are frighteningly broad. We have come to know that every individual lives, from one generation to the next, in some society; that he lives out a biography, and that he lives it out within some historical sequence. By the fact of his living he contributes, however minutely, to the shaping of this society and to the course of its history, even as he is made by society and by its historical push and shove.

The sociological imagination enables us to grasp history and biography and the relations between the two within society. That is its task and its promise. To recognize this task and this promise is the mark of the classic social analyst. It is characteristic of Herbert Spencer—turgid, polysyllabic, comprehensive; of E.A. Ross—graceful, muckraking, upright; of August Compte and Emile Durkheim; of the intricate and subtle Karl Mannheim. It is the quality of all that is intellectually excellent in Karl Marx; it is the clue to Thorstein Veblen's brilliant and ironic insight, to Joseph Schumpeter's many-sided constructions of reality; it is the basis of the psychological sweep of W.E.H. Lecky no less than of the profundity and clarity of Max Weber. And it is the signal of what is best in contemporary studies of man and society.

No social study that does not come back to the problems of biography, of history and of their intersections within a society has completed its intellectual journey. Whatever the specific problems of the classic social analysts, however limited or however broad the features of social reality they have examined, those who have been imaginatively aware of the promise of their work have consistently asked three sorts of questions:

1. What is the structure of this particular society as a whole? What are the essential components, and how are they related to one another? How does it differ from other varieties of social order? Within it, what is the meaning of any particular feature for its continuance and for its change?

2. Where does this society stand in human history? What are the mechanics by which it is changing? What is its place within and its meaning for the development of humanity as a whole? How does any particular feature we are examining affect, and how is it affected by, the historical period in which it moves? And this period—what are its essential features? How does it differ from other periods? What are its characteristic ways of history-making?

3. What varieties of men and women now prevail in this society and in this period? And what varieties are coming to prevail? In what ways are they selected and formed, liberated and repressed, made sensitive and blunted? What kinds of "human nature" are revealed in the conduct and character we observe in this society in this period? And what is the meaning for "human nature" of each and every feature of the society we are examining?

Whether the point of interest is a great power state or a minor literary mood, a family, a prison, a creed—these are the kinds of questions the best social analysts have asked. They are the intellectual pivots of classic studies of man in society—and they are the questions inevitably raised by any mind possessing the sociological imagination. For that imagination is the capacity to shift from one perspective to another—from the political to the psychological; from examination of a single family to comparative assessment of the national budgets of the world; from the theological school to the military establishment; from considerations of an oil industry to studies of contemporary poetry. It is the capacity to range from the most impersonal and remote transformations to the most intimate features of the human self—and to see the relations between the two. Back of its use there is always the urge to know the social and historical meaning of the individual in the society and in the period in which he has his quality and his being.

That, in brief, is why it is by means of the sociological imagination that men now hope to grasp what is going on in the world, and to understand what is happening in themselves as minute points of the intersections of biography and history within society. In large part, contemporary man's self-conscious view of himself as at least an outsider, if not a permanent stranger, rests upon an absorbed realization of social relativity and of the transformative power of history. The sociological imagination is the most fruitful form of self-consciousness. By its use men whose mentalities have swept only a series of limited orbits often come to feel as if suddenly awakened in a house with which they had only supposed themselves to be familiar. Correctly or incorrectly, they often come to feel that they can now provide themselves orientations. Older decisions that once appeared sound now seem to them products of a mind unaccountably dense. Their capacity for astonishment is made lively again. They acquire a new way of thinking, they experience a transvaluation of values; in a word, by their reflection and by their sensibility, they realize the cultural meaning of the social sciences.

II

Perhaps the most fruitful distinction with which the sociological imagination works is between "the personal troubles of milieu" and "the public issues of social structure." This distinction is an essential tool of the sociological imagination and a feature of all classic work in social science.

Troubles occur within the character of the individual and within the range of his immediate relations with others; they have to do with his self and with those limited areas of social life of which he is directly and personally aware. Accordingly, the statement and the resolution of troubles properly lie within the individual as a biographical entity and within the scope of his immediate milieu—the social setting that is directly open to his personal experience and to some extent his wilful activity. A trouble is a private matter: values cherished by an individual are felt by him to be threatened.

Issues have to do with matters that transcend these local environments of the individual and the range of his inner life. They have to do with the organization of many such milieux into the institutions of an historical society as a whole, with the ways in which various milieux overlap and interpenetrate to form the larger structure of social and historical life. An issue is a public matter: some value cherished by publics is felt to be threatened. Often there is a debate about what that value really is and about what it is that really threatens it. This debate is often without focus if only because it is the very nature of an issue, unlike even widespread trouble, that it cannot very well be defined in terms of the immediate and everyday environments of ordinary men. An issue, in fact, often involves a crisis in institutional arrangements, and often too it involves what Marxists call "contradictions" or "antagonisms."

In these terms, consider unemployment. When, in a city of 100,000, only one man is unemployed, that is his personal trouble, and for its relief we properly look to the character of the man, his skills, and his immediate opportunities. But when in a nation of 50 million employees, 15 million men are unemployed, that is an issue and we may not hope to find its solution within the range of opportunities open to any one individual. The very structure of opportunities has collapsed. Both the correct statement of the problem and the range of possible solutions require us to consider the economic and political institutions of the society, and not merely the personal situation and character of a scatter of individuals.

Consider war. The personal problem of war, when it occurs, may be how to survive it or how to die in it with honour; how to make money out of it; how to climb into the higher safety of the military apparatus; or how to contribute to the war's termination. In short, according to one's values, to find a set of milieux and within it to survive the war or make one's death in it meaningful. But the structural issues of war have to do with its causes; with what types of men it throws up into command; with its effects upon economic and political, family and religious institutions, with the unorganized irresponsibility of a world of nation-states.

Consider marriage. In a marriage a man and a woman may experience personal troubles, but when the divorce rate during the first four years of marriage is 250 out of every 1,000 attempts, this is an indication of a structural issue having to do with the institutions of marriage and the family and other institutions that bear upon them.

Or consider the metropolis—the horrible, beautiful, ugly, magnificent sprawl of the great city. For many upper-class people, the personal solution to "the problem of the city" is to have an apartment with private garage under it in the heart of the city, and forty miles out, a house by Henry Hill, garden by Garrett Eckbo, on a hundred acres of private land. In these two controlled environments—with a small staff at each end and a private helicopter connection—most people could solve many of the problems of personal milieux caused by the facts of the city. But all this, however splendid, does not solve the public issues that the structural fact of the city poses. What should be done with this wonderful monstrosity? Break it all up into scattered units, combining residence and work? Refurbish it as it stands? Or, after evacuation, dynamite it and build new cities according to new plans in new places? What should those plans be? And who is to decide and to accomplish whatever choice is made? These are structural issues; to confront them and to solve them requires us to consider political and economic issues that affect innumerable milieux.

In so far as an economy is arranged that slumps occur, the problem of unemployment becomes incapable of personal solution. In so far as war is inherent in the nation-state system and in the uneven industrialization of the world, the ordinary individual in his restricted milieu will be powerless—with or without psychiatric aid—to solve the troubles this system or lack of system imposes upon him. In so far as the family as an institution turns women into darling little slaves and men into their chief providers and unweaned dependants, the problem of a satisfactory marriage remains incapable of purely private solution. In so far as the overdeveloped megalopolis and the overdeveloped automobile are built-in features of the overdeveloped society, the issues of urban living will not be solved by personal ingenuity and private wealth.

What we experience in various and specific milieux, I have noted, is often caused by structural changes. Accordingly, to understand the changes of many personal milieux we are required to look beyond them. And the number and variety of such structural changes increase as the institutions within which we live become more embracing and more intricately connected with one another. To be aware of the idea of social structure and to use it with sensibility is to be capable of tracing such linkages among a great variety of milieux. To be able to do that is to possess the sociological imagination.

III

What are the major issues for publics and the key troubles of private individuals in our time? To formulate issues and troubles, we must ask what values are cherished yet threatened, and what values are cherished and supported, by the characterizing trends of our period. In the case both of threat and of support we must ask what salient contradictions of structure may be involved.

When people cherish some set of values and do not feel any threat to them, they experience well-being. When they cherish values but do feel them to be threatened, they experience a crisis—either as personal trouble or as public issue. And if all their values seem involved, they feel the total threat of panic.

But suppose people are neither aware of any cherished values nor experience any threat? That is the experience of indifference, which, if it seems to involve all their values, becomes apathy. Suppose, finally, they are unaware of any cherished values, but still are very much aware of a threat? That is the experience of uneasiness, of anxiety, which, if it is total enough, becomes a deadly unspecified malaise.

Ours is a time of uneasiness and indifference—not yet formulated in such ways as to permit the work of reason and the play of sensibility. Instead of troubles—defined in terms of values and threats—there is often the misery of vague uneasiness; instead of explicit issues there is often merely the beat feeling that all is somehow not right. Neither the values threatened nor whatever threatens them has been stated; in short, they have not been carried to the point of decision. Much less have they been formulated as problems of social science.

In the 'thirties there was little doubt—except among certain deluded business circles that there was an economic issue which was also a pack of personal troubles. In these arguments about "the crisis of capitalism," the formulations of Marx and the many unacknowledged re-formulations of his work probably set the leading terms of the issue, and some men came to understand their personal troubles in these terms. The values threatened were plain to see and cherished by all; the structural contradictions that threatened them also seemed plain. Both were widely and deeply experienced. It was a political age.

But the values threatened in the era after World War Two are often neither widely acknowledged as values nor widely felt to be threatened. Much private uneasiness goes unformulated; much public malaise and many decisions of enormous structural relevance never become public issues. For those who accept such inherited values as reason and freedom, it is the uneasiness itself that is the trouble; it is the indifference itself that is the issue. And it is this condition, of uneasiness and indifference, that is the signal feature of our period.

All this is so striking that it is often interpreted by observers as a shift in the very kinds of problems that need now to be formulated. We are frequently told that the problems of our decade, or even the crises of our period, have shifted from the external realm of economics and now have to do with the quality of individual life—in fact with the question of whether there is soon going to be anything that can properly be called individual life. Not child labour but comic books, not poverty but mass leisure, are at the centre of concern. Many great public issues as well as many private troubles are described in terms of "the psychiatric"—often, it seems, in a pathetic attempt to avoid the large issues and problems of modern society. Often this statement seems to rest upon a provincial narrowing of interest to the Western societies, or even to the United States—thus ignoring two-thirds of mankind; often, too, it arbitrarily divorces the individual life from the larger institutions within which that life is enacted, and which on occasion bear upon it more grievously than do the intimate environments of childhood.

Problems of leisure, for example, cannot even be stated without considering problems of work. Family troubles over comic books cannot be formulated as problems without considering the plight of the contemporary family in its new relations with the newer institutions of the social structure. Neither leisure nor its debilitating uses can be understood as problems without recognition of the extent to which malaise and indifference now form the social and personal climate of contemporary American society. In this climate, no problems of "the private life" can be stated and solved without recognition of the crisis of ambition that is part of the very career of men at work in the incorporated economy.

It is true, as psychoanalysts continually point out, that people do often have "the increasing sense of being moved by obscure forces within themselves which they are unable to define." But it is not true, as Ernest Jones asserted, that "man's chief enemy and danger is his own unruly nature and the dark forces pent up within him." On the contrary: "Man's chief danger" today lies in the unruly forces of contemporary society itself, with its alienating methods of production, its enveloping techniques of political domination, its international anarchy—in a word, its pervasive transformations of the very "nature" of man and the conditions and aims of his life.

It is now the social scientist's foremost political and intellectual task— for here the two coincide—to make clear the elements of contemporary uneasiness and indifference. It is the central demand made upon him by other cultural workmen—by physical scientists and artists, by the intellectual community in general. It is because of this task and these demands, I believe, that the social sciences are becoming the common denominator of our cultural period, and the sociological imagination our most needed quality of mind.

Social and Natural Phenomena and the Origin of Society

Emile Durkheim

In the first selection, "Social and Natural Phenomena," which is a brief excerpt from his book *Montesquieu and Rousseau*, Durkheim establishes the subject matter of the discipline of sociology—human society and its social nature. He argues that, historically, philosophers have aimed at discussing the best form of society rather than investigating social phenomena and the nature of society itself. Durkheim argues that social phenomena are "things, like all other things in nature," and calls for a science which can describe and explain their specific properties. But are social phenomena really like all other things in nature?

The second selection is from *The Division of Labor in Society*. I agree to do this and you agree to do that. That is, we agree to divide the labour. From such a contractual agreement, according to some theorists, society is born. But, on this essential point, Durkheim objects. He considers, rather, what makes agreement possible in the first place. That is, individuals do not just come together and agree to cooperate with one another, thus producing a society. The possibility of agreement presupposes an already-existing collective life. What is the nature of this pre-contractual solidarity that Durkheim argues is the primary fact of social and moral life?

SOCIAL AND NATURAL PHENOMENA

A discipline may be called a 'science' only if it has a definite field to explore. Science is concerned with things, realities. If it does not have a datum to describe and interpret, it exists in a vacuum. Separated from the description and interpretation of reality it can have no real function. Arithmetic is concerned with numbers, geometry with space and figures, the natural sciences with animate and inanimate bodies, and psychology

Sources: Emile Durkheim, "Social and Natural Phenomena," from Anthony Giddens, *Emile Durkheim: Selected Writings.* Cambridge University Press, 1972: 57–58. Reprinted with the permission of Cambridge University Press.

with the human mind. Before social science could begin to exist, it had first of all to be assigned a definite subject-matter.

At first sight, this problem presents no difficulty: the subject-matter of social science is social things: that is, laws, customs, religions, etc. Looking back in history, however, we find that no philosophers ever viewed matters in this way until quite recently. They thought that all such phenomena depended upon the human will and consequently failed to realise that they are actual things, like all other things in nature; they have their own specific properties, and these call for sciences which can describe and explain them. It seemed to them sufficient to ascertain what the human will should strive for and what it should avoid in established societies. Hence what they strove to discover was not the nature and origin of social phenomena, not what they actually are, but what they ought to be; their aim was not to offer us as valid a description of nature as possible, but to present us with the idea of a perfect society, a model to be imitated. Even Aristotle, who was far more concerned with empirical observation than Plato was, aimed at discovering, not the laws of social existence, but the best form of society. He begins by assuming that the sole objective of society should be to make its members happy through the practice of virtue, and that virtue lies in contemplation. He does not establish this principle as a law which societies actually follow, but as one which they should act upon in order that human beings may fulfil their specific nature. Certainly he does turn later to historical facts, but with little purpose other than to pass judgment upon them, and to show how his own principles could be adapted to various situations. The political thinkers who came after him on the whole followed his example. Whether they wholly disregard reality or pay a certain amount of attention to it, they all have a single purpose: to correct or transform it completely, rather than to know it. They take virtually no interest in the past and the present, but look to the future.

THE ORIGIN OF SOCIETY

[If society were based solely on social contracts]... the typical social relation would be the economic, stripped of all regulation and resulting from the entirely free initiative of the parties. In short, society would be solely the stage where individuals exchanged the products of their labor, without any action properly social coming to regulate this exchange.

Is this the character of societies whose unity is produced by the division of labor? If this were so, we could with justice doubt their stability. For if interest relates men, it is never for more than some few moments. It can create only an external link between them. In the fact of exchange, the various agents remain outside of each other, and when the business has been completed, each one retires and is left entirely on his own. Consciences are only superficially in contact; they neither penetrate each other, nor do they adhere. If we look further into the matter, we shall see that this total harmony of interests conceals a latent or deferred conflict. For where interest is the only ruling force each individual finds himself in a state of war with every other since nothing comes to mollify the egos, and any truce in this eternal antagonism would not be of long duration. There is nothing less constant than interest. Today, it unites me to you; tomorrow, it will make me your enemy. Such a cause can only give rise to transient relations and passing associations. We now understand how necessary it is to see if this is really the nature of organic solidarity.

To be sure, when men unite in a contract, it is because, through the division of labor, either simple or complex, they need each other. But in order for them to co-operate harmoniously, it is not enough that they enter into a relationship, nor even that they feel the state of mutual dependence in which they find themselves. It is still necessary that the conditions of this co-operation be fixed for the duration of their relations. The rights and duties of each must be defined, not only in view of the situation such as it presents itself at the moment when the contract is made, but with foresight for the circumstances which may arise to modify it. Otherwise, at every instant, there would be conflicts and endless difficulties. We must not forget that, if the division of labor makes interests solidary, it does not confound them; it keeps them distinct and opposite. Even as in the internal workings of the individual organism each organ is in conflict with others while co-operating with them, each of the contractants, while needing the other, seeks to obtain what he needs at the least expense; that is to say, to acquire as many rights as possible in exchange for the smallest possible obligations.

. . .

A corollary of all that has preceded is that the division of labor can be effectuated only among members of an already constituted society.

In effect, when competition places isolated and estranged individuals in opposition, it can only separate them more. If there is a lot of space at their disposal, they will flee; if they cannot go beyond certain boundaries, they will differentiate themselves, so as to become still more independent. No case can be cited where relations of pure hostility are transformed, without the intervention of any other factor, into social relations. Thus, as among individuals of the same animal or vegetable species, there is generally no bond, the war they wage has no other result than to diversify them, to give birth to dissimilar varieties which grow farther apart. It is this progressive disjunction that Darwin called the law of the divergence of characters. But the division of labor unites at the same time that it opposes; it makes the activities it differentiates converge; it brings together those it separates. Since competition cannot have determined this conciliation, it must have existed before. The individuals among whom the struggle is waged must already be solidary and feel so. That is to say, they must belong to the same society. . . .

To represent what the division of labor is suffices to make one understand that it cannot be otherwise. It consists in the sharing of functions up to that time common. But this sharing cannot be executed according to a preconceived plan. We cannot tell in advance where the line of demarcation between tasks will be found once they are separated; for it is not marked so evidently in the nature of things, but depends, on the contrary, upon a multitude of circumstances. The division of labor, then, must come about of itself and progressively. Consequently, under these conditions, for a function to be divided into two exactly complementary parts, as the nature of the division of labor demands, it is indispensable that the two specializing parts be in constant communication during all the time that this dissociation lasts. There is no other means for one to receive all the movement the other abandons, and which they adapt to each other. But in the same way that an animal colony whose members embody a continuity of tissue form one individual, every aggregate of individuals who are in continuous contact form a society. The division of labor can then be produced only in the midst of a pre-existing society. By that, we do not mean to say simply that individuals must adhere materially, but it is still necessary that there be moral links between them. First, material continuity by itself produces links of this kind, provided it is durable. But, moreover, they are directly necessary. If the relations becoming established in the period of groping were not subject to any rule, if no power moderated the conflict of individual interests, there would be chaos from which no new order could emerge. It is thought, it is true, that everything takes place through private conventions freely disputed. Thus, it seems that all social action is absent. But this is to forget that contracts are possible only where a juridical regulation, and, consequently, a society, already exists.

Hence, the claim sometimes advanced that in the division of labor lies the fundamental fact of all social life is wrong. Work is not divided among independent and already differentiated individuals who by uniting and associating bring together their different aptitudes. For it would be a miracle if differences thus born through chance circumstance could unite so perfectly as to form a coherent whole. Far from preceding collective life, they derive from it. They can be produced only in the midst of a society, and under the pressure of social sentiments and social needs. That is what makes them essentially harmonious. There is, then, a social life outside the whole division of labor, but which the latter presupposes. That is, indeed, what we have directly established in showing that there are societies whose cohesion is essentially due to a community of beliefs and sentiments, and it is from these societies that those whose unity is assured by the division of labor have emerged....

For a number of theorists, it is a self-evident truth that all society essentially consists of co-operation. Spencer has said that a society in the scientific sense of the word exists only when to the juxtaposition of individuals co-operation is added. We have just seen that this so-called axiom is contrary to the truth. Rather it is evident, as Auguste Comte points out, "that co-operation, far from having produced society, necessarily supposes, as preamble, its spontaneous existence." What bring men together are mechanical causes and impulsive forces, such as affinity of blood, attachment to the same soil, ancestral worship, community of habits, and so on. It is only when the group has been formed on these bases that co-operation is organized there.

Further, the only co-operation possible in the beginning is so intermittent and feeble that social life, if it had no other source, would be without force and without continuity. With stronger reason, the complex co-operation resulting from the division of labor is an ulterior and derived phenomenon. It results from internal movements which are developed in the midst of the mass, when the latter is constituted. It is true that once it appears it tightens the social bonds and makes a more perfect individuality of society. But this integration supposes another which it replaces. For social units to be able to be differentiated, they must first be attracted or grouped by virtue of the resemblances they present. This process of formation is observed, not only originally, but in each phase of evolution. We know, indeed, that higher societies result from the union of lower societies of the same type. It is necessary first that these latter be mingled in the midst of the same identical collective conscience for the process of differentiation to begin or recommence. It is thus that more complex organisms are formed by the repetition of more simple, similar organisms which are differentiated only if once associated. In short, association and co-operation are two distinct facts, and if the second, when developed, reacts on the first and transforms it, if human societies steadily become groups of co-operators, the duality of the two phenomena does not vanish for all that.

If this important truth has been disregarded by the utilitarians, it is an error rooted in the manner in which they conceive the genesis of society. They suppose originally isolated and independent individuals, who, consequently enter into relationships only to co-operate, for they have no other reason to clear the space separating them and to associate. But this theory, so widely held, postulates a veritable *creatio ex nihilo*.

It consists, indeed, in deducing society from the individual. But nothing we know authorizes us to believe in the possibility of such spontaneous generation. According to Spencer, for societies to be formed within this hypothesis, it is necessary that primitive units pass from the state of perfect independence to that of mutual dependence. But what can have determined such a complete transformation in them? Is it the prospect of the advantages presented by social life? But they are counterbalanced, perhaps more than counterbalanced, by the loss of independence, for, among individuals born for a free and solitary life, such a sacrifice is most intolerable. Add to this, that in the first social types social life is as absolute as possible, for nowhere is the individual more completely absorbed in the group. How would man, if he were born an individualist, as is supposed, be able to resign himself to an existence clashing violently with his fundamental inclination? How pale the problematical utility of co-operation must appear to him beside such a fall! With autonomous individualities, as are imagined, nothing can emerge save what is individual, and, consequently, co-operation itself, which is a social fact, submissive to social rules, cannot arise. Thus, the psychologist who starts by restricting himself to the ego cannot emerge to find the non-ego.

Collective life is not born from individual life, but it is, on the contrary, the second which is born from the first. It is on this condition alone that one can explain how the personal individuality of social units has been able to be formed and enlarged without disintegrating society. Indeed, as, in this case, it becomes elaborate in the midst of a pre-existing social environment, it necessarily bears its mark. It is made in a manner so as not to ruin this collective order with which it is solidary. It remains adapted to it while detaching itself. It has nothing anti-social about it because it is a product of society. It is not the absolute personality of the monad, which is sufficient unto itself, and could do without the rest of the world, but that of an organ or part of an organ having its determined function, but which cannot, without risking dissolution, separate itself from the rest of the organism. Under these conditions, co-operation becomes not only possible but necessary. Utilitarians thus reverse the natural order of facts, and nothing is more deceiving than this inversion. It is a particular illustration of the general truth that what is first in knowledge is last in reality. Precisely because co-operation is the most recent fact, it strikes sight first. If, then, one clings to appearance, as does common sense, it is inevitable that one see it in the primary fact of moral and social life.

But if it is not all of ethics, it is not necessary to put it outside ethics, as do certain moralists. As the utilitarians, the idealists have it consist exclusively in a system of economic relations, of private arrangements in which egotism is the only active power. In truth, the moral life traverses all the relations which constitute co-operation, since it would not be possible if social sentiments, and, consequently, moral sentiments, did not preside in its elaboration.

Types of Authority and Characteristics of Bureaucracy

Max Weber

Classical social theorists such as Weber continue to be read, not only because they elucidated some of the central principles and problems of the discipline of sociology, but also because they did so by asking enduring questions. Why do we generally comply with social expectations and the commands of others? One possibility is the use of force or coercion. Another is that we are persuaded to agree with an expectation. But neither explanation can adequately account for why and how members typically comply.

In the first selection, "Types of Authority," Weber explores the significant role of different kinds of authority, where authority is a form of power perceived to be legitimate by those who comply. In the second selection he outlines the characteristics of bureaucracy while examining how forms of organization such as bureaucracies promote compliance. But a key question remains: Why has rational-legal authority come to dominate life in the modern world in the way that it has?

TYPES OF AUTHORITY

All ruling powers, profane and religious, political and apolitical, may be considered as variations of, or approximations to, certain pure types. These types are constructed by searching for the basis of *legitimacy*, which the ruling power claims. Our modern 'associations,' above all the political ones, are of the type of 'legal' authority. That is the legitimacy of the power-holder to give commands rests upon rules that are rationally established by enactment, by agreement, or by imposition. The legitimation for establishing these rules rests, in turn, upon a rationally enacted or interpreted 'constitution.' Orders are given in the name of the impersonal norm, rather than in the name of a personal authority; and even the giving of a command constitutes obedience toward a norm rather than an arbitrary freedom, favor, or privilege.

Source: Max Weber, "Types of Authority," from Max Weber: Essays in Sociology (1973) edited by Gerth and Mills, pp. 224–229. By permission of Oxford University Press.

The 'official' is the holder of the power to command; he never exercises this power in his own right; he holds it as a trustee of the impersonal and 'compulsory institution.'[1] This institution is made up of the specific patterns of life of a plurality of men, definite or indefinite, yet specified according to rules. Their joint pattern of life is normatively governed by statutory regulations.

The 'area of jurisdiction' is a functionally delimited realm of possible objects for command and thus delimits the sphere of the official's legitimate power. A hierarchy of superiors, to which officials may appeal and complain in an order of rank, stands opposite the citizen or member of the association.. Today this situation also holds for the hierocratic association that is the church. The pastor or priest has his definitely limited 'jurisdiction,' which is fixed by rules. This also holds for the supreme head of the church. The present concept of [papal] 'infallibility' is a jurisdictional concept. Its inner meaning differs from that which preceded it, even up to the time of Innocent III.

The separation of the 'private sphere' from the 'official sphere' (in the case of infallibility: the *ex cathedra* definition) is carried through in the church in the same way as in political, or other, officialdoms. The legal separation of the official from the means of administration (either in natural or in pecuniary form) is carried through in the sphere of political and hierocratic associations in the same way as is the separation of the worker from the means of production in capitalist economy: it runs fully parallel to them.

No matter how many beginnings may be found in the remote past, in its full development all this is specifically modern. The past has known other bases for authority, bases which, incidentally, extend as survivals into the present. Here we wish merely to outline these bases of authority in a terminological way.

1. In the following discussion the term 'charisma' shall be understood to refer to an *extraordinary* quality of a person, regardless of whether this quality is actual, alleged, or presumed. 'Charismatic authority,' hence, shall refer to a rule over men, whether predominately external or predominantly internal, to which the governed submit because of their belief in the extraordinary quality of the specific *person*. The magical sorcerer, the prophet, the leader of hunting and booty expeditions, the warrior chieftain, the so-called 'Caesarist' ruler, and, under certain conditions, the personal head of a party are such types of rulers for their disciples, followings, enlisted troops, parties, et cetera. The legitimacy of their rule rests on the belief in and the devotion to the extraordinary, which is valued because it goes beyond the normal human qualities, and which was originally value as supernatural. The legitimacy of charismatic rule thus rests upon

the belief in magical powers, revelations and hero worship. The source of these beliefs is the 'proving' of the charismatic quality through miracles, through victories and other successes, that is, through the welfare of the governed. Such beliefs and the claimed authority resting on them therefore disappear, or threaten to disappear, as soon as proof is lacking and as soon as the charismatically qualified person appears to be devoid of his magical power or forsaken by his god. Charismatic rule is not managed according to general norms, either traditional or rational, but, in principle, according to concrete revelations and inspirations, and in this sense, charismatic authority is 'irrational.' It is 'revolutionary' in the sense of not being bound to the existing order: 'It is written — but I say unto you... !'

2. 'Traditionalism' in the following discussions shall refer to the psychic attitude-set for the habitual workaday and to the belief in the everyday routine as an inviolable norm of conduct. Domination that rests upon this basis, that is, upon piety for what actually, allegedly, or presumably has always existed, will be called 'traditionalist authority.'

   Patriarchalism is by far the most important type of domination the legitimacy of which rests upon tradition. Patriarchalism means the authority of the father, the husband, the senior of the house, the sib elder over the members of the household and sib; the rule of the master and patron over bondsmen, serfs, freed men; of the lord over the domestic servants and household officials; of the prince over house- and court-officials, nobles of office, clients, vassals; of the patrimonial lord and sovereign prince (*Landesvater*) over the 'subjects.'

   It is characteristic of patriarchical and of patrimonial authority, which represents a variety of the former, that the system of inviolable norms is considered sacred; an infraction of them would result in magical or religious evils. Side by side with this system there is a realm of free arbitrariness and favor of the lord, who in principle judges only in terms of 'personal,' not 'functional,' relations. In this sense, traditionalist authority is irrational.

3. Throughout early history, charismatic authority, which rests upon a belief in the sanctity or the value of the extraordinary, and traditionalist (patriarchical) domination, which rests upon a belief in the sanctity of everyday routines, divided the most important authoritative relations between them. The bearers of charisma, the oracles of prophets, or the edicts of charismatic war lords alone could integrate 'new' laws into the circle of what was upheld by tradition. Just as revelation and the sword were the two extraordinary powers,

so were they the two typical innovators. In typical fashion, however, both succumbed to routinzation as soon as their work was done.

With the death of the prophet or the war lord the question of successorship arises. This question can be solved by *Kürung*, which was originally not an 'election' but a selection in terms of charismatic qualification; or the question can be solved by the sacramental substantiation of charisma, the successor being designated by consecration, as is the case in hierocratic or apostolic succession; or the belief in the charismatic qualification of the charismatic leader's sib can lead to a belief in hereditary charisma, as represented by hereditary kingship and hereditary hierocracy. With these routinizations, *rules* in some form always come to govern. The prince or the hierocrat no longer rules by virtue of purely personal qualities, but by virtue of acquired or inherited qualities, or because he has been legitimized by an act of charismatic election. The process of routinzation, and thus traditionalization, has set in.

Perhaps it is even more important that when the organization of authority becomes permanent, the staff supporting the charismatic ruler becomes routinized. The ruler's disciples, apostles, and followers became priests, feudal vassals and, above all, officials. The original charismatic community lived communistically off donations, alms, and the booty of war: they were thus specifically alienated from the economic order. The community was transformed into a stratum of aids to the ruler and depended upon him for maintenance through the usufruct of land, office fees, income in kind, salaries, and hence, through prebends. The staff derived its legitimate power in greatly varying stages of appropriation, infeudation, conferment, and appointment. As a rule, this meant that princely prerogatives became *patrimonial* in nature. Patrimonialism can also develop from pure patriarchalism through the disintegration of the patriarchical master's strict authority. By virtue of conferment, the prebendary or the vassal has as a rule had a personal *right* to the office bestowed upon him. Like the artisan who possessed the economic means of production, the prebendary possessed the means of administration. He had to bear the costs of administration out of his office fees or other income, or he passed on to the lord only part of the taxes gathered from the subjects, retaining the rest. In the extreme case he could bequeath and alienate his office like other possessions. We wish to speak of *status* patrimonialism when the development by appropriation of prerogatory power has reached this stage, without regard to whether it developed from charismatic or patriarchical beginnings.

The development, however, has seldom stopped at this stage. We always meet with a *struggle* between the political or hierocratic lord and the owners or usurpers of prerogatives, which they have appropriated as status groups. The ruler attempts to expropriate the estates, and the estates attempt to expropriate the ruler. The more the ruler succeeds in attaching to himself a staff of officials who depend solely on him and whose interests are linked to his, the more this struggle is decided in favor of the ruler and the more the privilege-holding estates are gradually expropriated. In this connection, the prince acquires administrative means of his own and he keeps them firmly in his own hands. Thus we find political rulers in the Occident, and progressively from Innocent III to Johann XXII, also hierocratic rulers who have finances of their own, as well as secular rulers who have magazines and arsenals of their own for the provisioning of the army and the officials.

The *character* of the stratum of officials upon whose support the ruler has relied in the struggle for the expropriation of status prerogatives has varied greatly in history. In Asia and in the Occident during the early Middle Ages they were typically clerics; during the Oriental Middle Ages they were typically slaves and clients; for the Roman Principate, freed slaves to a limited extent were typical; humanist literati were typical for China; and finally, jurists have been typical for the modern Occident, in ecclesiastical as well as in political associations.

The triumph of princely power and the expropriation of particular prerogatives has everywhere signified at least the possibility, and often the actual introduction, of a rational administration. As we shall see, however, this rationalization has varied greatly in extent and meaning. One must, above all, distinguish between the *substantive* rationalization of administration and of judiciary by a patrimonial prince, and the *formal* rationalization carried out by trained jurists. The former bestows utilitarian and social ethical blessings upon his subjects, in the manner of the master of a large house upon the members of his household. The trained jurists have carried out the rule of general laws applying to all 'citizens of the state.' However fluid the difference has been—for instance, in Babylon or Byzantium, in the Sicily of the Hohenstaufen, or the England of the Stuarts, or the France of the Bourbons—in the final analysis, the difference between substantive and formal rationality has persisted. And, in the main, it has been the work of *jurists* to give birth to the modern Occidental 'state' as well as to the Occidental 'churches.' We shall not discuss at this point the source of their strength, the substantive ideas, and the technical means for this work.

With the triumph of *formalist* juristic rationalism, the legal type of domination appeared in the Occident at the side of the transmitted types of domination. Bureaucratic rule was not and is not the only variety of legal authority, but it is the purest. The modern state and municipal official, the modern Catholic priest and chaplain, the officials and employees of modern banks and of large capitalist enterprises represent, as we have already mentioned, the most important types of this structure of domination.

The following characteristic must be considered decisive for our terminology: in legal authority, submission does not rest upon the belief and devotion to charismatically gifted persons, like prophets and heroes, or upon sacred tradition, or upon piety toward a personal lord and master who is defined by an ordered tradition, or upon piety toward the possible incumbents of office fiefs and office prebends who are legitimized in their own right through privilege and conferment. Rather, submission under legal authority is based upon an *impersonal* bond to the generally defined and functional 'duty of office.' The official duty—like the corresponding right to exercise authority: the 'jurisdictional competency'—is fixed by *rationally established* norms, by enactments, decrees, and regulations, in such a manner that the legitimacy of the authority becomes the legality of the general rule, which is purposely thought out, enacted, and announced with formal correctness.

Reference
1. Anstalt.

CHARACTERISTICS OF BUREAUCRACY

Modern officaldom functions in the following specific manner:

I. There is the principle of fixed and official jurisdictional areas, which are generally ordered by rules, that is, by laws or administrative regulations.

1. The regular activities required for the purposes of the bureaucratically governed structure are distributed in a fixed way as official duties.
2. The authority to give the commands required for the discharge of these duties is distributed in a stable way and is strictly delimited by rules concerning the coercive means, physical, sacerdotal, or otherwise, which may be placed at the disposal of officials.
3. Methodical provision is made for the regular and continuous fulfilment of these duties and for the execution of the corresponding rights; only persons who have the generally regulated qualifications to serve are employed.

In public and lawful government these three elements constitute "bureaucratic authority[.]" In private economic domination, they constitute bureaucratic "management." Bureaucracy, thus   understood, is fully developed in political and ecclesiastical communities only in the modern state, and, in the private economy, only in the most advanced insitutions [institutions] of capitalism. Permanent and public office authority, with fixed jurisdiction, is not the historical rule but rather the exception. This is so even in large political structure such as those of the ancient Orient, the Germanic and Mongolian empires of conquest, or of many feudal structures of state. In all these cases, the ruler executes the most important measures through personal trustees, table-companions, or court-servants. Their commissions and authority are not precisely delimited and are temporarily called into being for each case.

II. The principles of office hierarchy and of levels of graded authority mean a firmly ordered system of super- and subordination in which there is a supervision of the lower offices by the higher ones. Such a system offers the governed the possibility of appealing the decision of a lower office to its higher authority, in a definitely regulated manner. With the full development of the bureaucratic type, the office hierarchy is monocratically organized. The principle of hierarchical office authority is found in all bureaucratic structures: in state and ecclesiastical structures as well as in large party organizations and private enterprises. It does not matter for the character of bureaucracy whether its authority is called "private" or "public."

When the principle of jurisdictional "competency" is fully carried through, hierarchical subordination—at least in public office—does not mean that the "higher" authority is simply authorized to take over the business of the "lower." Indeed, the opposite is the rule. Once established and having fulfilled its task, an office tends to continue in existence and be held by another incumbent.

III. The management of the modern office is based upon written documents ("the files"), which are preserved in their original or draught form. There is, therefore, a staff of subaltern officials and scribes of all sorts. The body of officials actively engaged in a "public" office, along with the respective apparatus of material implements and the files, make up a "bureau." In private enterprise, "the bureau" is often called "the office."

In principle, the modern organization of the civil service separates the bureau from the private domicile of the official, and, in general, bureaucracy segregates official activity as something distinct from the sphere of private life. Public monies and equipment are divorced from the private property of the official. This condition is everywhere the product of a long development. Nowadays, it is found in public as well as in private enterprises; in the latter, the principle extends even to the leading entrepreneur. In principle, the executive office is separated from the household, business from private correspondence, and business assets from private fortunes. The more consistently the modern type of business management has been carried through the more are these separations the case. The beginnings of this process are to be found as early as the Middle Ages.

It is the peculiarity of the modern entrepreneur that he conducts himself as the "first official" of his enterprise, in the very same way in which the ruler of a specifically modern bureaucratic state spoke of himself as "the first servant" of the state.[1] The idea that the bureau activities of the state are intrinsically different in character from the management of private economic offices is a continental European notion and, by way of contrast, is totally foreign to the American way.

IV. Office management, at least all specialized office management—and such management is distinctly modern—usually presupposes thorough and expert training. This increasingly holds for the modern executive and employee of private enterprises, in the same manner as it holds for the state official.

V. When the office is fully developed, official activity demands the full working capacity of the official, irrespective of the fact that his obligatory time in the bureau may be firmly delimited. In the normal case, this is only the product of a long development, in the public as well as in the private office. Formerly, in all cases, the normal state of affairs was reversed: official business was discharged as a secondary activity.

VI. The management of the office follows general rules, which are more or less stable, more or less exhaustive, and which can be learned. Knowledge of these rules represents a special technical learning which the officials possess. It involves jurisprudence, or administrative or business management.

The reduction of modern office management to rules is deeply embedded in its very nature. The theory of modern public administration, for instance, assumes that the authority to order certain matters by decree—which has been legally granted to public authorities—does not entitle the bureau to regulate the matter by commands given for each case, but only to regulate the matter abstractly. This stands in extreme contrast to the regulation of all relationships through individual privileges and bestowals of favor, which is absolutely dominant in patrimonialism, at least in so far as such relationships are not fixed by sacred tradition.

THE POSITION OF THE OFFICIAL

All this results in the following for the internal and external position of the official:

I. Office holding is a "vocation." This is shown, first, in the requirement of a firmly prescribed course of training, which demands the entire capacity of work for a long period of time, and in the generally prescribed and special examinations which are prerequisites of employment. Furthermore, the position of the official is in the nature of a duty. This determines the internal structure of his relations, in the following manner: Legally and actually, office holding is not considered a source to be exploited for rents or emoluments, as was normally the case during the Middle Ages and frequently up to the threshold of recent times. Nor is office holding considered a usual exchange of services for equivalents, as is the case with free labor contracts. Entrance into an office, including one in the private economy, is considered an acceptance of a specific obligation of faithful management in return for a secure existence. It is decisive for the specific nature of modern loyalty to an office that, in the pure type, it does not establish a relationship to a *person*, like the vassal's or disciple's faith in feudal or in patrimonial relations of authority. Modern loyalty is devoted to impersonal and functional purposes. Behind the functional purposes, of course, "ideas of culture-values" usually stand. These are *ersatz* for the earthly or supra-mundane personal master: ideas such as "state," "church," "community," "party," or "enterprise" are thought of as being realized in a community; they provide an ideological halo for the master.

The political official—at least in the fully developed modern state—is not considered the personal servant of a ruler. Today, the bishop, the priest, and the preacher are in fact no longer, as in early Christian times, holders of purely personal charisma. The supra-mundane and sacred values which they offer are given to everybody who seems to be worthy of them and who asks for them. In former times, such leaders acted upon the personal command of their master; in principle, they were responsible only to him. Nowadays, in spite of the partial survival of the old theory, such religious leaders are officials in the service of a functional purpose, which in the present-day "church" has become routinized and, in turn, ideologically hallowed.

Alienated Labour[*]

Karl Marx

There are few statements that capture the lived experience of a problem with such analytic force as Karl Marx's characterization of alienated labour, published in 1844. Marx's formulation of this ongoing problem of modern life centres on how work is generally organized and conducted in the world which emerged in the wake of industrialization and the rise of capitalism. What are the major features of alienated labour to which Marx directs our attention? Why is understanding the activity and structure of work so important for comprehending how men and women experience human society more generally?

We have proceeded from the premises of political economy. We have accepted its language and its laws. We presupposed private property, the separation of labour, capital and land, and of wages, profit of capital and rent of land—likewise division of labour, competition, the concept of exchange-value, etc. On the basis of political economy itself, in its own words, we have shown that the worker sinks to the level of a commodity and becomes indeed the most wretched of commodities; that the wretchedness of the worker is in inverse proportion to the power and magnitude of his production; that the necessary result of competition is the accumulation of capital in a few hands, and thus the restoration of monopoly in a more terrible form; that finally the distinction between capitalist and land-rentier, like that between the tiller of the soil and the factory-worker, disappears and that the whole of society must fall apart into the two classes—the property-*owners* and the propertyless *workers*.

Political economy proceeds from the fact of private property, but it does not explain it to us. It expresses in general, abstract formulae the *material* process through which private property actually passes, and these formulae it then takes for *laws*. It does not *comprehend* these laws— i.e., it does not demonstrate how they arise from the very nature of private property. Political economy does not disclose the source of the division between labour and capital, and between capital and land. When, for example, it defines the relationship of wages to profit, it takes the interest of the capitalists to be the ultimate cause; i.e., it takes for granted what it is supposed to evolve. Similarly, competition comes in everywhere. It is explained from external circumstances. As to how far these external and apparently fortuitous circumstances are but the expression of a necessary course of development, political economy teaches us nothing. We have seen how, to it, exchange itself appears to be a fortuitous fact. The only wheels which political economy sets in motion are *avarice* and the *war amongst the avaricious—competition.*

Precisely because political economy does not grasp the connections within the movement, it was possible to counterpose, for instance, the doctrine of competition to the doctrine of monopoly, the doctrine of craft-liberty to the doctrine of the corporation, the doctrine of the division of landed property to the doctrine of the big estate—for competition, craft-liberty and the division of landed property were explained and comprehended only as fortuitous, premeditated and violent consequences of monopoly, the corporation, and feudal property, not as their necessary, inevitable and natural consequences.

Now, therefore, we have to grasp the essential connection between private property, avarice, and the separation of labour, capital and landed property; between exchange and competition, value and the devaluation of men, monopoly and competition, etc.; the connection between this whole estrangement and the *money*-system.

Do not let us go back to a fictitious primordial condition as the political economist does, when he tries to explain. Such a primordial condition explains nothing. He merely pushes the question away into a grey nebulous distance. He assumes in the form of fact, of an event, what he is supposed to deduce—namely, the necessary relationship between two things—between, for example, division of labour and exchange. Theology in the same way explains the origin of evil by the fall of man: that is, it assumes as a fact, in historical form, what has to be explained.

We proceed from an *actual* economic fact.

The worker becomes all the poorer the more wealth he produces, the more his production increases in power and range. The worker becomes an ever cheaper commodity the more commodities he creates. With the *increasing value* of the world of things proceeds in direct proportion the *devaluation* of the world of men. Labour produces not only commodities; it produces itself and the worker as a *commodity*—and does so in the proportion in which it produces commodities generally.

This fact expresses merely that the object which labour produces—labour's product—confronts it as *something alien*, as a *power independent* of the producer. The product of labour is labour which has been congealed in an object, which has become material: it is the *objectification* of labour.

Labour's realization is its objectification. In the conditions dealt with by political economy this realization of labour appears as *loss of reality* for the workers; objectification as *loss of the object* and *object-bondage*; appropriation as *estrangement*, as *alienation*.[1]

So much does labour's realization appear as loss of reality that the worker loses reality to the point of starving to death. So much does objectification appear as loss of the object that the worker is robbed of the objects most necessary not only for his life but for his work. Indeed, labour itself becomes an object which he can get hold of only with the greatest effort and with the most irregular interruptions. So much does the appropriation of the object appear as estrangement that the more objects the worker produces the fewer can he possess and the more he falls under the dominion of his product, capital.

All these consequences are contained in the definition that the workers is related to the *product of his labour* as to an *alien* object. For on this premise it is clear that the more the worker spends himself, the more powerful the alien objective world becomes which he creates over-against himself, the poorer he himself—his inner world—becomes, the less belongs to him as his own. It is the same in religion. The more man puts into God, the less he retains in himself. The worker puts his life into the object; but now his life no longer belongs to him but to the object. Hence, the greater this activity, the greater is the worker's lack of objects. Whatever the product of his labour is, he is not. Therefore the greater this product, the less is he himself. The *alienation* of the worker in his product means not only that his labour becomes an object, an *external* existence, but that it exists *outside him*, independently, as something alien to him, and that it becomes a power of its own confronting him; it means that the life which he has conferred on the object confronts him as something hostile and alien.

Let us now look more closely at the *objectification*, at the production of the worker; and therein at the *estrangement*, the *loss* of the object, his product.

The worker can create nothing without *nature*, without the *sensuous external world*. It is the material on which is labor is manifested, in which it is active, from which and by means of which it produces.

But just as nature provides labor with the *means of life* in the sense that labour cannot *live* without objects on which to operate, on the other hand, it also provides the *means of life* in the more restricted sense—i.e., the means for the physical subsistence of the *worker* himself.

Thus the more the worker by his labour *appropriates* the external world, sensuous nature, the more he deprives himself of *means of life* in the double respect: first, that the sensuous external world more and more ceases to be an object belonging to his labour—to be his labour's *means of life*; and secondly, that it more and more ceases to be *means of life* in the immediate sense, means for the physical subsistence of the worker.

Thus in this double respect the worker becomes a slave of his object, first, in that he receives an *object of labour*, i.e., in that he receives *work*; and secondly, in that he received *means of subsistence*. Therefore, it enables him to exist, first, as a *worker*; and, second, as a *physical subject*. The extremity of this bondage is that it is only as a *worker* that he continues to maintain himself as a *physical subject*, and that it is only as a *physical subject* that he is a *worker*.

(The laws of political economy express the estrangement of the worker in his object thus: the more the worker produces, the less he has to consume; the more values he creates, the more valueless, the more unworthy he becomes; the better formed his product, the more deformed becomes the worker; the more civilized his object, the more barbarous becomes the worker; the mightier labour becomes, the more powerless becomes the worker; the more ingenious labour becomes, the duller becomes the worker and the more he becomes nature's bondsman.)

Political economy conceals the estrangement inherent in the nature of labour by not considering the direct relationship between the worker (labour) *and production.* It is true that labour produces for the rich wonderful things—but for the worker it produces privation. It produces palaces—but for the worker, hovels. It produces beauty—but for the worker, deformity. It replaces labour by machines—but some of the workers it throws back to a barbarous type of labour, and the other workers it turns into machines. It produces intelligence—but for the worker idiocy, cretinism.

The direct relationship of labour to its product is the relationship of the worker to the objects of his production. The relationship of the man of means to the objects of production and to production itself is only a *consequence* of this first relationship—and confirms it. We shall consider this other aspect later.

When we ask, then, what is the essential relationship of labour we are asking about the relationship of the *worker* to production.

Till now we have been considering the estrangement, the alienation of the worker only in one of its aspects, i.e., the worker's *relationship to the products of his labour*. But the estrangement is manifested not only in the result but in the *act of production*—within the *producing activity* itself. How would the worker come to face the product of his activity as a stranger, were it not that in the very act of production he was estranging himself from himself? The product is after all but the summary of the activity of production. If then the product of labour is alienation, production itself must be active alienation, the alienation of activity, the activity of alienation. In the estrangement of the object of labour is merely summarized the estrangement, the alienation, in the activity of labour itself.

What, then, constitutes the alienation of labour?

First, the fact that labour is *external* to the worker, i.e., it does not belong to his essential being; that in his work, therefore, he does not affirm himself but denies himself, does not feel content but unhappy, does not develop freely his physical and mental energy but mortifies his body and ruins his mind. The worker therefore only feels himself outside his work, and in his work feels outside himself. He is at home when he is not working, and when he is working he is not at home. His labour is therefore not voluntary, but coerced; it is *forced labour.* It is therefore not the satisfaction of a need; it is merely a *means* to satisfy needs external to it. Its alien character emerges clearly in the fact that as soon as no physical or other compulsion exists, labour is shunned like the plague. External labour, labour in which man alienates himself, is a labour of self-sacrifice, of mortification. Lastly, the external character of labour for the worker appears in the fact that it is not his own, but someone else's, that it does not belong to him, that in it he belongs, not to himself, but to another. Just as in religion the spontaneous activity of the human imagination, of the human brain and the human heart, operates independently of the individual—that is, operates on him as an alien, divine or diabolical activity—in the same way the worker's activity is not his spontaneous activity. It belongs to another; it is the loss of his self.

As a result, therefore, man (the worker) no longer feels himself to be freely active in any but his animal functions—eating, drinking, procreating, or at most in his dwelling and in dressing-up, etc.; and in his human functions he no longer feels himself to be anything but an animal. What is animal becomes human and what is human becomes animal.

Certainly eating, drinking, procreating, etc., are also genuinely human functions. But in the abstraction which separates them from the sphere of all other human activity and turns them into sole and ultimate ends, they are animal.

We have considered the act of estranging practical human activity, labour, in two of its aspects. (1) The relation of the worker to the *product of labour* as an alien object exercising power over him. This relation is at the same time the relation to the sensuous external world, to the objects of nature as an alien world antagonistically opposed to him. (2) The relation of labour to the *act of production* within the *labour* process. This relation is the relation of the worker to his own activity as an alien activity not belonging to him; it is activity as suffering, strength as weakness, begetting as emasculating, the worker's *own* physical and mental energy, his personal life or what is life other than activity—as an activity which is turned against him, neither depends on nor belongs to him. Here we have *self-estrangement*, as we had previously the estrangement of the *thing.*

We have yet a third aspect of *estranged labour* to deduce from the two already considered.

Man is a species being, not only because in practice and in theory he adopts the species as his object (his own as well as those of other things), but—and this is only another way of expressing it—but also because he treats himself as the actual, living species; because he treats himself as a *universal* and therefore a free being.

The life of the species, both in man and in animals, consists physically in the fact that man (like the animal) lives on inorganic nature; and the more universal man is compared with an animal, the more universal is the sphere of inorganic nature on which he lives. Just as plants, animals, stones, the air, light, etc., constitute a part of human consciousness in the realm of theory, partly as objects of natural science, partly as objects of art—his spiritual inorganic nature, spiritual nourishment which he must first prepare to make it palatable and digestible—so too in the realm of practice they constitute a part of human life and human activity. Physically man lives only on these products of nature, whether they appear in the form of food, heating, clothes, a dwelling, or whatever it may be. The universality of man is in practice manifested precisely in the universality which makes all nature his *inorganic* body—both inasmuch as nature is (1) his direct means of life, and (2) the material, the object, and the instrument of his life-activity. Nature is man's *inorganic body*—nature, that is, in so far as it is not itself the human body. Man *lives* on nature—means that nature is his *body*, with which he must remain in continuous intercourse if he is not to die. That man's physical and spiritual life is linked to nature means simply that nature is linked to itself, for man is a part of nature.

In estranging from man (1) nature, and (2) himself, his own active functions, his life-activity, estranged labour estranges the *species* from man. It turns for him the *life of the species* into a means of individual life. First it estranges the life of the species and individual life, and secondly it makes individual life in its abstract form the purpose of the life of the species, likewise in its abstract and estranged form.

For in the first place labour, *life-activity, productive life* itself, appears to man merely as a *means* of satisfying a need—the need to maintain the physical existence. Yet the productive life is the life of the species. It is life-engendering life. The whole character of a species—its species character—is contained in the character of its life-activity; and free, conscious activity is man's species character. Life itself appears only as *a means to life.*

The animal is immediately identical with its life-activity. It does not distinguish itself from it. It is *its life-activity.* Man makes his life-activity itself the object of his will and of his consciousness. He has conscious life-activity. It is not a determination with which he directly merges. Conscious life-activity directly distinguishes man from animal life-activity. It is just because of this that he is a species being. Or it is only because he is a species being that he is a Conscious Being, i.e., that his own life is an object for him. Only because of that is his activity free activity. Estranged labour reverses this relationship, so that it is just because man is a conscious being that he makes his life-activity, his *essential* being, a mere means to his *existence*.

NOTES

1. "Alienation"—*Entäusserung*
* Tucker translates the title as "Estranged Labour." We have chosen to use the more familiar translation of "Alienated Labour."

The Methodological Position of Symbolic Interactionism

Herbert Blumer

Symbolic interactionism is one of the principal theoretical approaches to the study of social life. In contrast to functionalism's emphasis on group cohesion or stability, or conflict theory's image of society as a dynamic negotiation among competing interest groups, symbolic interactionism investigates how individuals are continually creating and re-creating society from the ground up as they go about their everyday affairs. Drawing on Max Weber's definition of action as behaviour that is subjectively meaningful to the acting individual, Blumer examines the meaning of "symbolic interactionism" and discusses three main premises of this approach to studying human society.

Blumer's attention to meaning highlights the interpretive work done by individuals as they interact with one another and the world. Yet his discussion assumes, rather than directly addresses, the intersubjective nature of social interaction. That is, that meanings are, or can be, shared by participants in the interaction. So the question remains: How exactly is this possible?

The term "symbolic interactionism" has come into use as a label for a relatively distinctive approach to the study of human group life and human conduct.[1] The scholars who have used the approach or contributed to its intellectual foundation are many, and include such notable American figures as George Herbert Mead, John Dewey, W. L. Thomas, Robert E. Park, William James, Charles Horton Cooley, Florian Znaniecki, James Mark Baldwin, Robert Redfield, and Louis Wirth. Despite significant differences in the thought of such scholars, there is a great similarity in the general way in which they viewed and studied human group life. . . .

THE NATURE OF SYMBOLIC INTERACTIONISM

Symbolic interactionism rests in the last analysis on three simple premises. The first premise is that human beings act toward things on the basis of the meanings that the things have for them. Such things include

Source: Blumer, Herbert, SYMBOLIC INTERACTIONISM: PERSPECTIVE & METHOD, © 1969, pp. 1–5. Reprinted by permission of Pearson Education, Inc., Upper Saddle River, NJ.

everything that the human being may note in his world—physical objects, such as trees or chairs; other human beings, such as a mother or a store clerk; categories of human beings, such as friends or enemies; institutions, as a school or a government; guiding ideals, such as individual independence or honesty; activities of others, such as their commands or requests; and such situations as an individual encounters in his daily life. The second premise is that the meaning of such things is derived from, or arises out of, the social interaction that one has with one's fellows. The third premise is that these meanings are handled in, and modified through, an interpretative process used by the person in dealing with the things he encounters. I wish to discuss briefly each of these three fundamental premises.

It would seem that few scholars would see anything wrong with the first premise—that human beings act toward things on the basis of the meanings which these things have for them. Yet, oddly enough, this simple view is ignored or played down in practically all of the thought and work in contemporary social science and psychological science. Meaning is either taken for granted and thus pushed aside as unimportant or it is regarded as a mere neutral link between the factors responsible for human behavior and this behavior as the product of such factors. We can see this clearly in the predominant posture of psychological and social science today. Common to both of these fields is the tendency to treat human behavior as the product of various factors that play upon human beings; concern is with the behavior and with the factors regarded as producing them. Thus, psychologists turn to such factors as stimuli, attitudes, conscious or unconscious motives, various kinds of psychological inputs, perception and cognition, and various features of personal organization to account for given forms or instances of human conduct. In a similar fashion sociologists rely on such factors as social position, status demands, social roles, cultural prescriptions, norms and values, social pressures, and group affiliation to provide such explanations. In both such typical psychological and sociological explanations the meanings of things for the human beings who are acting are either bypassed or swallowed up in the factors used to account for their behavior. If one declares that the given kinds of behavior are the result of the particular factors regarded as producing them, there is no need to concern oneself with the meaning of the things toward which human beings act; one merely identifies the initiating factors and the resulting behavior. Or one may, if pressed, seek to accommodate the element of meaning by lodging it in the initiating factors or by regarding it as a neutral link intervening between the initiating factors and the behavior they are alleged to produce. In the first of these latter cases the meaning disappears by being merged into the initiating or causative factors; in the second case meaning becomes a mere transmission link that can be ignored in favor of the initiating factors.

The position of symbolic interactionism, in contrast, is that the meanings that things have for human beings are central in their own right. To ignore the meaning of the things toward which people act is seen as falsifying the behavior under study. To bypass the meaning in favor of factors alleged to produce the behavior is seen as a grievous neglect of the role of meaning in the formation of behavior.

The simple premise that human beings act toward things on the basis of the meaning of such things is much too simple in itself to differentiate symbolic interactionism—there are several other approaches that share this premise. A major line of difference between them and symbolic interactionism is set by the second premise, which refers to the source of meaning. There are two well-known traditional ways of accounting for the origin of meaning. One of them is to regard meaning as being intrinsic to the thing that has it, as being a natural part of the objective makeup of the thing. Thus, a chair is clearly a chair in itself, a cow a cow, a cloud a cloud, a rebellion a rebellion, and so forth. Being inherent in the thing that has it, meaning needs merely to be disengaged by observing the objective thing that has the meaning. The meaning emanates, so to speak, from the thing and as such there is no process involved in its formation; all that is necessary is to recognize the meaning that is there in the thing. It should be immediately apparent that this view reflects the traditional position of "realism" in philosophy—a position that is widely held and deeply entrenched in the social and psychological sciences. The other major traditional view regards "meaning" as a psychical accretion brought to the thing by the person for whom the thing has meaning. This psychical accretion is treated as being an expression of constituent elements of the person's psyche, mind, or psychological organization. The constituent elements are such things as sensations, feelings, ideas, memories, motives, and attitudes. The meaning of a thing is but the expression of the given psychological elements that are brought into play in connection with the perception of the thing; thus one seeks to explain the meaning of a thing by isolating the particular psychological elements that produce the meaning. One sees this in the somewhat ancient and classical psychological practice of analyzing the meaning of an object by identifying the sensations that enter into perception of that object; or in the contemporary practice of tracing the meaning of a thing, such as let us say prostitution, to the attitude of the person who views it. This lodging of the meaning of things in psychological elements limits the processes of the formation of meaning to whatever processes are involved in arousing and bringing together the given psychological elements that produce the meaning. Such processes are psychological in nature, and include perception, cognition, repression, transfer of feelings, and association of ideas.

Symbolic interactionism views meaning as having a different source than those held by the two dominant views just considered. It does not regard meaning as emanating from the intrinsic makeup of the thing that has meaning, nor does it see meaning as arising through a coalescence of psychological elements in the person. Instead, it sees meaning as arising in the process of interaction between people. The meaning of a thing for a person grows out of the ways in which other persons act toward the person with regard to the thing. Their actions operate to define the thing for the person. Thus, symbolic interactionism sees meanings as social products, as creations that are formed in and through defining activities of people as they interact. This point of view gives symbolic inter-actionism a very distinctive position, with profound implications that will be discussed later.

The third premise mentioned above further differentiates symbolic interactionism. While the meaning of things is formed in the context of social interaction and is derived by the person from that interaction, it is a mistake to think that the use of meaning by a person is but an application of the meaning so derived. This mistake seriously mars the work of many scholars who otherwise follow the symbolic interactionist approach. They fail to see that the use of meanings by a person in his action involves an interpretive process. In this respect they are similar to the adherents of the two dominant views spoken of above—to those who lodge meaning in the objective makeup of the thing that has it and those who regard it as an expression of psychological elements. All three are alike in viewing the use of meaning by the human being in his action as being no more than an arousing and application of already established meanings. As such, all three fail to see that the use of meanings by the actor occurs through *a process of interpretation.* This process has two distinct steps. First, the actor indicates to himself the things toward which he is acting; he has to point out to himself the things that have meaning. The making of such indications is an internalized social process in that the actor is interacting with himself. This interaction with himself is something other than an interplay of psychological elements; it is an instance of the person engaging in a process of communication with himself. Second, by virtue of this process of communicating with himself, interpretation becomes a matter of handling meanings. The actor selects, checks, suspends, regroups, and transforms the meanings in the light of the situation in which he is placed and the direction of his action. Accordingly, inter-pretation should not be regarded as a mere automatic application of established meanings but as a formative process in which meanings are used and revised as instruments for the guidance and formation of action. It is necessary to see that meanings play their part in action through a process of self-interaction.

NOTE

1. The term "symbolic interactionism" is a somewhat barbaric neologism that I coined in an offhand way in an article written in *Man and Society* (Emerson P. Schmidt, ed. New York: Prentice-Hall, 1937). The term somehow caught on and is now in general use.

The Body of the Condemned

Michel Foucault

The nature of social control in modern Western societies has changed over the last two hundred years. Social control refers to the various institutions and procedures developed within a society to create suffering for those who violate societal norms and values and, possibly, to bring them back into line with expected, approved conduct. As this excerpt from Foucault's book *Discipline and Punish* indicates, the change is most dramatically seen in the shift from the spectacle of punishment focused on the body to the disciplined environment of the prison with its focus on improving the "soul." But what does Foucault mean by the "soul" and "the soul is the prison of the body?" And how might his theorizing about these changes apply to other examples of social control in modern society besides the penal system?

I. THE BODY OF THE CONDEMNED

On 2 March 1757 Damiens the regicide was condemned 'to make the *amende honourable* before the main door of the Church of Paris', where he was to be 'taken and conveyed in a cart, wearing nothing but a shirt, holding a torch of burning wax weighing two pounds'; then, 'in the said cart, to the Place de Grève, where, on a scaffold that will be erected there, the flesh will be torn from his breasts, arms, thighs and calves with red-hot pincers, his right hand, holding the knife with which he committed the said parricide, burnt with sulphur, and, on those places where the flesh will be torn away, poured molten lead, boiling oil, burning resin, wax and sulphur melted together and then his body drawn and quartered by four horses and his limbs and body consumed by fire, reduced to ashes and his ashes thrown to the winds' (*Pièces originales...*, 372–4).

. . .

Bouton, an officer of the watch, left us his account: 'The sulphur was lit, but the flame was so poor that only the top skin of the hand was burnt,

Source: DISCIPLINE AND PUNISH by Michel Foucault, English Translation copyright © 1977 by Alan Sheridan (New York: Pantheon). Originally published in French as Surveiller et Punir Copyright © 1975 by Editions Gallimard. Reprinted by permission of Georges Borchardt, Inc., for Editions Gallimard.

and that only slightly. Then the executioner, his sleeves rolled up, took the steel pincers, which has been especially made for the occasion, and which were about a foot and a half long, and pulled first at the calf of the right leg, then at the thigh, and from there at the two fleshy parts of the right arm; then at the breasts. Though a strong, sturdy fellow, this executioner found it so difficult to tear away the pieces of flesh that he set about the same spot two or three times, twisting the pincers as he did so, and what he took away formed at each part a wound about the size of a six-pound crown piece.

'After these tearings with the pincers, Damiens, who cried out profusely, though without swearing, raised his head and looked at himself; the same executioner dipped an iron spoon in the pot containing the boiling potion, which he poured liberally over each wound. Then the ropes that were to be harnessed to the horses were attached with cords to the patient's body; the horses were then harnessed and placed alongside the arms and legs, one at each limb...' (Zevaes, 201).

Eighty years later, Léon Faucher drew up his rules 'for the House of young prisoners in Paris':

'Art. 17. The prisoners' day will begin at six in the morning in winter and at five in summer. They will work for nine hours a day throughout the year. Two hours a day will be devoted to instruction. Work and the day will end at nine o'clock in winter and at eight in summer.

Art. 18. *Rising.* At the first drum-roll, the prisoners must rise and dress in silence, as the supervisor opens the cell doors. At the second drum-roll, they must be dressed and make their beds. At the third, they must line up and proceed to the chapel for morning prayer. There is a five-minute interval between each drum-roll.

Art. 19. The prayers are conducted by the chaplain and followed by a moral or religious reading. This exercise must not last more than half an hour.

Art. 20. *Work.* At a quarter to six in the summer, a quarter to seven in winter, the prisoners go down into the courtyard where they must wash their hands and faces, and receive their first ration of bread. Immediately afterwards, they form into work-teams and go off to work, which must begin at six in summer and seven in winter.

Art. 21. *Meal.* At ten o'clock the prisoners leave their work and go to the refectory; they wash their hands in their courtyards and assemble in divisions. After the dinner, there is recreation until twenty minutes to eleven.

Art. 22. *School.* At twenty minutes to eleven, at the drum-roll, the prisoners form into ranks, and proceed in divisions to the school. The class lasts two hours and consists alternately of reading, writing, drawing and arithmetic.

Art. 23. At twenty minutes to one, the prisoners leave the school, in divisions, and return to their courtyards for recreation. At five minutes to one, at the drum-roll, they form into work-teams.

Art. 24. At one o'clock they must be back in the workshops: they work until four o'clock.

Art. 25. At four o'clock the prisoners leave their workshops and go into the courtyards where they wash their hands and form into divisions for the refectory.

Art. 26. Supper and the recreation that follows it last until five o'clock: the prisoners then return to the workshops.

Art. 27. At seven o'clock in the summer, at eight in winter, work stops; bread is distributed for the last time in the workshops. For a quarter of an hour one of the prisoners or supervisors reads a passage from some instructive or uplifting work. This is followed by evening prayer.

Art. 28. At half-past seven in summer, half-past eight in winter, the prisoners must be back in their cells after the washing of hands and the inspection of clothes in the courtyard; at the first drum-roll, they must undress, and at the second get into bed. The cell doors are closed and the supervisors go the rounds in the corridors, to ensure order and silence' (Faucher, 274-82).

We have, then, a public execution and a time-table. They do not punish the same crimes or the same type of delinquent. But they each define a certain penal style. Less than a century separates them. It was a time when, in Europe and in the United States, the entire economy of punishment was redistributed. It was a time of great 'scandals' for traditional justice, a time of innumerable projects for reform. It saw a new theory of law and crime, a new moral or political justification of the right to punish; old laws were abolished, old customs died out. 'Modern' codes were planned or drawn up: Russia, 1769; Prussia, 1780; Pennsylvania and Tuscany, 1786; Austria, 1788; France, 1791, Year IV, 1808 and 1810. It was a new age for penal justice.

Among so many changes, I shall consider one: the disappearance of torture as a public spectacle. Today we are rather inclined to ignore it; perhaps, in its time, it gave rise to too much inflated rhetoric; perhaps it has been attributed too readily and too emphatically to a process of 'humanization', thus dispensing with the need for further analysis. And, in any case, how important is such a change, when compared with the great institutional transformations, the formulation of explicit, general codes and unified rules of procedure; with the almost universal adoption of the jury system, the definition of the essentially corrective character of the penalty and the tendency, which has become increasingly marked since the nineteenth century, to adapt punishment to the individual offender? Punishment of a less immediately physical kind, a certain discretion in the art of inflicting pain, a combination of more subtle, more subdued sufferings, deprived of their visible display, should not all this be treated as a special case, an incidental effect of deeper changes? And yet the fact remains that a few decades saw the disappearance of the tortured, dismembered, amputated body, symbolically branded on face or shoulder, exposed alive or dead to public view. The body as the major target of penal repression disappeared.

By the end of the eighteenth and the beginning of the nineteenth century, the gloomy festival of punishment was dying out, though there and there it flickered momentarily into life. In this transformation, two processes were at work. They did not have quite the same chronology or the same *raison d'être*. The first was the disappearance of punishment as a spectacle. The ceremonial of punishment tended to decline; it survived only as a new legal or administrative practice.... It was as if the punishment was thought to equal, if not to exceed, in savagery the crime itself, to accustom the spectators to a ferocity from which one wished to divert them, to show them the frequency of crime, to make the executioner resemble a criminal, judges murderers, to reverse roles at the last moment, to make the tortured criminal an object of pity or admiration. As early as 1764, Beccaria remarked: 'The murder that is depicted as a horrible crime is repeated in cold blood, remorselessly' (Beccaria, 101). The public execution is now seen as a hearth in which violence bursts again into flame.

Punishment, then, will tend to become the most hidden part of the penal process. This has several consequences: it leaves the domain of more or less everyday perception and enters that of abstract consciousness; its effectiveness is seen as resulting from its inevitability, not from its visible intensity; it is the certainty of being punished and not the horrifying spectacle of public punishment that must discourage crime; the exemplary mechanics of punishment changes its mechanisms. As a result, justice no longer takes public responsibility for the violence that is bound up with its practice. If it too strikes, if it took kills, it is not as a glorification of its strength, but as an element of itself that it is obliged to tolerate, that it finds difficult to account for. The apportioning of blame is redistributed: in punishment-as-spectacle a confused horror spread from the scaffold; it enveloped both executioner and condemned; and, although it was always ready to invert the shame inflicted on the victim into pity or glory, it often turned the legal violence of the executioner into shame. Now the scandal and the light are to be distributed differently; it is the conviction itself that marks the offender with the unequivocally negative sign: the publicity has shifted to the trial, and to the sentence; the execution itself is like an additional shame that justice is ashamed to impose on the condemned man; so it keeps its distance from the act, tending always to entrust it to others, under the seal of secrecy. It is ugly to be punishable, but there is no glory in punishing. Hence that double system of protection that justice has set up between itself and the punishment it imposes. Those who carry out the penalty tend to become an autonomous sector; justice is relieved of responsibility for it by a bureaucratic concealment of the penalty itself. It is typical that in France the administration of the prisons should for so long have been the responsibility of the Ministry of the Interior, while responsibility for the *bagnes*, for penal servitude in the convict ships and penal settlements, lay with the Ministry of the Navy or the Ministry of the Colonies. And beyond this distribution of roles operates a theoretical disavowal: do not imagine that the sentences that we judges pass are activated by a desire to punish; they are intended to correct, reclaim, 'cure'; a technique of improvement represses, in the penalty, the strict expiation of evil-doing, and relieves the magistrates of the demeaning task of punishing. In modern justice and on the part of those who dispense it there is a shame in punishing, which does not always preclude zeal. This sense of shame is constantly growing: the psychologists and the minor civil servants of moral orthopaedics proliferate on the wound it leaves.

The disappearance of public executions marks therefore the decline of the spectacle; but it also marks a slackening of the hold on the body.... [G]enerally speaking, punitive practices had become more reticent. One no longer touched the body, or at least as little as possible, and then only to reach something other than the body itself. It might be objected that imprisonment, confinement, forced labour, penal servitude, prohibition from entering certain areas, deportation—which have occupied so important a place in modern penal systems—are 'physical' penalties: unlike fines, for example, they directly affect the body. But the punishment–body relation is not the same as it was in the torture during public executions. The body now serves as an instrument or intermediary: if one intervenes upon it to imprison it, or to make it work, it is in order to deprive the individual of a liberty that is regarded both as a right and as property. The body, according to this penalty, is caught up in a system of constraints and privations, obligations and prohibitions. Physical pain, the pain of the body itself, is no longer the constituent element of the penalty. From being an art of unbearable sensations punishment has become an economy of suspended rights. If it is still necessary for the law to reach and manipulate the body of the convict, it will be at a distance, in the proper way, according to strict rules, and with a much 'higher' aim. As a result of this new restraint, a whole army of technicians took over from the executioner, the immediate anatomist of pain: warders, doctors, chaplains, psychiatrists, psychologists, educationalists; by their very presence near the prisoner, they sing the praises that the law needs: they reassure it that the body and pain are not the ultimate objects of its punitive action. Today a doctor must watch over those condemned to death, right up to the last moment—thus juxtaposing himself as the agent of welfare, as the alleviator of pain, with the official whose task it is to end life. This is worth thinking about. When the moment of execution approaches, the patients are injected with tranquillizers. A utopia of judicial reticence: take away life, but prevent the patient from feeling it; deprive the prisoner of all rights, but do not inflict pain; impose penalties free of all pain. Recourse to psycho-pharmacology and to various physiological 'disconnectors', even if it is temporary, is a logical consequence of this 'non-corporal' penalty.

The modern rituals of execution attest to this double process: the disappearance of the spectacle and the elimination of pain. The same movement has affected the various European legal systems, each at its own rate: the same death for all—the execution no longer bears the specific mark of the crime or the social status of the criminal; a death that lasts only a moment—no torture must be added to it in advance, no further actions performed upon the corpse; an execution that affects life rather than the body.

. . .

Punishment had no doubt ceased to be centred on torture as a technique of pain; it assumed as its principal object loss of wealth or rights. But a punishment like forced labour or even imprisonment—mere loss of liberty—has never functioned without a certain additional element of punishment that certainly concerns the body itself: rationing of food, sexual deprivation, corporal punishment, solitary confinement. Are these the unintentional, but inevitable, consequence of imprisonment? In fact, in its most explicit practices, imprisonment has always involved a certain degree of physical pain. The criticism that was often levelled at the penitentiary system in the early nineteenth century (imprisonment is not a sufficient punishment: prisoners are less hungry, less cold, less deprived in general than many poor people or even workers) suggests a postulate that was never explicitly denied: it is just that a condemned man should suffer physically more than other men. It is difficult to dissociate punishment from additional physical pain. What would a non-corporal punishment be?

There remains, therefore, a trace of 'torture' in the modern mechanisms of criminal justice—a trace that has not been entirely overcome, but which is enveloped, increasingly, by the non-corporal nature of the penal system.

The reduction in penal severity in the last 200 years is a phenomenon with which legal historians are well acquainted. But, for a long time, it has been regarded in an overall way as a quantitative phenomenon: less cruelty, less pain, more kindness, more respect, more 'humanity'. In fact, these changes are accompanied by a displacement in the very object of the punitive operation. Is there a diminution of intensity? Perhaps. There is certainly a change of objective.

If the penalty in its most severe forms no longer addresses itself to the body, on what does it lay hold? The answer of the theoreticians—those who, about 1760, opened up a new period that is not yet at an end—is simple, almost obvious. It seems to be contained in the question itself: since it is no longer the body, it must be the soul. The expiation that once rained down upon the body must be replaced by a punishment that acts in depth on the heart, the thoughts, the will, the inclinations. Mably formulated the principle once and for all: 'Punishment, if I may so put it, should strike the soul rather than the body' (Mably, 326).

...

During the 150 or 200 years that Europe has been setting up its new penal systems, the judges have gradually, by means of a process that goes back very far indeed, taken to judging something other than crimes, namely, the 'soul' of the criminal.

And, by that very fact, they have begun to do something other than pass judgement. Or, to be more precise, within the very judicial modality of judgement, other types of assessment have slipped in, profoundly altering its rules of elaboration. Ever since the Middle Ages slowly and painfully built up the great procedure of investigation, to judge was to establish the truth of a crime, it was to determine its author and to apply a legal punishment. Knowledge of the offence, knowledge of the offender, knowledge of the law: these three conditions made it possible to ground a judgement in truth. But now a quite different question of truth is inscribed in the course of the penal judgement. The question is no longer simply: 'Has the act been established and is it punishable?' But also: 'What *is* this act, what *is* this act of violence or this murder? To what level or to what field of reality does it belong? Is it a phantasy, a psychotic reaction, a delusional episode, a perverse action?' It is no longer simply: 'Who committed it?' But: 'How can we assign the causal process that produced it? Where did it originate in the author himself? Instinct, unconscious, environment, heredity?' It is no longer simply: 'What law punishes this offence?' But: 'What would be the most appropriate measures to take? How do we see the future development of the offender? What would be the best way of rehabilitating him?' A whole set of assessing, diagnostic, prognostic, normative judgements concerning the criminal have become lodged in the framework of penal judgement. Another truth has penetrated the truth that was required by the legal machinery; a truth which, entangled with the first, has turned the assertion of guilt into a strange scientifico-juridical complex. A significant fact is the way in which the question of madness has evolved in penal practice.... [A]lready the reform of 1832, introducing attenuating circumstances, made it possible to modify the sentence according to the supposed degrees of an illness or the forms of semi-insanity. And the practice of calling on psychiatric expertise, which is widespread in the assize courts and sometimes extended to courts of summary jurisdiction, means that the sentence, even if it is always formulated in terms of legal punishment, implies, more or less obscurely, judgements of normality, attributions of causality, assessments of possible changes, anticipations as to the offender's future.... And the sentence that condemns or acquits is not simply a judgement of guilt, a legal decision that lays down punishment; it bears within it an assessment of normality and a technical prescription for a possible normalization. Today the judge—magistrate or juror—certainly does more than 'judge'.

And he is not alone in judging. Throughout the penal procedure and the implementation of the sentence their swarms a whole series of subsidiary authorities. Small-scale legal systems and parallel judges have multiplied around the principal judgement: psychiatric or psychological experts, magistrates concerned with the implementation of sentences, educationalists, members of the prison service, all fragment the legal power to punish; it might be objected that none of them really shares the right to judge; that some, after sentence is passed, have no other right than to implement the punishment laid down by the court and, above all, that others—the experts—intervene before the sentence not to pass judgement, but to assist the judges in their decision....

To sum up, ever since the new penal system—that defined by the great codes of the eighteenth and nineteenth centuries—has been in operation, a general process has led judges to judge something other than crimes; they have been led in their sentences to do something other than judge; and the power of judging has been transferred, in part, to other authorities than the judges of the offence. The whole penal operation has taken on extra-juridical elements and personnel. It will be said that there is nothing extraordinary in this, that it is part of the destiny of the law to absorb little by little elements that are alien to it. But what is odd about modern criminal justice is that, although it has taken on so many extra-juridical elements, it has done so not in order to be able to define them juridically and gradually to integrate them into the actual power to punish: on the contrary, it has done so in order to make them function within the penal operation as non-juridical elements; in order to stop this operation being simply a legal punishment; in order to exculpate the judge from being purely and simply he who punishes. 'Of course, we pass sentence, but this sentence is not in direct relation to the crime. It is quite clear that for us it functions as a way of treating a criminal. We punish, but this is a way of saying that we wish to obtain a cure.' Today, criminal justice functions and justifies itself only by this perpetual reference to something other than itself, by this unceasing reinscription in non-juridical systems....

But we can surely accept the general proposition that, in our societies, the systems of punishment are to be situated in a certain 'political economy' of the body: even if they do not make use of violent or bloody punishment, even when they use 'lenient' methods involving confinement or correction, it is always the body that is at issue—the body and its forces, their utility and their docility, their distribution and their submission. It is certainly legitimate to write a history of punishment against the background of moral ideas or legal structures. But can one write such a history against the background of a history of bodies, when such systems of punishment claim to have only the secret souls of criminals as their objective?

Historians long ago began to write the history of the body. They have studied the body in the field of historical demography or pathology; they have considered it as the seat of needs and appetites, as the locus of physiological processes and metabolisms, as a target for the attacks of germs or viruses; they have shown to what extent historical processes were involved in what might seem to be thet purely biological base of existence; and what place should be given in the history of society to biological 'events' such as the circulation of bacilli, or the extension of the life-span (cf. Le Roy-Ladurie). But the body is also directly involved in a political field; power relations have an immediate hold upon it; they invest it, mark it, train it, torture it, force it to carry out tasks, to perform ceremonies, to emit signs. This political investment of the body is bound up, in accordance with complex reciprocal relations, with its economic use; it is largely as a force of production that the body is invested with relations of power and domination; but, on the other hand, its constitution as labour power is possible only if it is caught up in a system of subjection (in which need is also a political instrument meticulously prepared, calculated and used); the body becomes a useful force only if it is both a productive body and a subjected body. This subjection is not only obtained by the instruments of violence or ideology; it can also be direct, physical, pitting force against force, bearing on material elements, and yet without involving violence; it may be calculated, organized, technically thought out; it may be subtle, make use neither of weapons nor of terror and yet remain of a physical order. That is to say, there may be a 'knowledge' of the body that is not exactly the science of its functioning, and a mastery of its forces that is more than the ability to conquer them: this knowledge and this mastery constitute what might be called the political technology of the body. Of course, this technology is diffuse, rarely formulated in continuous, systematic discourse; it is often made up of bits and pieces; it implements a disparate set of tools or methods. In spite of the coherence of its results, it is generally no more than a multiform instrumentation. Moreover, it cannot be localized in a particular type of institution or state apparatus. For they have recourse to it; they use, select or impose certain of its methods. But, in its mechanisms and its effects, it is situated at a quite different level. What the apparatuses and institutions operate is, in a sense, a micro-physics of power, whose field of validity is situated in a sense between these great functionings and the bodies themselves with their materiality and their forces.

. . .

[O]ne might imagine a political 'anatomy'. This would not be the study of a state in terms of a 'body' (with its elements, its resources and its forces), nor would it be the study of the body and its surroundings in terms of a small state. One would be concerned with the 'body politic', as a set of material elements and techniques that serve as weapons, relays, communication routes and supports for the power and knowledge relations that invest human bodies and subjugate them by turning them into objects of knowledge.

It is a question of situating the techniques of punishment—whether they seize the body in the ritual of public torture and execution or whether they are addressed to the soul—in the history of this body politic; of considering penal practices less as a consequence of legal theories than as a chapter of political anatomy.

. . .

It would be wrong to say that the soul is an illusion, or an ideological effect. On the contrary, it exists, it has a reality, it is produced permanently around, on, within the body by the functioning of a power that is exercised on those punished—and, in a more general way, on those one supervises, trains and corrects, over madmen, children at home and at school, the colonized, over those who are stuck at a machine and supervised for the rest of their lives. This is the historical reality of this soul, which, unlike the soul represented by Christian theology, is not born in sin and subject to punishment, but is born rather out of methods of punishment, supervision and constraint. This real, non-corporal soul is not a substance; it is the element in which are articulated the effects of a certain type of power and the reference of a certain type of knowledge, the machinery by which the power relations give rise to a possible corpus of knowledge, and knowledge extends and reinforces the effects of this power. On this reality-reference, various concepts have been constructed and domains of analysis carved out: psyche, subjectivity, personality, consciousness, etc.; on it have been built scientific techniques and discourses, and the moral claims of humanism. But let there be no misunderstanding: it is not that a real man, the object of knowledge, philosophical reflection or technical intervention, has been substituted for the soul, the illusion of the theologians. The man described for us, whom we are invited to free, is already in himself the effect of a subjection much more profound than himself. A 'soul' inhabits him and brings him to existence, which is itself a factor in the mastery that power exercises over the body. The soul is the effect and instrument of a political anatomy; the soul is the prison of the body.

That punishment in general and the prison in particular belong to a political technology of the body is a lesson that I have learnt not so much from history as from the present. In recent years, prison revolts have occurred throughout the world. There was certainly something paradoxical about their aims, their slogans and the way they took place. They were revolts against an entire state of physical misery that is over a century old: against cold, suffocation and overcrowding, against decrepit walls, hunger, physical maltreatment. But they were also revolts against model prisons, tranquillizers, isolation, the medical or educational services. Were they revolts whose aims were merely material? Or contradictory revolts: against the obsolete, but also against comfort; against the warders, but also against the psychiatrists? In fact, all these movements— and the innumerable discourses that the prison has given rise to since the early nineteenth century—have been about the body and material things. What has sustained these discourses, these memories and invectives are indeed those minute material details. One may, if one is so disposed see them as no more than blind demands or suspect the existence behind them of alien strategies. In fact, they were revolts, at the level of the body, against the very body of the prison. What was at issue was not whether the prison environment was too harsh or too aseptic, too primitive or too efficient, but its very materiality as an instrument and vector of power; it is this whole technology of power over the body that the technology of the 'soul'—that of the educationalists, psychologists and psychiatrists— fails either to conceal or to compensate, for the simple reason that it is one of its tools.

References

Beccaria, C. de, *Traité des délits et des peines.* 1764, ed. 1856.
Faucher, L, *De la réforme des prisons.*1838.
Le Roy-Ladurie, E., *Contrepointe.*1973.
Le Roy-Ladurie, E., "L'histoire immobile", *Annales,* May-June, 1974.
Mably, G. de, *De la législation,* Oeuvres complètes, IX, 1789.
Pièces originales et procédures du procès fait à Robert-François Damiens, III, 1757.
Zevaes, A. L., *Damiens le régicide,*1937.

Women's Experience as a Radical Critique of Sociology

Dorothy Smith

In this reading Dorothy Smith begins her critical inquiry into "the conceptual practices of power" within society by starting with, and reflecting on, where she is—doing and teaching sociology in the university. In particular Smith wants to consider how sociology participates in the way governing in the broadest sense occurs in our society, and how students are customarily taught to think and do as they learn sociology. What exactly is Smith most critical of regarding mainstream sociology? What are her recommendations for an alternative approach to thinking about, and experiencing, the social world?

Although Smith is challenging how sociologists have customarily studied and participated in society, are her criticisms also relevant to the profession, trade, or occupation you are preparing for, or already participating in, including the occupation of student?

RELATIONS OF RULING AND OBJECTIFIED KNOWLEDGE

When I speak here of governing or ruling I mean something more general than the notion of government as political organization. I refer rather to that total complex of activities, differentiated into many spheres, by which our kind of society is ruled, managed, and administered. It includes what the business world calls *management*, it includes the professions, it includes government and the activities of those who are selecting, training, and indoctrinating those who will be its governors. The last includes those who provide and elaborate the procedures by which it is governed and develop methods for accounting for how it is done— namely, the business schools, the sociologists, the economists. These are the institutions through which we are ruled and through which we, and I emphasize this *we*, participate in ruling.

Source: Dorothy Smith, "Women's Experience as a Radical Critique of Sociology," from *The Conceptual Practices of Power: A Feminist Sociology of Knowledge.* © Dorothy Smith, 1990. University of Toronto Press, 1990. Reprinted with permission of the publisher.

Sociology, then, I conceive as much more than a gloss on the enterprise that justifies and rationalizes it, and at the same time as much less than "science." The governing of our kind of society is done in abstract concepts and symbols, and sociology helps create them by transposing the actualities of people's lives and experience into the conceptual currency with which they can be governed.

Thus the relevances of sociology are organized in terms of a perspective on the world, a view from the top that takes for granted the pragmatic procedures of governing as those that frame and identify its subject matter. Issues are formulated because they are administratively relevant, not because they are significant first in the experience of those who live them. The kinds of facts and events that matter to sociologists have already been shaped and given their character and substance by the methods and practice of governing. Mental illness, crimes, riots, violence, work satisfaction, neighbors and neighborhoods, motivation, and so on— these are the constructs of the practice of government. Many of these constructs, such as mental illness, crimes, or neighborhoods, are constituted as discrete phenomena in the institutional contexts of ruling; others arise as problems in relation to the actual practice of government or management (for example, concepts of violence, motivation, or work satisfaction).

. . .

As . . . students . . . we learn to think sociology as it is thought and to practice it as it is practiced. We learn that some topics are relevant and others are not. We learn to discard our personal experience as a source of reliable information about the character of the world and to confine and focus our insights within the conceptual frameworks and relevances of the discipline. Should we think other kinds of thoughts or experience the world in a different way or with horizons that pass beyond the conceptual, we must discard them or find some way to sneak them in. We learn a way of thinking about the world that is recognizable to its practitioners as the sociological way of thinking.

. . .

An important set of procedures that serve to separate the discipline's body of knowledge from its practitioners is known as *objectivity*. The ethic of objectivity and the methods used in its practice are concerned primarily with the separation of knowers from what they know and in particular with the separation of what is known from knowers' interests, "biases," and so forth, that are not authorized by the discipline. In the social sciences the pursuit of objectivity makes it possible for people to be paid to pursue a knowledge to which they are otherwise indifferent. What they feel and think about society can be kept out of what they are professionally or academically interested in. Correlatively, if they are interested in exploring a topic sociologically, they must find ways of converting their private interest into an objectified, unbiased form.

Sociologists, when they go to work, enter into the conceptually ordered society they are investigating. They observe, analyze, explain, and examine that world as if there were no problem in how it becomes observable to them. They move among the doings of organizations, governmental processes, and bureaucracies as people who are at home in that medium. The nature of that world itself, how it is known to them, the conditions of its existence, and their relation to it are not called into question. Their methods of observation and inquiry extend into it as procedures that are essentially of the same order as those that bring about the phenomena they are concerned with. Their perspectives and interests may differ, but the substance is the same. They work with facts and information that have been worked up from actualities and appear in the form of documents that are themselves the product of organizational processes, whether their own or those of some other agency. They fit that information back into a framework of entities and organizational processes which they take for granted as known, without asking how it is that they know them or by what social processes the actual events—what people do or utter—are construed as the phenomena known.

Where a traditional gender division of labor prevails, men enter the conceptually organized world of governing without a sense of transition. The male sociologist in these circumstances passes beyond his particular and immediate setting (the office he writes in, the libraries he consults, the streets he travels, the home he returns to) without attending to the shift in consciousness. He works in the very medium he studies.

But, of course, like everyone else, he exists in the body in the place in which it is. This is also then the place of his sensory organization of immediate experience; the place where his coordinates of here and now, before and after, are organized around himself as center; the place where he confronts people face to face in the physical mode in which he expresses himself to them and they to him as more and other than either can speak. This is the place where things smell, where the irrelevant birds fly away in front of the window, where he has indigestion, where he dies. Into this space must come as actual material events—whether as sounds of speech, scratchings on the surface of paper, which he constitutes as text, or directly—anything he knows of the world. It has to happen here somehow if he is to experience it at all.

Entering the governing mode of our kind of society lifts actors out of the immediate, local, and particular place in which we are in the body. What becomes present to us in the governing mode is a means of passing beyond the local into the conceptual order. This mode of governing creates, at least potentially, a bifurcation of consciousness. It establishes two modes of knowing and experiencing and doing, one located in the body and in the space it occupies and moves in, the other passing beyond it. Sociology is written in and aims at the latter mode of action. Robert Bierstedt writes, "Sociology can liberate the mind from time and space themselves and remove it to a new and transcendental realm where it no longer depends upon these Aristotelian categories."[1] Even observational work aims at description in the categories and hence conceptual forms of the "transcendental realm." Yet the local and particular site of knowing that is the other side of the bifurcated consciousness has not been a site for the development of systematic knowledge.

The suppression of the local and particular as a site of knowledge has been and remains gender organized. The domestic sites of women's work, traditionally identified with women, are outside and subservient to this structure. Men have functioned as subjects in the mode of governing; women have been anchored in the local and particular phase of the bifurcated world. It has been a condition of a man's being able to enter and become absorbed in the conceptual mode, and to forget the dependence of his being in that mode upon his bodily existence, that he does not have to focus his activities and interests upon his bodily existence. Full participation in the abstract mode of action requires liberation from attending to needs in the concrete and particular. The organization of work in managerial and professional circles depends upon the alienation of subjects from their bodily and local existence. The structure of work and the structure of career take for granted that these matters have been provided for in such a way that they will not interfere with a man's action and participation in that world. Under the traditional gender regime, providing for a man's liberation from Bierstedt's Aristotelian categories is a woman who keeps house for him, bears and cares for his children, washes his clothes, looks after him when he is sick, and generally provides for the logistics of his bodily existence.

Women's work in and around professional and managerial settings performs analogous functions. Women's work mediates between the abstracted and conceptual and the material form in which it must travel to communicate. Women do the clerical work, the word processing, the interviewing for the survey; they take messages, handle the mail, make appointments, and care for patients. At almost every point women mediate for men at work the relationship between the conceptual mode of action and the actual concrete forms in which it is and must be realized, and the actual material conditions upon which it depends.

Marx's concept of alienation is applicable here in a modified form. The simplest formulation of alienation posits a relation between the work individuals do and an external order oppressing them in which their work contributes to the strength of the order that oppresses them. This is the situation of women in this relation. The more successful women are in mediating the world of concrete particulars so that men do not have to become engaged with (and therefore conscious of) that world as a condition to their abstract activities, the more complete men's absorption in it and the more effective its authority. The dichotomy between the two worlds organized on the basis of gender separates the dual forms of consciousness; the governing consciousness dominates the primary world of a locally situated consciousness but cannot cancel it; the latter is a subordinated, suppressed, absent, but absolutely essential ground of the governing consciousness. The gendered organization of subjectivity dichotomizes the two worlds, estranges them, and silences the locally situated consciousness by silencing women.

. . .

KNOWING A SOCIETY FROM WITHIN: A WOMAN'S PERSPECTIVE

An alternative sociological approach must somehow transcend this contradiction without reentering Bierstedt's "transcendental realm." Women's standpoint, as I am analyzing it here, discredits sociology's claim to constitute an objective knowledge independent of the sociologist's situation. Sociology's conceptual procedures, methods, and relevances organize its subject matter from a determinate position in society. This critical disclosure is the basis of an alternative way of thinking sociology. If sociology cannot avoid being situated, then it should take that as its beginning and build it into its methodological and theoretical strategies. As it is now, these strategies separate a sociologically constructed world from that of direct experience; it is precisely that separation that must be undone.

I am not proposing an immediate and radical transformation of the subject matter and methods of the discipline nor the junking of everything that has gone before. What I am suggesting is more in the nature of a reorganization of the relationship of sociologists to the object of our knowledge and of our problematic. This reorganization involves first placing sociologists where we are actually situated, namely, at the beginning of those acts by which we know or will come to know, and second, making our direct embodied experience of the everyday world the primary ground of our knowledge.

A sociology worked on in this way would not have as its objective a body of knowledge subsisting in and of itself; inquiry would not be justified by its contribution to the heaping up of such a body. We would reject a sociology aimed primarily at itself. We would not be interested in contributing to a body of knowledge whose uses are articulated to relations of ruling in which women participate only marginally, if at all. The professional sociologist is trained to think in the objectified modes of sociological discourse, to think sociology as it has been and is thought; that training and practice has to be discarded. Rather, as sociologists we would be constrained by the actualities of how things come about in people's direct experience, including our own. A sociology for women would offer a knowledge of the social organization and determinations of the properties and events of our directly experienced world.[2] Its analyses would become part of our ordinary interpretations of the experienced world, just as our experience of the sun's sinking below the horizon is transformed by our knowledge that the world turns away from a sun that seems to sink.

The only way of knowing a socially constructed world is knowing it from within. We can never stand outside it. A relation in which sociological phenomena are objectified and presented as external to and independent of the observer is itself a special social practice also known from within. The relation of observer and object of observation, of sociologist to "subject," is a specialized social relationship. Even to be a stranger is to enter a world constituted from within as strange. The strangeness itself is the mode in which it is experienced.

When Jean Briggs[3] made her ethnographic study of the ways in which an Eskimo people structure and express emotion, what she learned emerged for her in the context of the actual developing relations between her and the family with whom she lived and other members of the group. Her account situates her knowledge in the context of those relationships and in the actual sites in which the work of family subsistence was done. Affections, tensions, and quarrels, in some of which she was implicated, were the living texture in which she learned what she describes. She makes it clear how this context structured her learning and how what she learned and can speak of became observable to her.

Briggs tells us what is normally discarded in the anthropological or sociological telling. Although sociological inquiry is necessarily a social relation, we have learned to dissociate our own part in it. We recover only the object of our knowledge as if it stood all by itself. Sociology does not provide for seeing that there are always two terms to this relation. An alternative sociology must preserve in it the presence, concerns, and experience of the sociologist as knower and discover.

To begin from direct experience and to return to it as a constraint or "test" of the adequacy of a systematic knowledge is to begin from where we are located bodily. The actualities of our everyday world are already socially organized. Settings, equipment, environment, schedules, occasions, and so forth, as well as our enterprises and routines, are socially produced and concretely and symbolically organized prior to the moment at which we enter and at which inquiry begins. By taking up a standpoint in our original and immediate knowledge of the world, sociologists can make their discipline's socially organized properties first observable and then problematic.

When I speak of *experience* I do not use the term as a synonym for *perspective*. Nor in proposing a sociology grounded in the sociologist's actual experience am I recommending the self-indulgence of inner exploration or any other enterprise with self as sole focus and object. Such subjectivist interpretations of *experience* are themselves an aspect of that organization of consciousness that suppresses the locally situated side of the bifurcated consciousness and transports us straight into mind country, stashing away the concrete conditions and practices upon which it depends. We can never escape the circles of our own heads if we accept that as our territory. Rather, sociologists' investigation of our directly experienced world as a problem is a mode of discovering or rediscovering the society from within. We begin from our own original but tacit knowledge and from within the acts by which we bring it into our grasp in making it observable and in understanding how it works. We aim not at a reiteration of what we already (tacitly) know, but at an exploration of what passes beyond that knowledge and is deeply implicated in how it is.

SOCIOLOGY AS STRUCTURING RELATIONS BETWEEN SUBJECT AND OBJECT

Our knowledge of the world is given to us in the modes by which we enter into relations with the object of knowledge. But in this case the object of our knowledge is or originates in the co-ordering of activities among "subjects." The constitution of an objective sociology as an authoritative version of how things [are] done from a position in and as part of the practices of ruling in our kind of society. Our training as sociologists teaches us to ignore the uneasiness at the junctures where multiple and diverse experiences are transformed into objectified forms. That junction shows in the ordinary problems respondents have of fitting their experience of the world to the questions in the interview schedule. The sociologist who is a woman finds it hard to preserve this exclusion, for she discovers, if she will, precisely that uneasiness in her relation to her discipline as a whole. The persistence of the privileged sociological version (or versions) relies upon a substructure that has already discredited and deprived of authority to speak the voices of those who know the society differently. The objectivity of a sociological version depends upon a special relationship with others that makes it easy for sociologists to remain outside the others' experience and does not require them to recognize that experience as a valid contention.

Riding a train not long ago in Ontario I saw a family of Indians— woman, man, and three children—standing together on a spur above a river watching the train go by. I realized that I could tell this incident—the train, those five people seen on the other side of the glass—as it was, but that my description was built on my position and my interpretations. I have called them "Indians" and a family; I have said they were watching the train. My understanding has already subsumed theirs. Everything may have been quite different for them. My description is privileged to stand as what actually happened because theirs is not heard in the contexts in which I may speak. If we begin from the world as we actually experience it, it is at least possible to see that we are indeed located and that what we know of the other is conditional upon that location. There are and must be different experiences of the world and different bases of experience. We must not do away with them by taking advantage of our privileged speaking to construct a sociological version that we then impose upon them as their reality. We may not rewrite the other's world or impose upon it a conceptual framework that extracts from it what fits with ours. Their reality, their varieties of experience, must be an unconditional datum. It is the place from which inquiry begins.

    . . .

The aim of an alternative sociology would be to explore and unfold the relations beyond our direct experience that shape and determine it. An alternative sociology would be a means to anyone of understanding how the world comes about for us and how it is organized so that it happens to us as it does in our experience. An alternative sociology, from the standpoint of women, makes the everyday world its problematic.

THE STANDPOINT OF WOMEN AS A PLACE TO START

The standpoint of women situates the inquirer in the site of her bodily existence and in the local actualities of her working world. It is a standpoint that positions inquiry but has no specific content. Those who undertake inquiry from this standpoint begin always from women's experience as it is for women. We are the authoritative speakers of our experience. The standpoint of women situates the sociological subject prior to the entry into the abstracted conceptual mode, vested in texts, that is the order of the relations of ruling. From this standpoint, we know the everyday world through the particularities of our local practices and activities, in the actual places of our work and the actual time it takes. In making the everyday world problematic we also problematize the everyday localized practices of the objectified forms of knowledge organizing our everyday worlds.

A bifurcated consciousness is an effect of the actual social relations in which we participate as part of a daily work life. Entry as subject into the social relations of an objectified consciousness is itself an organization of actual everyday practices. The sociology that objectifies society and social relations and transforms the actualities of people's experience into the synthetic objects of its discourse is an organization of actual practices and activities. We know and use practices of thinking and inquiring sociologically that sever our knowledge of society from the society we know as we live and practice it. The conceptual practices of an alienated knowledge of society are also in and of the everyday world. In and through its conceptual practices and its everyday practices of reading and writing, we enter a mode of consciousness outside the everyday site of our bodily existence and experiencing. The standpoint of women, or at least, *this* standpoint of women at work, in the traditional ways women have worked and continue to work, exposes the alienated knowledge of the relations of ruling as the everyday practices of actual individuals. Thus, though an alienated knowledge also alienates others who are not members of the dominant white male minority, the standpoint of women distinctively opens up for exploration the conceptual practices and activities of the extralocal, objectified relations of ruling as what actual people do.

NOTES

1. Robert Bierstedt, "Sociology and general education," in *Sociology and contemporary education*, ed. Charles H. Page (New York: Random House, 1966).
2. Dorothy E. Smith, *The everyday world as problematic: A feminist sociology* (Boston: Northeastern University Press, 1987).
3. Jean Briggs, *Never in anger* (Cambridge: Harvard University Press, 1970).

Postmodernity?

Steve Bruce

Modern societies are by their nature characterized by dramatic change since they, unlike traditional forms of society, are organized around the principle of change often valorized as progress. But is the "postmodern world" really a newly emerging one that requires new modes of theorizing, or is it a further extension of the modern world as the modernization process embraces more and more societies around the world?

Although scholars differ in the weight they give to different causes in [their] account of modernization, there is widespread agreement that industrial societies are fundamentally unlike the agrarian societies that preceded them. For most of the twentieth century, there was a further argument about which features of our world were a consequence of industrialization and which were a feature of the *capitalist* form of economy in which our industrialization had taken shape. The class structures, gender relations, patterns of religious observation, and crime rates of capitalist democracies were compared with those of the Communist bloc states. The complete collapse of Communism in the 1980s ended those debates. The attempt to see which parts of our past were somehow 'essential' and which were accidental has now shifted to comparison between our past and the present development of Third World countries. It is, for example, now possible to understand better our own history when we see how the development of the nation state, representative politics, and industrialization proceed in the very different context of Singapore, Japan, Korea, and China. The existence of such comparators has certainly contributed to the decline of the confidence of Western sociologists of the 1950s who believed that the history of the West provided a universal template for modernization.

There has also been an important shift in depictions of the West as many scholars (interestingly often philosophers and social theorists rather than sociologists) have argued that, though the above account of

Source: Steve Bruce, "Postmodernity?" from *Sociology: A Very Short Introduction*, Oxford University Press, 2000, pp. 76–79. By permission of Oxford University Press, Inc.

modernization is reasonably accurate for the nineteenth and early twentieth centuries, we have now moved into another epoch: the *postmodern* world. Although there are many strands to 'postmodernism' (an art style before it became a social theory), the basic idea is that individual freedom has combined with increased geographical mobility and better communication to create a world in which 'consumers' select elements of culture from a global cafeteria. Economies based on the production and distribution of things have been superseded by economies based on the production and distribution of ideas and images. Idiosyncratic preference, taste, and choice have extended to the degree that it makes little sense to talk of social formations such as class. An obvious illustration can be found in the matter of accents. Before the 1970s there was a clear association in such societies as the British and American between a certain accent and social prestige. Mass-media broadcasters spoke in the accents of the upper classes. It used to be possible to guess the political party of a politician by accent. British Tories spoke like members of the royal family; Labour politicians spoke in the regional tones of the working class. Such typing is now vastly more difficult. Well-educated middle-class children listen to 'gangsta rap' and other musical styles associated with the black poor of the inner cities arid borrow vocabulary and accent, as well as borrowing dress and posture styles.

In politics, it is no longer possible to 'read off' people's preferences from their position in the class structure. Instead we find a variety of consciously created interest groups: radical student movements, environmental movements, animal rights campaigns, gay rights groups, and women's groups.

The nation state has become impotent. The globalization of trade and finance has radically reduced the ability of states to control their economies. Modern digital technologies of communication have radically reduced the ability of states to control their citizens. States are becoming increasingly subordinate to supra-national entities such as the European Union.

Even the certainties of birth, sex, and death have been blown away by transplant and reproductive innovations. Genetic engineering has given us the power fundamentally to alter the biological bases of identity. We have now cloned a sheep. We will soon clone people. In the postmodern world, nothing is solid. All is flux.

While there is something in such a description, it is grossly exaggerated. It is always useful to remind the intellectuals of London, Paris, and New York that much of the life in the provinces goes on little changed. Satellite TV might be a novel form of communication, but the soap operas we watch are little different from the novels of Dickens. Cheap international travel is now possible, but it takes as long to cross London now as it did when Sherlock Holmes solved the crimes of Victorian

London. The heavy industries of the Ruhr and the Clyde have disappeared, but workers are still organized in trade unions and occupational class still affects people's attitudes, beliefs, and political behaviour. More significantly for the postmodern image of the autonomous consumer, the hard facts of people's lives, their health and longevity, remain heavily determined by class. To give just one example, at the start of the twentieth century working-class boys in London and Glasgow were on average 2.5 inches shorter than their middle-class counterparts. At the end of the century that is still the case. The popularity of divorce and remarriage has made the structure of the modern family more complex than that of its nineteenth-century predecessor, but the family remains the primary unit of reproduction and socialization and, for most of us, it remains a source of great satisfaction and psychic stability. While technological and social innovations have threatened the family, they have also provided resources for retaining the old habits in new times. Cheap high-speed travel allows us to be further away from each other, but it also allows us to regroup frequently. As we see when states limit trade, set quotas on immigration, and haggle over which will play what part in the 'international community's' response to this or that political crisis, the announcement of the passing of the nation state is, to say the least, premature.

It may well be that the modern societies that have preoccupied sociology were so dependent for their character on industrial manufacturing that a shift to an economy based on technological expertise and exchange will bring about such far-reaching changes to society and culture that we will at the start of the twenty-first century be justified in claiming a new epoch. But at present such designation seems premature and masks the fact that many of the changes heralded as 'post-modern' are only extensions of the features of the modern world that fascinated Marx, Weber, and Durkheim.

PART THREE

Being and Becoming Social: Self, Social Action, and Culture

In the introduction to Section II we highlighted some differences between social problems and fundamental sociological problems. These kinds of sociological problems are puzzling and enduring in that they are inherent in the very existence of human societies.

One of the most enduring questions is the nature of human sociality itself. While we routinely point out that humans are social beings, live in groups, and are interdependent creatures, understanding humans' social nature, and what it means to act and interact socially remains, in many ways, a mystery. For example, if social interaction presumes relations between or among selves, two fundamental questions arise: What is a "self"? How does it emerge?

While in ordinary language we use the term in a variety of ways, in sociology *self* takes on a particular meaning. It is bound up with one's awareness of oneself as distinct from others, as well as one's experience of being in the world with and for others. In line with G. H. Mead's well-known formulation, it is not an amorphous "thing" either pervading one's body or located behind one's forehead, but rather is an ongoing socially situated, symbolic process involving the abilities to think and act toward oneself in terms of others. The self is a socially constructed presence in the world which involves a reflective loop, and it is a presence which is conscious—at least on some level—of how one's well-being is bound up with the well-being of the group as a whole.

Understanding human sociality also requires that we address the subject of social interaction and one's participation in groups as social actors. Thinking of individuals as social actors and human behaviour in terms of social action is derived in large part from the work of Max Weber and is reflected in his classic statement:

> In action is included all human behaviour when and in so far as the acting individual attaches subjective meaning to it ... Action is social in so far as, by virtue of the subjective meaning attached to it by the acting individual (or individuals), it takes account of others and is thereby oriented in its course.[1]

Contemporary sociological approaches have drawn heavily on Weber's concept of social action. Treating individuals as *social actors* means regarding them as

- thinking, feeling beings who attach subjective meaning to situations and who act on the basis of that interpretation;
- volitional, historical beings who create, invent and participate in interactions rather than as passive recipients of external and internal stimuli who then respond directly;
- moral beings who are oriented to some value and employ standards of good and bad, right and wrong in their decisions and actions;
- social beings who take others into account along with their anticipated interpretations and reactions;
- symbolic beings whose actions and interactions involve representations of ourselves, others, and the situation itself, and who perform roles and have the capacity to take the role of the other.

Finally, one of the most significant aspects of group life that both enables and is produced by interaction is culture. Why do human groups everywhere have some form of culture? That is, what deep need or requirement of human existence does culture both express and respond to?

If we are to have a human existence—if we are to live as people rather than merely as animals—there must be limits. What we call culture both expresses and represents a collective response to the deep human need for meaningful limits. What is it to act rather than simply behave, to live in a society rather than in a state of nature? Culture provides the resources to distinguish ourselves from others and to define ourselves and, in so doing, reflects a people's quest for value and for meaningful standards for action and choice.

Culture serves not only to morally regulate those living in a society, but to provide "collective re-presentations" and a collective memory of the world and of those within it. These re-presentations precede people's births and survive their deaths and, in so doing, make the past meaningful and the future possible.

A culture locates us in space by providing us with a place (i.e., a meaningful relation to the space we inhabit). It locates us in time by situating us at a temporal point between past and future. And it provides us collectively with an identity. Together, these enable us to belong in a "world" or "life world" and not just exist in nature. Culture limits the realm of human possibilities such that we are something and someone in particular and not just anything and everything that might have existed anywhere at any time.

Insofar as culture represents a collective response to the problem of who, what, when, and where we are, it becomes more understandable why cultural clashes are so often at the heart of severe conflicts. In this age of colliding cultural diversity, migration within and among nation-states, and such rapid, dramatic changes in our worlds, much is at stake.

NOTE

1. Max Weber, *The Theory of Social and Economic Organization,* Free Press, 1947, p. 88.

The Issue of Educational Opportunity

Stephen Richer

As a primary site and instrument of socialization, our education system allegedly provides an equal opportunity for all members of society to fulfill their potential. But is this actually the case? In this excerpt Richer argues that the pedagogy and the curriculum content, particularly at an elementary level, effectively reproduce inequalities persistent in the social order. The uniformity of the organizational structure and overt curriculum content across schools, classrooms and teachers masks a "hidden curriculum" that, says Richer, contributes to the reproduction of the inequality between men and women and the lower and middle classes. Students entering school encounter this hidden curriculum with differing amounts and types of cultural capital, and this, he argues, has a significant effect on their scholastic achievement as well as shaping their life chances generally. Are Richer's findings generalizable to the educational process you have been through?

INTRODUCTION

The issue of equality of opportunity has been a dominant theme in the writings of liberal philosophers since the onset of the industrial age. With the Industrial Revolution came the recognition that not only a nation's natural endowments but also her people could be viewed as part of her resource pool, to be developed to the maximum....

The connection between these ideas and equality of opportunity lies in the notion of emphasizing the potential of all societal members. Only if all individuals had equal access to public schooling could a society's talent pool be fully tapped and maximum societal productivity be attained. Equality of opportunity was thus initially defined in terms of the availability of public education to all, regardless of ascriptive factors such as an individual's race, sex or family of origin....

Source: Stephen Richer, "Equality to Benefit from Schooling: The Issue of Educational Opportunity," from Dennis Forcese and Stephen Richer (eds.) *Social Issues: Sociological Views of Canada* (2nd ed.) Scarborough: Prentice-Hall, 1988: 271–7; 279. Reprinted with permission from the author.

Despite the undoubted centrality of the equality of opportunity value in U.S. and Canadian culture, however, its realization was quite another matter.

Societies vary in the extent to which dominant values are institutionalized in social structure. One set of reasons for this in North America has to do with the persistence of what was essentially a myth about the non-existence of social class....

Inequality for those embracing a conflict perspective results from the ability and desire of certain groups (notably class and status groups) to transmit to their offspring a raft of social skills and values which will ensure them an advantage in the competitive process. This they ostensibly do at home via family socialization. As well, they ensure that the schools will reflect these values through their influential positions as school board members, members of the elite in general, and through active parental intervention in the process of schooling itself. Within this perspective, universalism is typically viewed as an ideology which legitimizes the outcome of the competition, ensuring that the "losers" (members of the lower classes) accept their fate and do not overthrow the system.

[This] framework... can be tested empirically.... One area ... concerns the role of ascriptive factors.... . [A] conflict perspective would predict that their impact would remain fairly constant or even increase over time. Data from both the U.S. and Canada are consistent with the latter position....

An important clue to how one might begin to understand [this is to recall that] both lower-class students and women in nonmonitored segments of the labor market receive lower incomes for the same educational level compared to middle-class students and males. One can classify the possible explanations as emphasizing extra school or school-related factors. Regarding the former, one might argue that family socialization differentially prepares individuals for using their education. That is, middle-class students, through parents and other well-placed contacts, may be better equipped to know both how to present themselves to prospective employers and exactly where in the labor market to display their particular educational credentials. Further, males in general might receive better family training in this regard than females.

It may also be the case, however, that there are salient differences among relevant groups in the nature of formal schooling. That is, there may be qualitative differences in the educational experience by sex and social class which are not captured by the traditional measure of years of schooling attained. One such difference which leaps to mind is the greater probability of the offspring of the upper classes to attend private schools....

Another aspect of schooling ... lies in the idea that within the public sector itself schools vary with respect to the kind of education to which their students are exposed.... [H]owever, support for this contention is quite weak. There are very few studies which speak directly to the issue.

Further, those which do exist point to the similarity (at least at the elementary level) of pedagogical structure and content across various schools rather than to any disparity. Private schools excluded, the nature of teacher-student relationships, and of the formal and "hidden curriculum" appear very similar no matter what kind of school one is considering (Sharp and Green, 1975; Martin, 1976; Brophy and Good, 1974)....

Do we therefore dismiss the notion of the school as reproducer of the social order? The answer is no, but in order to analyze the role of the school in this regard we must be sensitive to the dynamics of *classroom interaction*, particularly at the early elementary level. My contention is that it is precisely the *lack* of interschool and intraschool variation in pedagogical structure and curriculum content, particularly at the elementary level, which is partially responsible for the data alluded to above on inequality of returns to education. I shall argue that *it is the uniformity of schooling juxtaposed against the variability in children which is salient.* I shall take the position that there is an inequality to benefit from the schooling experience due to the exposure of different children to the *same* educational experience.[1] The reasons for focusing on early elementary schooling arises out of the increasing conviction of many researchers in education that it is in these initial years that basic processes are set in motion which to a large extent determine educational and perhaps even occupational mobility. It is at this level that children are the most malleable and hence most vulnerable to the school as an agent of socialization....

CULTURAL CAPITAL AND THE HIDDEN CURRICULUM

Given this desired focus, how can we begin to investigate equality to benefit from education in elementary schools? Two useful concepts are those of cultural capital and the notion of the hidden curriculum. The former, most fully developed by Bourdieu (1964, 1966, 1970) connotes the idea of a differential distribution in society of cultural trappings which are essential for success. Kennett (1973), in summarizing the thrust of Bourdieu's work, explicates five postulates underlying it:

1. Society is essentially a repressive system.
2. There is diffused within society a cultural capital "transmitted by inheritance and invested in order to be cultivated."
3. The education system functions to "discriminate in favour of those who are the inheritors of this cultural capital."
4. The notion of school failure as due to lack of talents, or of groups lacking certain characteristics which makes then unfit for success, is "a mystification, an ideology of the dominant group."
5. Culture has a political function.

There are thus two sides of the cultural capital coin:

1. People vary with respect to their possession of cultural capital.
2. Schools operate within the assumptions underlying this cultural capital.

The latter point leads us to the so-called hidden curriculum, the name given to the bundle of values and norms implicitly transmitted in the schools. To quote Giroux and Penna (1977), the concept refers to "... those unstated norms, values and beliefs that are transmitted to students through the underlying structure of classrooms, as opposed to formally recognized and sanctioned dimensions of classroom experience."

I am convinced that in order for students to be successful in school they would be better to master the hidden rather than the formal curriculum. The point is, of course, that certain students (those already imbued with the cultural capital underlying the hidden curriculum) have a greater capacity to learn its subtleties than other children. In the rest of this paper I shall do two things:

1. Outline the content of the hidden curriculum.
2. Develop links between school success and certain types of children based on compatibility with the demands of the hidden curriculum.

I rely heavily for our discussion on my longitudinal study of Ontario kindergarten children.

CONTENT OF THE HIDDEN CURRICULUM

The study just alluded to was carried out in Ottawa and involved four years of observation in six kindergarten classrooms. With the aid of video tapes and a team of observers, it was possible to collect detailed data on the organization and daily workings of such classrooms. A thematic analysis of the tapes and researchers' diaries led to the following description of what has been termed the "hidden curriculum." Basically, we identified two major aspects of the hidden curriculum: the cognitive (i.e., the way in which school knowledge is organized), and the social, which we discuss at two levels—the societal level and the level of schooling as an institution.

COGNITIVE DIMENSIONS OF THE HIDDEN CURRICULUM

To understand the way knowledge is organized in our society, it is helpful to begin with a typical kindergarten activity schedule.

Figure 1 presents such a schedule for one particular class (although it is very similar to that of other classes observed).[2] The important implicit aspect of this curriculum structure is the relatively clear demarcation

Figure 1
A typical kindergarten schedule (Ottawa school, 1978)

Activity	Space	Time
Good Morning Time (Series of songs, e.g., Good Morning, If You're Happy & You Know It)	Piano area	9:00 – 9:10 am
Demonstration by teacher of how to cut and paste a fire engine	Piano area	9:10 – 9:20 am
Coordination exercises	Piano area	9:20 – 9:25 am
Game (Simon Says) Area in front of doll's house		9:25 – 9:35 am
Walking across hall to French	Hall	9:35 – 9:38 am
French	French teacher's room across the hall	9:38 – 9:50 am
1/2 class—construction of fire engine	Work-table area	9:50 – 10:15 am
Others—Free Play	Doll's house area, Jungle Gym, Block area	
Rotation of above		10:15 – 10:40 am
Snack	Piano area	10:40 – 10:55 am
Show and Tell (children speak a little about materials brought from home)	Piano area	10:55 – 11:00 am
Story (read by teacher)	Piano area	11:00 – 11:15 am
Prepare for going home	Counter area	11:15 am

which exists among subjects, even at this very early level of schooling. As Figure 1 indicates, the day is divided into clearly defined time-space-activity blocks which have virtually no linkages or connections among them. That is, "knowledge" is presented to the child as a set of relatively discrete, self-contained subjects. In Bernstein's words, this exemplifies a strong "classification" type of knowledge organization. Classification refers to "the nature of the differentiation" among curriculum contents—"where classification is strong, contents are well insulated from each other by strong boundaries. Where classification is weak, there is reduced insulation between contents for the boundaries between contents are weak and blurred." (Bernstein, 1971: 49) Such a knowledge code clearly reflects wider trends in industrial societies towards increased specialization in the division of labor (see also Esland, 1971; Young 1971).

The point is that children have to learn in a way that forces them to think in terms of relatively fragmented units of information, as opposed to thinking styles which preserve the gestalt or interconnectedness of the social and/or physical world. We shall return to this issue at a later point.

SOCIAL DIMENSIONS OF THE HIDDEN CURRICULUM

This aspect of the hidden curriculum consists, I suggest, of two major levels of information—information about the society the children will eventually be entering, and information about public schooling as an institution.

The Societal Level

Talcott Parsons, writing about the school class as a social system, delineated the function played by formal schooling in preparing children for life in an industrial society.... [H]e argues convincingly, albeit rather abstractly, that the school weans the child away from the particularism and affectivity characteristic of family life, gradually replacing these with the values of universalism and affective neutrality. The school experience is thus the child's first encounter with the kind of roles he or she will have to engage in when out in the work world (Parsons, 1959). My own study of several kindergarten classes found Parsons' arguments still relevant. Underlying the various classroom activities, even at this initial stage of schooling, can be discerned a set of general values characteristics of the wider society. The major themes involve

1. the ethic of interindividual competition;
2. an emphasis on materialism;
3. the primacy of work over play; and
4. the submission of self.

First, the child from the first day of formal schooling finds himself competing with other children. Differentiation of the children in terms of success of failure is evident in the games played, early printing exercises, proper school comportment and achieving attention from the teacher. Some children do better than others in motor coordination events and are rewarded accordingly. Games such as Simon Says and Cross the river produced a winner—the first, the best in attentiveness and reflex, and the second, the best jumper in the class. By the middle of the second month in the classes observed, stars were allotted for especially neat printing, usually accompanied by verbal praise. Show and Tell, a period where children talked briefly about items brought from home, became a period where they sough to impress the teacher with their favourite doll, toy soldier or truck, this inevitably at the expense of their peers. At a more covert level, the children were placed in the position of competing throughout the day for both teacher approval and attention. Regarding the former, children adhering to the teacher's conception of proper school behaviour were clearly treated differently from those behaving otherwise. While there were children in the class who rejected this competition for a while and refused to participate, by the end of the first two months of school all the children were actively seeking the teacher's approval. This was accomplished by the teacher through various types of rewards and punishments. Regarding competition for teacher attention, Jackson has given a good description of the competition that constantly exists for this commodity. Given the situation of 30 or so children in a teacher-centered communication structure, "delay, denial and interruption" are no doubt inevitable and frequent occurrences (Jackson, 1968).

The institution of Show and Tell, we argue, became a competition along the axis of material possessions. This period, which occurred every day in the classes studied (and in virtually all elementary classes I have observed), was ostensibly established in schools to provide children the opportunity to speak before their peers about an object or objects familiar to them. Confidence in front of others and verbal skills are assumedly enhanced in the process. While these certainly may occur, an unanticipated consequence of Show and Tell would appear to reinforce interindividual competition and simultaneously to inculcate the values of materialism and private property. In a typical session, a child stands before the group exhibiting a toy or watch, or perhaps a new article of clothing. The teacher usually comments positively on its "niceness" and asks various questions about the object; for example, "Where did you get this?" "Who gave it to you?" "What is it supposed to do?" and "Does anyone else here have something like this?" For the child who has many toys and games at home, this activity becomes an exciting one. He or she proudly produces possessions day after day. Rewarded for bringing them by teacher as well as peer attention, he or she cannot help but see the value of material possessions.

The activities for which tangible rewards are allotted provide a clue as to which types of activities are valued in a group. In kindergarten, rewards such as paper stars, animal picture stamps or colored check marks were distributed only for "3 R" type activities; that is, letter and number printing and various puzzle worksheets. Play activities, including games and songs, produced occasional praise if well done but no further recognition. In short, those we would term school work activities were associated with tangible rewards, while play activities were not. This, along with the physical centrality of the teacher's desk, the blackboard, and the children's work area, served, I would argue, to convey to the child the primacy of work over play, a primacy to be reinforced from that point on in his/her life.

Perhaps the least "hidden" aspect of the hidden curriculum is the transmission of what we would call submission of self. The theme of "man" as essentially wild, self-interested and aggressive in the pursuit of his interests appears in the writings of many philosophers and social scientists. Parsons, for example, in *The Structure of Social Action* wrestles with the contradiction between these attributes of man and the existence of order, eventually coming to his solution of the internalization of norms regulating social action (Parsons, 1939). The kindergarten classroom, although at first glance an unlikely arena for the acting out of the Parsonian solution, nevertheless evoked for me time and again the struggle between individual voluntarism and societal constraint. What the child learns here, sometimes painfully, is that the collectivity has precedence over his or her own desires and wishes.

In Parsons' terms, there is a move from a self to a collectivistic orientation (Parsons, 1951). The child learns that he/she must put aside his or her own wishes in favor of the wishes of a recognized authority figure in adult society, in this case the teacher. He/she is to accept as natural and right that people above him/her in a status hierarchy can dictate his or her own behavior, an acceptance that is to be generalized to other organizations he or she will encounter.

From the first day of school, the teachers observed made this their primary task. Convinced, as are the large majority of teachers I have observed, that "You cannot teach them a thing if you can't control them," a great deal of effort was expended on achieving classroom order. Through an elaborate set of rewards and punishments, orderly behavior was reinforced and its opposite punished. The concern of the teachers with their initial inability to effect control was also evident in the teacher interviews conducted regularly after each observation session. Two standard questions were asked during these interviews: "What were your impressions today?" and "Which children stood out today?" The questions were purposely phrased very generally so that teachers could raise anything they wished. Nevertheless, in the first several weeks, both questions were answered with regard to the presence or absence of control on that particular day:

> The class is too big to handle. They don't know enough to sit in groups. We have to have some conformity. While some are sitting at the piano others are wandering around. (September 5)
>
> It was better today, there was a little more control over the children. They didn't abuse the free playtime. The end of the day is a very disappointing time. I just can't control them anymore. You cannot teach them a thing if you can't control them. In all, they are responding to me better as a person of authority and resource. (September 6)
>
> Monday mornings always seem to be good. It's a good teaching day. I'm refreshed from the weekend and the children are too. It gets worse up to the end of the week and by Friday they're right up there. (September 10)

The question "Which children stood out today?" might theoretically have resulted in discussions of particularly "bright" children, of conspicuous articles of clothing, or of children with various physical attributes. Instead, the children who "stood out" for the teacher were those she perceived as behavior problems: "There are three problem boys, John, Mark, Phillip." (September 5) or "Billy—I can't get through to him." (September 6) or "Robert—I can't figure him out. I can't reason with him." (September 7) or "Billy needs constant discipline. I believe things would be better with a smaller class." (September 12)[3]

The relationships between the teacher and six or seven of the children observed in the first month or so thus consisted largely of a contest of wills. The following exchanges, one from September 5 and another from September 8, provide vivid illustrations.

(September 5) Teacher: "Billy, join the group please."
Billy: "I don't feel like coming."
Teacher: "Yes, you do."
Billy: "No, I don't. You might think I do, but I don't."

(September 8) Teacher: "You're not doing what everyone else is doing."
Robert: "I know, I don't want to."
Teacher: "But I want you to. Come on."

It is clear, then, that for the child there is no alternative but to eventually submit. In terms of resources, the contest is inherently unequal, the teacher possessing greater age and physical size, not to mention legitimate authority. The latter is manifested both in the support of fellow professionals as well as in the tacit backing of the larger community who, in effect, grant a mandate to teachers to socialize their sons and daughters.

The outcome, then, undoubtedly functional from a societal perspective, is that the children learn to accept external authority and concomitantly to repress their own egocentric interests, a requisite for later participation in large-scale bureaucratic organizations.

The Institution of Public Schooling

From the first day of kindergarten, the child is presented with a set of values about appropriate school behavior which he/she will encounter time and again through his/her formal education. These are in large part school-specific counterparts of the societal level values alluded to above. School is presented as a place where children compete with one another along the work axis for various rewards which are meted out by an adult to whom they are expected continually to defer. This last aspect, the submission of self, is expressed in schools in terms of a set of normative expectations clustered into the role of student. These expectations are transmitted by the teacher and involve attentiveness, ways of receiving attention and ways of overall comportment.

First, children are expected to listen when the teacher speaks. One of the teachers, for example, experimented with two major techniques of obtaining attention. In the first week, she told the children that when she called "All hands up," they were to stop whatever they were doing, raise their hands and "close their mouths." This signified her desire to address the class. This was abandoned, however, as it proved unsuccessful after the third or fourth attempt. The second technique, eventually adopted, was to require certain types of behavior in particular part of the classroom. We refer here to the association between spatial area and appropriate activity, exemplified in Figure 1. The most relevant area was the piano area. When then children were gathered here, the teacher expected the exhibition of "piano manners." These entailed being quiet, sitting up straight, and listening when others were talking. An excerpt from October 29 is a typical illustration of the way in which the teacher sought to bring about the acceptance of the manners idea:

> Teacher: "Can we have piano manners? (pause) What are piano manners?"
>
> Two or three children: "Sitting up straight and listening."
>
> Teacher: "That is right. If I have people who don't give me piano manners, they'll put their hands on top of their heads."

By the end of the first week in November, piano manners were virtually automatically displayed by the children in this area of the room, and the teacher was able to drop explanatory repetitions of the kind quoted above.

Children were thus expected to give attention on demand to the teacher. They were also expected to conform to a set of rules concerning the *seeking* of attention. As early as September 5, in all the classes the children were told to raise their hands if they wished to speak and not to interrupt when someone else was speaking. It should be pointed out that most of the children were quite familiar with the hand-raising phenomenon before the teacher held forth on it, which implies some anticipatory socialization in the home *vis-à-vis* student behavior.

As for general comportment, the children were to learn that school was a place where one behaves differently than one does away from school. This difference is expressed in one teacher's distinction between "inside" and "outside" voices − school is a place for the former; that is, quiet talk, hushed voices. A set of expected behaviors called "hall manners" by the same teacher are also a good illustration. When the children left the classroom to go to French or Gym, or to a school assembly, they were expected to line up in single file, place their two hands on their head, and walk directly to the appropriate area. In short, as a student one is attentive to the teacher, one seeks recognition through appropriate channels, and one generally behaves docilely.

STUDENT CULTURAL CAPITAL
AND THE HIDDEN CURRICULUM

The argument to be made here is that certain children, because of the cultural capital which they carry, are more compatible than others with the above aspects of schooling. Specifically, it is my contention that middle-class children and females are better "matched" to the demands of schooling than their lower-class and male counterparts. This accounts to a large extent, I suggest, for their differential success in the schools.

There has been much work done in the U.S. on cultural differences among the social classes (classic works are Hyman, 1954; Kohn, 1963, 1969; Kluckhohn and Strodbeck, 1961; Reissman, 1962). Despite some inconsistencies among studies, there is general agreement in this literature that one can differentiate the American lower and middle classes along four lines: attitudes towards achievement; future versus present orientation; extent to which competition versus cooperation characterizes life style; and extent to which materialism is a salient value. With some caution, one can produce a fair amount of evidence consistent with the following: members of the lower as opposed to the middle class are less motivated to achieve, more likely to stress the present as opposed to the future as the major source of rewards, more inclined to value cooperation as opposed to competition, more inclined to respond to material as opposed to symbolic rewards and, related to this latter point, more inclined to settle disputes through physical rather than verbal means.

The relevant research in Canada, although less abundant than that in the U.S., is nevertheless consistent with these typifications. (For general summaries see Elkin and Handel, 1972; Elkin, 1964; Pike and Zureik, 1975; Jones and Selby, 1972.) ...

By way of summarizing the above social class discussion, I am suggesting that certain middle-class characteristics, notably the tendency to defer gratification, the greater concern with interindividual competition, the tendency towards an analytical-goal orientation, the greater reliance on verbal rather than physical skills, and the greater likelihood of responding to nonmaterial rewards, ensure a rather distinct advantage from the first days of formal schooling.

NOTES

1. Clearly, at the secondary level one finds greater differentiation of schools by curriculum content and perhaps teaching style. As I shall be arguing, however, it is the focus on elementary schooling (particularly the early grades) which is most salient.

2. Much of the data reported in this paper came from one classroom, the central one in our study. The findings, however, hold without exception for all the classes observed.

3. Observations made in other classes as well as the literature on classroom research (Jackson, 1968) indicate that the concern with order is by no means unique to this study. Indeed, it seems to pervade the atmosphere in most, if not all, traditional schools.

References

Bernstein, B. 1971 "On the Classification and Framing of Educational Knowledge." In M.F.D.Young, ed., *Knowledge and Control*. London: Collier-MacMillan.

Bourdieu, P. 1966 "L'Ecole conservatrice: Les inégalités devant l'école et devant la culture," *Revue Française de Sociologie, 7*: 325–347.

——. 1970 *La Reproduction*. Paris: Editions de Minuit.

Bourdieu, P. and J. C. Passeron 1964 *Les Héritiers*. Paris: Editions de Minuit.

Brophy, J. E. and T. L. Good 1974 *Teacher-Student Relationships*. New York: Holt, Rinehart and Winston.

Elkin, F. 1964 *The Family in Canada*. Ottawa: Canadian Conference on the Family.

Elkin, F. and G. Handel 1972 *The Child and Society: The Process of Socialization*. New York: Random House.

Esland, G. M. 1971 "Teaching and Learning as the Organization of Knowledge." In MichaelYoung, ed., *Knowledge and Control*.

Giroux, H. and A. Penna 1977 "Social Relations in the Classroom: The Dialectic of the Hidden Curriculum," *Edcentric*, 40–41: 39–46.

Hyman, H. H. 1954 "The Value-System of Different Classes." In R. Bendix and S. Lipset, eds., *Class Status and Power*. London: Routledge and Kegan Paul.

Jackson, P. W. 1968 *Life in the Classroom*. New York: Holt, Rinehart and Winston.

Jones, F. and J. Selby 1972 "School Performance and Social Class." In T. J. Ryan, *Poverty and the Child*. Toronto: McGraw-Hill Ryerson.

Kennett, J. 1973 "The Sociology of Pierre Bourdieu," *Educational Review*, 25.

Kluckholn, F. R. and F. L. Strodtbeck 1961 *Variations in Value Orientations*. Chicago: Row Peterson.

Kohn, M. 1963 "Social Class and Parent-Child Relationships: An Interpretation," *American Journal of Sociology, 68*: 471–480.

Martin, W. 1976 *The Negotiated Order of the School*. Toronto: MacMillan.

Parsons, T. 1939 *The Structure of Social Action*. Glencoe: The Free Press.

——. 1951 *The Social System*. Glencoe: The Free Press.

——. 1959 "The School Class as a Social System," *Harvard Educational Review, 29*: 297–318.

Pike, R. M. and E. Zureik, eds. 1975 *Socialization and Values in Canadian Society, Vol. 2*. Toronto: McClelland and Stewart.

Riessman, F. 1962 *The Culturally Deprived Child*. New York: Harper and Row.

Sharp, R. and A. Green 1975 *Educational and Social Control*. London: Routledge and Kegan Paul.

Young, M. 1971 "Curricula, Teaching and Learning as the Organization of Knowledge." In M. Young, ed., *Knowledge and Control*. London: Collier-MacMillan.

"Even If I Don't Know What I'm Doing I Can Make It Look Like I Know What I'm Doing": Becoming a Doctor in the 1990s

Brenda L. Beagan

There are two themes in this article, the meaning of competence and learning to play a role. The training of doctors has been a research topic for sociologists for almost half a century. Beagan revisited medical schools to find out what hasn't changed over the decades and what has, given that many students now are female, working class, gay and/or of diverse racial and ethnic backgrounds. One thing that hasn't changed is the emphasis on competence. This competence is found to reside less in a technical mastery of scientific methods or knowledge of "objective" facts, than in an ability to appear competent, accomplished especially by demonstrating an understanding of what is expected even if one is unsure of what one is doing. But why does the appearance of competence continue to be of such overriding significance in the training of qualified doctors? And what has changed over the decades?

Socialization does not cease after childhood or adolescence, but is also part of the world of work. The socialization of adults is examined within the context of a formal organization, namely medical school. Beagan neatly outlines the requirements for, and the process of, learning to play a role. But the question remains: What is the relationship of playing a role to the self? That is, what is the relationship of what you do to who you (really) are?

Source: Brenda Beagan, "'Even if I don't know what I'm doing I can make it look like I know what I'm doing': Becoming a doctor in the 1990s," *Canadian Review of Sociology and Anthropology*, 38 (3) 2001: 275–292. Reprinted with permission.

> *For most medical students, a remarkable and*
> *important transformation occurs from the time they*
> *enter medical school to the time they leave.... They*
> *become immersed in the culture, environment and*
> *lifestyle of the school. They slowly lose their initial*
> *identity and become redefined by the new situation.*
> *Medical students have to look for something to hang*
> *on to. And that something is provided: their new*
> *identity as 'doctor.'*
> Shapiro (1987: 27)

When students enter medical school they are lay people with some science background. When they leave four years later they have become physicians; they have acquired specialized knowledge and taken on a new identity of medical professional. What happens in those four years? What processes of socialization go into the making of a doctor?

Most of what we know about how students come to identify as future-physicians derives from research conducted when students were almost exclusively male, white, middle- or upper-class, young and single—for example, the classics *Boys in White* (Becker, Greer, Strauss, & Hughes, 1961) and *Student Physician* (Merton, Reader, & Kendall, 1957). When women and students of colour were present in this research it was in token numbers. Even when women and non-traditional students were present, as in Sinclair's (1997) recent ethnography, their impact on processes of professional identity formation and the potentially distinct impact of professional socialization on these students have been largely unanalysed. What does becoming a doctor look like in a medical school of the late 1990s, where many students are female, are of diverse racial and cultural backgrounds, are working-class, gay and/or parents?

This study draws on survey and interview data from students and faculty at one Canadian medical school to examine the processes of professional identity formation and how they are experienced by diverse undergraduate medical students in the late 1990s. As the results will show, the processes are remarkably unchanged from the processes documented 40 years ago.

. . .

FIRST EXPERIENCES BECOME COMMONPLACE

When identifying how they came to think of themselves as medical students, participants described a process whereby what feels artificial and unnatural initially comes to feel natural simply through repetition. For many students, a series of "first times" were transformative moments.

> Denise:[1] I think there are sort of seminal experiences. The first cut in anatomy, the first time you see a patient die, first time you see a treatment that was really aggressive and didn't work.... First few procedures that I conducted myself, first time I realized that I really did have somebody's life in my hands.... It seems like a whole lot of first times. The first time you take a history, the first time you actually hear the murmur. There are a lot of "Ah-ha!" sort of experiences.

Part of the novelty is the experience of being entitled—even required—to violate conventional social norms, touching patients' bodies, inquiring about bodily functions, probing emotional states: "You have to master a sense that you're invading somebody, and to feel like it's all right to do that, to invade their personal space...."

CONSTRUCTING A PROFESSIONAL APPEARANCE

Students are quite explicitly socialized to adopt a professional appearance: "When people started to relax the dress code a letter was sent to everybody's mailbox, commenting that we were not to show up in jeans, and a tie is appropriate for men." Most students, however, do not require such reminders; they have internalized the requisite standards.

Dressing neatly and appropriately is important to convey respect to patients, other medical staff, and the profession. It probably also helps in patients taking students seriously (survey comment).

Asked whether or not they ever worry about their appearance or dress at the hospital, 41% of the survey respondents said they do not, while 59% said they do.

There were no statistically significant differences by gender, class background or "minority" status, yet gendered patterns emerged when students detailed their concerns in an open-ended question. Most of the men satisfied their concerns about professional appearance with a shave and a collared shirt, perhaps adding a tie: "I do make sure that I am dressed appropriately when I see patients i.e. well-groomed, collared shirt (but no tie)." Women, on the other hand, struggled with the complex messages conveyed by their clothing, trying to look well-dressed yet not convey sexual messages. For women, "dressed up" normally means feminine while a professional image is intended to convey competence. Striking a balance at the intersection can be difficult: "Is it professional enough? Competent looking? ... I do not want to appear 'sexy' on the job." As one student noted, while both men and women sometimes violate standards of professional dress, men's violations tend to involve being too informal; women's may involve dressing too provocatively, thereby sexualizing a doctor-patient encounter.

CHANGES IN LANGUAGE, THINKING AND COMMUNICATION SKILLS

Acquiring a huge vocabulary of new words and old words with new meanings—what one student called "medical-ese"—is one of the central tasks facing medical students, and one of the major bases for examining them (Sinclair, 1997). Students were well aware of adopting the formal language of medicine.

> Dawna: All of a sudden all I can think of is this lingo that people won't understand. My brother told me the other day, "Sometimes I just don't understand what you are talking about anymore." I don't realize it! I'll use technical terms that I didn't think that other people wouldn't know.

The language of medicine is the basis for constructing a new social reality. Even as it allows communication, language constructs "zones of meaning that are linguistically circumscribed" (Berger & Luckmann, 1966: 39). Medical language encapsulates and constructs a worldview wherein reducing a person to body parts, tissues, organs and systems becomes normal, natural, "the only reasonable way to think" (Good & Good, 1993: 98-9). Students described this as learning to pare away "extraneous" information about a patient's life to focus on what is clinically relevant.

> Becky: I see how it happens.... The first day of medicine were just people. We relate by asking everything about a person, just like you'd have a conversation with anybody. And then that sort of changes and you become focussed on the disease ... because right now there's just too much. It's overwhelming. I'm hoping that as I learn more and become more comfortable with what I know and I can apply it without having to consciously go through every step in my mind, that I'll be able to focus on the *person* again.

In part through the language of medicine students learn a scientific gaze that reduces patients to bodies, allowing them to concentrate on what is medically important—disease, procedures, and techniques (Haas & Shaffir, 1987).

Not surprisingly, students may simultaneously lose the communication abilities they had upon entering medical school.

Dr. W.: Their ability to talk to people becomes corrupted by the educational process. They learn the language of medicine but they give up some of the knowledge that they brought in . . . . The knowledge of how to listen to somebody, how to be humble, how to hear somebody else's words . . . . It gets overtaken by the agenda of medical interviewing.

Another faculty member noted that students' communication skills improved significantly during their first term of first year, but "by the end of fourth year they were worse than they had been before medical school."

LEARNING THE HIERARCHY

Key to becoming a medical student is learning to negotiate the complex hierarchy within medicine, with students positioned at the bottom. A few faculty saw this hierarchy as a fine and important tradition facilitating students' learning.

Dr. U.: You're always taught by the person above you. Third-year medical students taught by the fourth-year student . . . . Fourth-year student depends on the resident to go over his stuff. Resident depends on maybe the senior or the chief resident or the staff person. So they all get this hierarchy which is wonderful for learning because the attendings can't deal with everybody.

Students, and most faculty, were far less accepting of this traditional hierarchy—particularly of students' place in it.

Both faculty and students pointed out the compliance the hierarchical structure inculcates in students, discouraging them from questioning those above them.

Dr. G.: If they don't appear compliant and so on they will get evaluated poorly. And if you get evaluated poorly then you might not get a good residency position. There's that sort of thing over their shoulders all of the time . . . the fear.

For students being a "good medical student" means not challenging clinicians.

Valerie: If I ever saw something blatantly sexist or racist or wrong I hope that I would say something. But you get so caught up in basically clamming up, shutting up, and just taking it . . . . Is it going to ruin my career, am I going to end up known as the fink, am I going to not get the [residency] spot that I want because I told?

Though virtually every student described seeing things on the wards that they disagreed with, as long as there was no direct harm to a patient they stayed silent and simply filed away the incident in their collection of "things not to do when I am a doctor."

Other researchers have noted that medical students develop an approach geared to getting along with faculty, pleasing them whatever their demands (Becker et al., 1961: 281; Bloom, 1973: 20; Sinclair, 1997: 29). Some students, however, had *internalized* the norm of not criticizing clinicians, adopting an unspoken "code of silence" not just to appease faculty, but as part of being a good physician. In particular, one should never critique a colleague in front of patients.

> Mark: As students we all critique the professors and our attendings.... But I don't think we'd ever do that in front of a patient. It's never been told to us not to. But most of us wouldn't do that. Even if a patient describes something their doctor has prescribed to them or a treatment they've recommended which you know is totally wrong, maybe even harmful, I think most of us, unless it was really harmful, would tend to ignore it and just accept, "This is the doctor and his patient. What happens between them is okay."

These students had developed a sense of alliance with other members of the profession rather than with lay people and patients—a key to professional socialization. Several faculty referred to good medical students as "good team players' (cf. Sinclair, 1997), invoking a notion of belonging.

> Dr. M.: That sense of belonging, I think, is a sense of belonging to the profession.... You're part of the process of health care.... I mean, you haven't a lot of the responsibility, but at least you're connected with the team.

For some students, too, the desire to present a united front for patients was expressed as being a good team player: "You have to go along with some things ... in front of the patient. For teams it wouldn't be good to have the ranks arguing amongst themselves about the best approach for patient care." To remain good team players, many students, residents and physicians learn to say nothing even when they see colleagues and superiors violating the ethics and standards of the profession; such violations are disregarded as matters of personal style (Light, 1988).

RELATIONSHIP TO PATIENTS

As students are learning their place in the hierarchy within medicine, they are simultaneously learning an appropriate relationship to patients. Within the medical hierarchy students feel powerless at the bottom. Yet in relation to patients even students hold a certain amount of power. In the interviews there were widely diverging views on the degree of professional authority physicians and student-physicians should display.

Some faculty drew a very clear connection between professionalism and the "emotional distancing" Fox documented in medicine in 1957, describing students developing a "hard shell" as a "way of dealing with feelings" to prevent over-identifying with patients. Emotional involvement and over-identification are seen as dangerous; students must strike a balance between empathy and objectivity, learning to overcome or master their emotions (Conrad, 1988; Haas & Shaffir, 1987): "I only become of use if I can create some distance so that I can function."

> Dr. E.: Within the professional job that you have to do, one can be very nice to patients but there's a distancing that says you're not their friend, you're their doctor.

In contrast, several faculty members rejected the "emotional distancing" approach to medicine in favour of one based in egalitarian connection.

> Dr. V.: I reject that way of dealing with it.... When I'm seeing a patient I have to try to get into understanding what's bothering them. And in fact it's a harder job, I mean I need to understand well enough so I can help them to understand. 'Cause the process of healing is self-understanding.

These faculty members talked about recognizing and levelling power or sharing power. They saw professional distancing as the loss of humanitarianism, the adoption of a position of superiority, aloofness, emphasizing that clinicians need to know their patients as something more than a diagnosis. Women were slightly over-represented among those expressing the egalitarian perspective, but several male clinicians also advocated this position.

PLAYING A ROLE GRADUALLY BECOMES REAL

Along with emotional distancing, Fox (1957) identified "training for uncertainty" as key to medical socialization, including the uncertainty arising from not knowing everything, and not knowing enough. Alongside gathering the knowledge and experience that gradually reduces feelings of uncertainty, students also grow to simply tolerate high levels of uncertainty. At the same time they face routine expectations of certainty—from patients who expect them "to know it all" and faculty who often expect them to know far more than they do and who evaluate the students' competence (Haas & Shaffir, 1987). Students quickly learn it is risky to display lack of certainty; impression management becomes a central feature of clinical learning (Conrad, 1988). Haas and Shaffir (1987: 110) conclude that the process of professionalization involves *above all* the successful adoption of a cloak of competence such that audiences are convinced of the legitimacy of claims to competence.

Robert Coombs argues that medical professional socialization is partly a matter of *playing* the role of doctor, complete with the props of white coat, stethoscope, name tag, and clipboard (1978: 222). The symbols mark medical students off as distinct from lay people and other hospital staff, differentiating between We and They. Students spoke of "taking on a role" that initially made them feel like "total frauds," "impostors."

> Erin: It was really role-playing. You were doing all these examinations on these patients which were not going to go into their charts, were not going to ever be read by anybody who was treating the people so it really was just practice. Just play-acting.

They affirmed the importance of the props to successful accomplishment of their role play—even as it enhanced the feeling of artifice: "During third year when we got to put the little white coat on and carry some instruments around the hospital, have a name tag ... it definitely felt like role-playing."

Despite feeling fraudulent, the role play allows students to meet a crucial objective: demonstrating to faculty, clinical instructors, nurses and patients that they know something. They quickly learn to at least look competent.

> Nancy: Even if I don't know what I'm doing I can make it look like I know what I'm doing.... It was my acting in high school.... I get the trust of the patient....

RESPONSES FROM OTHERS

The more students are treated by others as if they really were doctors the more they feel like doctors (cf. Coombs, 1978). In particular, the response from other hospital personnel and patients can help confirm the student's emerging medical professional identity.

> Rina: The more the staff treats you as someone who actually belongs there, that definitely adds to your feeling like you do belong there.... It's like, "Wow! This nurse is paging me and wants to know my opinion on why this patient has no urine output?!"

For many students patients were the single most important source of confirmation for their emerging identity as physicians. With doctors and nurses, students feel they can easily be caught out for what they don't know; with patients they feel fairly certain they can pull off a convincing performance, and they often realize they *do* know more than the average person.

One response from others that has tremendous impact is simply being called doctor by others (Konner, 1987; Shapiro, 1987).... *Not* being called doctor—especially when your peers *are*—can be equally significant. In previous accounts, being white and being male have greatly improved a medical student's chances of being taken for a doctor (Dickstein, 1993; Gamble, 1990; Kirk, 1994; Lenhart, 1993). In this study, although social class background, minority status and first language made no difference, significantly more men than women were *regularly* called doctor and significantly more women had *never* been called doctor.[2]

These data suggest a lingering societal assumption that the doctor is a man. According to the interviews, women medical students and physicians are still often mistaken for nurses. Two of the male students suggested the dominant assumption that a doctor is a man facilitates their establishing rapport with patients and may ease their relationships with those above them in the medical hierarchy: "I've often felt because I fit like a stereotypical white male, that patients might see me as a bit more trustworthy. A bit more what they'd like to see. Who they want to see." Goffman notes that the part of a social performance intended to impress others, which he calls the "front," and which includes clothing, gender, appearance and manner, is predetermined: "When an actor takes on an established social role, usually he finds that a particular front has already been established for it" (1959: 27). In this case it appears that the role doctor, or medical student, still carries an attached assumption of maleness.

SECONDARY SOCIALIZATION: SUBSUMING THE FORMER SELF?

The fact that roles carry with them established expectations heightens the potential for clashes with the identity characteristics of new incumbents. Education processes, inevitably process of secondary socialization, must always contend with individuals' already formed and persistent selves, selves established through primary socialization. As Berger and Luckmann (1966: 129) note, "Whatever new contents are now to be internalized must somehow be superimposed upon this already present reality."

In his study of how medical students put together identities as spouses, parents, and so on with their developing identities as physicians, Broadhead (1983) stresses the need for individuals to "articulate" their various identities to one another, sorting out convergences and divergences of attitudes, assumptions, activities and perspectives that accompany different subject positions.

In this research, most students indicated that medicine had largely taken over the rest of their lives, diminishing their performance of other responsibilities. While 55% of survey respondents thought they were doing a good job of being a medical student, many thought they were doing a poor to very poor job of being a spouse (26%) or family member (37%); 46% gave themselves failing grades as friends. Fewer than a quarter of respondents thought they were doing a good job of being an informed citizen (18%) or member of their religion, if they had one (17%).

What emerged from most interviews and from the survey was a picture of medical school dominating all other aspects of daily life. Overwhelmingly students talked about sacrifice.

> Lew: You just sacrifice so much. I don't know about people who don't have children, but I value my family more than anything, and, and I cannot–I didn't know you had to sacrifice that much.

Many students have given things up, at least temporarily: musical instruments, art, writing, sports activities, volunteer activities. Some students spoke of putting themselves on hold, taking on new medical-student identities by subsuming former identities.

This sacrifice of self-identity can be quite serious. Several faculty and students suggested students from non-Western, non-Caucasian cultural backgrounds need to assimilate: "Students from other cultures leave behind a lot of their culture in order to succeed. There's a trade-off." Similarly, faculty and students suggested gay and lesbian students frequently become more "closeted" as they proceed through undergraduate training. One clinician said of a lesbian fourth-year student, "Now all of a sudden her hair's cut very business-like and the clothes are different. . . . She's fitting into medicine. Medicine isn't becoming a component of her, she's becoming a component of the machine." Some faculty suggested women in medicine may need to relinquish their identity as women in order to fit in as physicians.

> Dr. Q.: The women who are in those positions are white men. You just have to look at the way they dress. They're wearing power suits often with ties, you know, they're really trying to fit the image. [One of the women here] recently retired and in the elevator in the hospital they talked about her as one of the boys. So that's the perception of the men is that this is not a woman, this is one of the boys.

Women, they argued, become more-or-less men during medical training, "almost hyper-masculine in their interactions," "much more like men in terms of thought processes and interactions with people."

In addition to letting go of gender identity, sexual identity and cultural identity, some students described losing connections to their families and old friends after entering medical school. Often this was due to time constraints and diverging interest, but for some there was also a growing social distance as they moved into a new social status and education level. Lance was disconnecting from his working-class family:

> Lance: My family actually were very unsupportive [when I got into medicine]. They didn't even know what I was doing. And there's still this huge gap between them and myself because they don't want to understand what's going on in my world, and their world seems quite simple, simplistic to me .... I see that gap getting larger over time.

Relationships with family, friends from outside of medicine, and anyone else who cannot relate to what students are doing every day are put "on the back burner." Intimate relationships are frequent casualties of medical school.

Thus some students do not or cannot integrate their medical student identities with their former sense of self; rather they let go of parts of themselves, bury them, abandon them, or put them aside, at least for a while. Another option for students who experience incongruities between their medical-student identities and other aspects of themselves is to segregate their lives. Because human beings have the ability to reflect on our own actions, it becomes possible to experience a segment of the self as distinct, to "detach a part of the self and its concomitant reality as relevant only to the role-specific situation in question" (Berger & Luckmann, 1966: 131). In this research 31% of survey respondents felt they are one person at school and another with friends and family. Perhaps as a consequence, many students maintain quite separate groups of friends, within medicine and outside medicine. Indeed, some faculty stressed the importance of maintaining strong outside connections to make it through medical school without losing part of yourself.[3]

DIFFERENCE AS A BASIS FOR RESISTANCE

Elsewhere I have argued that intentional and unintentional homogenizing influences in medical education neutralize the impact of social differences students bring into medicine (Beagan, 2000). Students come to believe that the social class, "race," ethnicity, gender and sexual orientation of a physician is not—and should not be relevant during physician-patient interactions. Nonetheless, at the same time those social differences can provide a basis for critique of and resistance to aspects of medical professional socialization. A study of medical residents found that those most able to resist socialization pressures minimized contact and interaction with others in medicine; maintained outside relationships that supported an alternative orientation to the program; and entered their programs with a "relatively strong and well-defined orientation" (Shapiro & Jones,

1979: 243). Complete resocialization requires "an intense concentration of all significant interaction within the [new social] group" (Berger & Luckmann, 1966: 145); it is also facilitated by minimal contradictions between the previous social world and the new world.

In this research, age played a clear role in students' ability to resist some aspects of professional socialization. Older students usually had careers before medicine, which helped put medical school in a different perspective....

The strongest basis for resisting professional socialization, however, came from having a working-class or impoverished family background. Most of the working-class students said they are not seen as particularly praiseworthy within their families—if anything they are somewhat suspect. They expressed a sustained anti-elitism that keeps them from fully identifying with other medical professionals....

Although the number of working class students was small, the data showed quite clearly that they tended to be among the least compliant with the processes of secondary socialization encountered in medical school.

> Lance: I think I'm very much different from my classmates ... more outspoken, definitely.... Other people tend to say the right thing because they're a little afraid of the consequences. I don't care.... It comes from my background, you know, fishing. I've seen these tough, hard guys, think they're pretty something, but they're puking their guts out being seasick. It kind of reduces to the common denominator.

CONCLUSION

What is perhaps most remarkable about these findings is how little has changed since the publication of *Boys in White* (Becker et al., 1961) and *Student Physician* (Merton et al., 1957), despite the passage of 40 years and the influx of a very different student population. The basic processes of socializing new members into the profession of medicine remain remarkably similar, as students encounter new social norms, a new language, new thought processes, and a new world view that will eventually enable them to become full-fledged members of "the team" taking the expected role in the medical hierarchy.

Yet, with the differences in the 1990s student population, there are also some important differences in experiences. The role of medical student continues to carry with it certain expectations of its occupant. At a time when medical students were almost exclusively white, heterosexually identified, upper- or middle-class men, the identity may have "fit" more easily than it does for students who are women, who are from minority racial groups, who identify as gay or lesbian or working-class. If role-playing competence and being reflected back to yourself as "doctor" are as central to medical socialization as Haas and Shaffir (1987) suggest, what does it mean that women students are less likely than their male peers to be called doctor? This research has indicated the presence of a lingering societal assumption that Doctor = Man. Women students struggle to construct a professional appearance that male students find a straightforward accomplishment. Women search for ways to be in relationship with their patients that are unmarked by gender. Despite the fact that they make up half of all medical students in Canada, women's experiences of medical school remain different. In this research, almost half (6 of 14) of the women students interviewed indicated that they do not identify themselves as medical students in casual social settings outside school lest they be seen as putting on airs; none of the male students indicated this. It remains for future research to determine whether gender differences in the "fit" of the physician role make a difference to medical practice.

Interestingly, it is commonly assumed that the source of change in medical education will be the next generation of physicians—in other words, the current crop of medical students and residents (cf. Sinclair, 1997: 323–24). Over and over again I heard the refrain, "Surely the new generation of doctors will do things differently." This was the response to the hierarchy that stifles questioning or dissent through fear; to the inhumane hours expected of student interns and residents; to the need to show deference to superiors; to the need to pretend competence and confidence; to the need to sacrifice family, friends and outside interests to succeed in medicine. Yet, there have been many new generations of doctors in the past 40 years ... with remarkably little change. Why should we expect change now? Students, residents and junior physicians have very little power in the hierarchy to bring about change. Moreover, if they have been well socialized, why would we expect them to facilitate change? As one physician suggested, those who fit in well in medical school, who thrive on the competition and succeed, those are the students who return as physicians to join the faculty of the medical school. The ones who did not fit in, the ones who hated medical school, the ones who barely made it through—they are unlikely to be involved enough in medical education to bring about change.

Medical training has not always been good for patients (see Beagan, 2000). Nor has it been particularly good for medical students in many ways. Yet efforts at change on a structural level seem to have made little overall difference. In fact medical schools have a history of revision and reform without change (Bloom, 1988). Sinclair suggests moves toward entire new educational processes, such as the move to problem-based learning in medical schools throughout North America, simply "realign existing elements in the traditional training" (1997: 325). Furthermore, additions of new and very different components of the curriculum—such as classes on social and cultural aspects of health and illness, communication courses, and courses critiquing the social relations of the medical profession—are often seriously undermined in clinical teaching (Sinclair, 1997). Again, further empirical research should investigate the impact of such curriculum changes on professional socialization.

NOTES

In order to gain access to the research site, it was agreed that the medical school would remain unnamed. The school in question was in a large Canadian city with a racially and ethnically diverse population. It followed a traditional undergraduate curriculum.

At this medical school classes have been 40%–50% female for about 15 years (Association of Canadian Medical Colleges, 1996: 16); the class studied here was 48% female. Using subjective assessment of club photos, over the past 15 years about 30% of each class would be considered "visible minority" students, mainly of Asian and South Asian heritage.

1. All names are pseudonyms.
2. Never been called doctor, 14% of women, 0% of men; occasionally or regularly, 57% of women, 78% of men (Cramer's V = 0.32).
3. All of the gay/lesbian faculty and students described themselves as leading highly segregated lives during medical school.

References

Association of Canadian Medical Colleges. 1996. *Canadian medical education statistics*, Vol. 18.

Beagan, B. L. 2000. Neutralizing differences: Producing neutral doctors for (almost) neutral patients. *Social Science & Medicine*, 51(8): 1253–65.

Becker, H. S., B. Greer, A. L. Strauss, and E. C. Hughes. 1961. *Boys in white: Student culture in medical school*. Chicago: University of Chicago Press.

Berger, P. L., and T. Luckmann. 1966. *The social construction of reality: A treatise in the sociology of knowledge*. New York: Doubleday and Co.

Bloom, S. W. 1973. *Power and dissent in the medical school*. New York: The Free Press.

——. 1988. Structure and ideology in medical education: An analysis of resistance to change. *Journal of Health and Social Behavior*, 29: 294–306.

Broadhead, R. 1983. *The private lives and professional identities of medical students*. New Brunswick, N.J.: Transaction.

Conrad, E. 1988. Learning to doctor: Reflections on recent accounts of the medical school years. *Journal of Health and Social Behavior*, 29: 323–32.

Cooley, C. H. 1964. *Human nature and the social order*. New York: Schocken.

Dickstein, L. J. 1993. Gender bias in medical education: Twenty vignettes and recommended responses. *Journal of the American Medical Women's Association*, 48(5): 152–62.

Fox, R. C. 1957. Training for uncertainty. In *The student-physician: Introduction studies in the sociology of medical education*, eds. R. K. Merton, G. G. Reader, and E. L. Kendall, 207–44. Cambridge, Mass.: Harvard University Press.

Gamble, V. N. 1990. On becoming a physician: A dream not deferred. In *The black women's health book: Speaking for ourselves*, ed. E. C. White, 52–64. Seattle: Seal Press.

Goffman, E. 1959. *The presentation of self in everyday life.* New York: Doubleday.

Good, B. J., and M. J. DelVecchio Good. 1993. "Learning medicine." The constructing of medical knowledge at Harvard medical school. In *Knowledge, power, and practice: The anthropology of medicine and everyday life*, eds. S. Lindbaum, and M. Lock, 81–107. Berkeley: University of California Press.

Haas, J. and W. Shaffir. 1987. *Becoming doctors: The adoption of a cloak of competence.* Greenwich, Conn.: JAI Press.

Kirk, J. 1994. A feminist analysis of women in medical schools. In *Health, illness, and health care in Canada*, 2nd ed., eds. B. S.Bolaria, and H. D. Dickenson, 158–82. Toronto: Harcourt Brace.

Konner, M. 1987. *Becoming a doctor: A journey of initiation in medical school.* New York: Viking.

Lenhart, S. 1993. Gender discrimination: A health and career development problem for women physicians. *Journal of the American Medical Women's Association*, 48(5): 155–59.

Light, D. W. 1988. Towards a new sociology of medical education. *Journal of Health and Social Behavior*, 29: 307–22.

Mead, G. H. 1934. *Mind, self, and society: From the standpoint of a social behaviorist.* Chicago: University of Chicago Press.

Merton, R. K., G. G. Reader, and P. L. Kendall 1957. *The student physician: Introductory studies in the sociology of medical education.* Cambridge, Mass.: Harvard University Press.

Shapiro, E. C., and A. B. Jones. 1979. Women physicians and the exercise of power and authority in health care. In *Becoming a physician: Development of values and attitudes in medicine*, eds., E. Shapiro, and L. Lowenstein, 237–45. Cambridge: Bellinger.

Shapiro, M. 1987. *Getting doctored: Critical reflections on becoming a physician.* Toronto: Between the Lines.

Sinclair, S. 1997. *Making doctors: An institutional apprenticeship.* New York: Berg.

Goffman: Rituals of Interaction

Saul Geiser

Erving Goffman's studies of everyday life develop the idea of social action as performance. In this article Geiser highlights several key concepts in Goffman's research on social interaction and describes the special intensity of, and the risks inherent in, the presentation of self and everyday encounters with others. While the art of managing impressions, utilizing strategies, and engaging in ritual interactions may have an immediate practical significance for those involved, Geiser asks us to recognize that they simultaneously serve a broader social function for the group and its members as a whole. But what is at risk in an individual's mutually managed encounter with others? Why is there a special intensity to face-to-face interaction?

THE EPISODIC NATURE OF THE MICRO-ORDER

The micro-order may be conceived of as made up of millions of minute and transient episodes of social life. Even where people have long-standing relationships over many years, the actual time they are in communication consists of relatively brief encounters and occasions. In this sense "society" is not an abstraction—it is made up of very specific activities and communications, many of which are fleeting and precarious. To some degree, society as it is really lived is continuously coming into being and passing out of existence.

> A sociology of occasions is here advocated. Social organization is the central theme, but what is organized is the co-mingling of persons and the temporary interactional enterprises that can arise therefrom. A normatively stabilized structure is at issue, a 'social gathering,' but this is a shifting entity, necessarily evanescent, created by arrivals and killed by departures (1967: 2).

Source: Saul Geiser, "Goffman: Rituals of Interaction," in Leonard Broom and Phillip Selznick, *Essentials of Sociology* (2nd ed), Harper and Row, 1979: 28–33. Reprinted with permission.

Being episodic, the micro-order must be created anew at each successive encounter. Re-creation is accomplished through an exchange of cues and gestures by which the participants indicate to each other their own intended roles in the situation as well as what they expect the others' roles to be. This working consensus will differ from one interaction setting to another.

> Thus, between two friends at lunch, a reciprocal show of affection, respect, and concern for the other is maintained. In service occupations, on the other hand, the specialist often maintains an image of disinterested involvement in the problem of the client, while the client responds with a show of respect for the competence and integrity of the specialist (1959: 10).

INTERACTION AS THEATRE

In an important sense, therefore, social interaction is much like theatre. There is an expressive, dramatized element designed to project a definition of reality as much as to carry out practical tasks. Shakespeare's metaphor "all the world's a stage" can be developed into a dramaturgical model of the micro-order, showing how everyday life is pervaded by features of a theatrical performance. Thus, many social establishments are divided into "front-stage" and "backstage" regions. In front-stage areas, such as living rooms and food counters, an idealized display of decorum and cleanliness is affected whenever outsiders are present; backstage, in bedrooms and kitchens, performers can relax in guarded secrecy. Social performances are often staged by teams, such as the husband and wife hosting a dinner party or the doctor and nurse showing spotless clinical efficiency in the presence of patients.

THE MANAGEMENT OF IMPRESSIONS

If interaction is like theatre, then individuals must be like actors. The individual as effective "actor" in social encounters must be skilled in the art of "impression management"—controlling his or her image in the eyes of others so as to create a favourable definition of the situation. "For example, in American society we find that eight-year-old children claim lack of interest in the television programs that are directed to five- and six-year-olds, but sometimes surreptitiously watch them. We also find that middle-class housewives may leave *The Saturday Evening Post* on their living room end table but keep a copy of *True Romance* ('It's something the cleaning woman must have left around') concealed in their bedroom . . . ."

> In this capacity as performers, individuals will be concerned with maintaining the impression that they are living up to the many

standards by which they and their products are judged. But qua performers, individuals are concerned not with the moral issue of realizing these standards, but with the amoral issue of engineering a convincing impression that these standards are being realized (1959: 42, 251).

THE HAZARDS OF IMPRESSION MANAGEMENT

Once the individual has projected an impression of himself or herself to others, the others will expect this impression to be maintained throughout the remainder of the encounter as well as in any subsequent encounters.

> The expressive coherence that is required in performances points out a crucial discrepancy between our all-too-human selves and our socialized selves. As human beings we are presumably creatures of variable impulse with moods and energies that change from one moment to the next. As characters put on for an audience, however, we must not be subject to ups and downs. A certain bureaucratization of the spirit is expected so that we can be relied upon to give a perfectly homogeneous performance at every appointed time (1959: 56).

However, such expressive coherence is often difficult to sustain. The individual may make *faux pas* and boners that give away the act; the audience may come into possession of past information about the individual which is inconsistent with the character presently portrayed; outsiders may accidentally enter backstage regions, catching a team in the midst of activity at odds with their front-stage image. A fundamental theme that runs through Goffman's work is the ever-present danger that someone will see through, contradict, or otherwise disrupt a performance.

> When these disruptive events occur, the interaction itself may come to a confused and embarrassed halt. At such moments the individual whose presentation has been discredited may feel ashamed while the others present may feel hostile, and all the participants may come to feel ill at ease, non-plussed, out of countenance, embarrassed, experiencing the kind of anomy that is generated when the minute social system of face-to-face interaction breaks down.
>
> While the likelihood of disruption will vary widely from interaction to interaction.... There is no interaction in which the participants do not take an appreciable chance of being slightly embarrassed or a slight chance of being deeply humiliated. Life may not be much of a gamble, but interaction is (1959: 12, 243).

THE SPECIAL INTENSITY OF FACE-TO-FACE INTERACTION

Face-to-face interaction has a kind of multiplier effect, serving to raise both the emotional intensity as well as the hazards of social interaction. By being in the physical presence of others, the individual gives off information not only by verbal expressions but by clothing, gestures, and physical demeanour; as a result, face-to-face interaction is potentially more threatening to projected definitions of self than interaction mediated by telephone or writing, since more information about the individual is available to the observer.

> Each individual can *see* that he is being experienced in some way, and he will guide at least some of his conduct according to the perceived identity and initial response of his audience. Further, he can be seen to be seeing this, and can see that he has been seen seeing this. Ordinarily, then, to use our naked senses is to use them nakedly and to be made naked by their use. Copresence renders persons uniquely accessible, available, and subject to one another (1963: 16, 22).

INTERPERSONAL RITUALS

The main foundation of social order lies in the minute interpersonal rituals—hellos, goodbyes, courtesies, compliments, apologies, and handshakes—that punctuate everyday interaction. "The gestures which we sometimes call empty are perhaps in fact the fullest things of all" (1967: 91).

In the context of religion, "ritual" denotes standardized conduct through which an individual shows respect and regard to an object of ultimate value (usually a supernatural being) or to its stand-in (for example, an idol or a priest). In other words, rituals have mostly ceremonial, but little practical value. Nevertheless, as Durkheim—to whom Goffman owes a theoretical debt—pointed out, ritual and ceremony play an essential role in holding society together; through ritual worship of a common totem, members of primitive tribes reaffirmed their mutual commitment and collective solidarity.

Religious rituals have declined in importance in modern, secular societies, but despite their passing, many kinds of interpersonal rituals remain in force and perform the same function.

There are several types of interpersonal rituals. *Presentation* rituals include such acts as salutations, invitations, and compliments, by which the actor depicts appreciation of the recipient.

When members of the [hospital] ward passed by each other, salutations would ordinarily b[e] exchanged, the length of the salutation depending upon the period that had elapsed since the last salutation and the period that seemed likely before the next. At table, when eyes met, a brief smile of recognition would be exchanged; when someone left for the weekend, a farewell involving a pause in on-going activity and a brief exchange of words would be involved. In any case, there was the understanding that when members of the ward were in a physical position to enter into eye-to-eye contact of some kind, this contact would be effected. It seemed that anything less would not have shown proper respect for the state of relatedness that existed among members of the ward (1956/1967: 71).

Avoidance rituals, on the other hand, are practices in which the actors respect the privacy of others by such distancing behaviours as limiting eye contact between persons who do not know each other.

In performing this courtesy the eyes of the looker may pass over the eyes of the other, but no "recognition" is typically allowed. Where the courtesy is performed between two persons passing on the street, civil inattention may take the special form of eyeing the other up to eight feet, during which the sides of the street are apportioned by gesture, and then casting the eyes down as the other passes—a kind of dimming of the lights. In any case, we have here what is perhaps the slightest of interpersonal rituals, yet one that constantly regulates the social intercourse of persons in our society (1963: 84).

Maintenance rituals reaffirm the well-being of a relationship, for example, where persons with a long-standing relationship who have not seen one another for a time arrange an encounter: "It is as if the strength of a bond slowly deteriorates if nothing is done to celebrate it, and so at least occasionally a little invigoration is called for" (1971: 73).

Ratification rituals, such as congratulations at marriage and commiserations at divorce, mark the passage of an individual from one status to another:

Ratificatory rituals express that the performer is alive to the situation of the one who has sustained change, that he will continue his relationship to him, that support will be maintained, that in fact things are what they were in spite of the acknowledged change (1971: 67).

Access rituals, such as greetings and farewells, are commonly employed to mark the transition of persons to and from a state of increased access to one another:

The enthusiasm of greetings compensates for the weakening of the relationship caused by the absence just terminated, while the enthusiasm of farewells compensates the relationship for the harm that is about to be done to it by separation (1955/1967: 41).

FUNCTIONS OF INTERPERSONAL RITUALS

Different rituals serve different specific functions, but all have one essential feature in common: They are a conventionalized means by which the actor portrays ceremonial respect and regard for the self of another. Interpersonal rituals are important for three main reasons: First, they may be likened to traffic signs that serve to keep the flow of inter-action moving smoothly and direct it away from areas that could prove dangerous. Like "do not enter" signs, avoidance rituals prevent us from being conversationally accosted in public places and so leave us free to go about our business. Like a green light, invitations and greetings tell us when our interaction is welcome and appropriate. Without the con-versational rituals of speaking and listening in turn, social interaction would degenerate into chaotic babble.

A second basic function of interpersonal ritual is to ensure that one's self will escape relatively unscathed when one enters interactional traffic. Each new interactional episode poses a potential threat to the self image the individual attempts to project; and the self is therefore a "ritually delicate object," ever alert to offences and slights that would reflect unfavourably upon it. Even having one's remarks ignored in conversation can be taken as a sign that one's self is somehow deficient; conversational etiquette requires that one's remarks, however trivial, be acknowledged.

Third, the principle of reciprocity is built into the very structure of interpersonal ritual. It has a "dialogistic" character; that is, it typically involves a standardized exchange of moves and countermoves, or dialogue, between two or more actors: "Hi, how are you?" "Fine, thanks. And you?" Together these moves make up a little ceremony in which both selves receive ritual support; it is like holding hands in a circle, in which one gets back in the left hand what he or she gives with the right. Because such ritual exchanges occur so often in everyday interaction, they provide repeated opportunities for actors mutually to ratify their projected identities and thereby to sustain a workable, if idealized, definition of the situation. Interpersonal ritual thus involves a tacit teamwork, allowing each individual room in which to construct and uphold his or her own chosen identity. It is a curious twist. Though individuals are selfishly concerned to sustain a favourable impression of themselves, their efforts must be ritually expressed as altruistic regard for the identities of others: "His aim is to save face; his effect is to save the situation" (1955/1967: 39).

THE CONSERVATIVE NATURE OF THE MICRO-ORDER

Because of its episodic and fleeting character, the micro-order may seem to be rather flimsy and unstable, lacking a consistent influence on behaviour. In some senses this is true. Since the ritual order depends primarily on informal sanctions to effect conformity, anyone with enough self-assurance can override attempted sanctions simply by not allowing himself or herself to be emotionally affected. Ethnomethodologists profess to see the seeds of a "revolutionary" viewpoint in this analysis: The micro-order exists only because people *believe* it exists; social order is really quite precarious.

Yet Goffman's work has a conservative ring. He is at pains to show how interaction constrains people and why such constraints are necessary if people are to create a shared and consistent definition of reality:

> By entering a situation in which he is given a face to maintain, a person takes on the responsibility of standing guard over the flow of events as they pass before him. He must ensure that a particular *expressive order* is sustained—an order that regulated the flow of events, large or small, so that anything that appears to be expressed by them will be consistent with his face.... While his social face can be his most personal possession and the centre of his security and pleasure, it is only on loan to him from society; it will be withdrawn unless he conducts himself in a way that is worthy of it. Approved attributes and their relation to face make of every man his own jailer; this is a fundamental social constraint even though each man may like his cell (1955/1967: 9-10).

Perhaps the most conservative element in Goffman's account for the micro-order is his emphasis on shared values as the social glue by which society is held together. Unlike theorists of a more radical persuasion who see society as marked by continual conflict and held together by force, he locates the basis of social order in the values people hold in common. Ritual is important insofar as it is a means of ceremonially reaffirming these values. Yet unlike earlier conservative theorists such as Durkheim, who emphasized abstractly shared values and ritual, Goffman shows how ritual performances are a pervasive feature of everyday life:

> To the degree that a performance highlights the common official values of the society in which it occurs, we may look upon it ... as a ceremony—as an expressive ... reaffirmation of the moral values of the community. Furthermore, insofar as the expressive bias of performances comes to be accepted as reality, then that which is

accepted at the moment as reality will have some of the characteristics of a celebration. To stay in one's room away from the
place where the party is given, or away from where the practitioner
attends his client, is to stay away from where reality is being
performed. The world, in truth, is a wedding (1959: 35-36).

References

A summary and interpretation of works on the micro-order by Erving Goffman. His
analyses appear in the following volumes: *The Presentation of Self in Everyday Life*
(Garden City, NY: Doubleday Anchor Books, 1959); *Encounters: Two Studies in the
Sociology of Interaction* (Indianapolis: Bobbs-Merrill, 1961), *Behaviour in Public Places:
Notes on the Social Organization of Gatherings* (New York: The Free Press of Glencoe,
1963); *Interaction Ritual: Essays on Face-to-Face Behaviour* (Garden City, NY:
Doubleday Anchor Books, 1967); *Relations in Public: Microstudies of the Public Order*
(New York: Harper Colophon Books, 1971). Quoted material, some of which is
abridged, is used by permission of the author and copyright holder. This adaptation
was prepared by Saul Geiser.

The Dissolution of the Self

Kenneth Gergen

A social scientific understanding of socialization and the development of the self, with its focus on the influence of the social environment, generally implies that the character of the self will change as the character of society changes historically. In this selection Gergen discusses whether or not we are undergoing a radical transformation into a new, postmodern form of society with new cultural sensibilities, and considers what the implications of this form of society will be for the kind of self that develops within it.

The postmodern individual, he argues, is related to a multiplicity of identities. We show different sides of our self, depending on the situation we encounter. But where, or of what, is the essential self composed? The sheer diversity and rapid, fleeting nature of the changing environments the individual encounters in contemporary society make it difficult, if not impossible, to sustain the fiction of an authentic self of personal depth and singular reasoning ability. Gergen argues that, as we come to know more about others and how the world works, we change and are changed. But what is the nature of this change? If there is only a fictional essential self how does the self change as a person learns? What are our responsibilities and obligations towards others and ourselves given Gergen's conception of self?

... Cultural life in the twentieth century has been dominated by two major vocabularies of the self. Largely from the nineteenth century, we have inherited a romanticist view of the self, one that attributes to each person characteristics of personal depth: passion, soul, creativity, and moral fiber. This vocabulary is essential to the formation of deeply committed relations, dedicated friendships, and life purposes. But since the rise of the *modernist* worldview beginning in the early twentieth century, the romantic vocabulary has been threatened. For modernists, the chef [chief] characteristics of the self reside not in the domain of depth, but rather in our ability to reason—in our beliefs, opinions, and conscious intentions. In the modernist idiom, normal persons are predictable, honest, and sincere. Modernists believe in educational systems, a stable family life, moral training, and rational choice of marriage partners.

Yet, as I shall argue, both the romantic and the modern beliefs about the self are falling into disuse, and the social arrangements that they support are eroding. This is largely a result of the forces of social saturation. Emerging technologies saturate us with the voices of humankind—both harmonious and alien. As we absorb their varied rhymes and reasons, they become part of us and we of them. Social saturation furnishes us with a multiplicity of incoherent and unrelated languages of the self. For everything we "know to be true" about ourselves, other voices within respond with doubt and even derision. This fragmentation of self-conceptions corresponds to a multiplicity of incoherent and disconnected relationships. These relationships pull us in myriad directions, inviting us to play such a variety of roles and the very concept of an "authentic self" with knowable characteristics recedes from view. The fully saturated self becomes no self at all . . . .

I . . . equate the saturating of self with the condition of *postmodernism*. As we enter the postmodern era, all previous beliefs about the self are placed in jeopardy, and with them the patterns of action they sustain. Postmodernism does not bring with it a new vocabulary for understanding ourselves, new traits or characteristics to be discovered or explored. Its impact is more apocalyptic than that: the very concept of personal essences is thrown into doubt. Selves as possessors of real and identifiable characteristics—such as rationality, emotion, inspiration, and will—are dismantled . . . .

THE PROCESS OF SOCIAL SATURATION

A century ago, social relationships were largely confined to the distance of an easy walk. Most were conducted in person, within small communities: family, neighbors, townspeople. Yes, the horse and carriage made longer trips possible, but even a trip of thirty miles could take all day. The railroad could speed one away, but the cost and availability limited such travel. If one moved from the community, relationships were likely to end. From birth to death, one could depend on relatively even-textured social surroundings. Words, faces, gestures, and possibilities were relatively consistent, coherent, and slow to change.

For much of the world's population, especially the industrialized West, the small, face-to-face community is vanishing into the pages of history. We go to country inns for weekend outings, we decorate condominium interiors with clapboards and brass beds, and we dream of old age in a rural cottage. But as a result of the technological developments just described, contemporary life is a swirling sea of social relations. Words thunder in by radio, television, newspaper, mail, radio, telephone, fax, wire service, electronic mail, billboards, Federal Express, and more. Waves of new faces are everywhere—in town for a day, visiting for the weekend, at the Rotary lunch, at the church social—and incessantly and incandescently on television. Long weeks in a single community are unusual; a full day within a single neighborhood is becoming rare. We travel casually across town, into the countryside, to neighboring towns, cities, states; one might go thirty miles for coffee and conversation.

Through the technologies of the century, the number and variety of relationships in which we are engaged, potential frequency of contact, expressed intensity of relationship, and endurance through time all are steadily increasing. As this increase becomes extreme, we reach a state of social saturation.

In the face-to-face community, the cast of others remained relatively stable. There were changes by virtue of births and deaths, but moving from one town—much less state or country—to another was difficult. The number of relationships commonly maintained in today's world stands in stark contrast. Counting one's family, the morning television news, the car radio, colleagues on the train, and the local newspaper, the typical commuter may confront as many different persons (in terms of views or images) in the first two hours of a day as the community-based predecessor did in a month. The morning calls in a business office may connect one to a dozen different locales in a given city, often across the continent, and very possibly across national boundaries. A single hour of prime-time melodrama immerses one in the lives of a score of individuals. In an evening of television, hundreds of engaging faces insinuate themselves into our lives. It is not only the immediate community that occupies our thoughts and feelings, but a constantly changing cast of characters spread across the globe . . . .

POPULATING THE SELF

Consider the moments:

- Over lunch with friends, you discuss Northern Ireland. Although you have never spoken a word on the subject, you find yourself heatedly defending British policies.
- You work as an executive in the investments department of a bank. In the evenings, you smoke marijuana and listen to the Grateful Dead.
- You sit in a café and wonder what it would be like to have an intimate relationship with various strangers walking past.
- You are a lawyer in a prestigious midtown firm. On the weekends, you work on a novel about romance with a terrorist.
- You go to a Moroccan restaurant and afterward take in the latest show at a country-and-western bar.

In each case, individuals harbor a sense of coherent identity or self-sameness, only to find themselves suddenly propelled by alternative impulses. They seem securely to be one sort of person, but yet another comes bursting to the surface—in a suddenly voiced opinion, a fantasy, a turn of interests, or a private activity. Such experiences with variation and self-contradiction may be viewed as preliminary effects of social saturation. They may signal a *populating of the self*, the acquisition of multiple and disparate potentials for beginning. It is this process of self-population that begins to undermine the traditional commitments to both romanticist and modernist forms of being. It is of pivotal importance in setting the stage for the postmodern turn. Let us explore.

The technologies of social saturation expose us to an enormous range of persons, new forms of relationship, unique circumstances and opportunities, and special intensities of feeling. One can scarcely remain unaffected by such exposure. As child-development specialists now agree, the process of socialization is lifelong. We continue to incorporate information from the environment throughout our lives. When exposed to other persons, we change in two major ways. We increase our capacities for *knowing that* and for knowing how. In the first case, through exposure to others, we learn myriad details about their words, actions, dress, mannerisms, and so on. We ingest enormous amounts of information about patterns of interchange. Thus, for example, from an hour on a city street, we are informed of the clothing styles of blacks, whites, upper class, lower class, and more. We may learn the ways of Japanese businessmen, bag ladies, Sikhs, Hare Krishnas, or flute players from Chile. We see how relationships are carried out between mothers and daughters, business executives, teenage friends, and construction workers. An hour in a business office may expose us to the political views of a Texas oilman, a Chicago lawyer, and a gay activist from San Francisco. Radio commentators espouse views on boxing, pollution, and child abuse; pop music may advocate machoism, racial bigotry, and suicide. Paperback books cause hearts to race over the unjustly treated, those who strive against impossible odds, those who are brave or brilliant. And this is to say nothing of television input. Via television, myriad figures are allowed into the home who would never otherwise trespass. Millions watch as talk-show guests—murders, rapists, women prisoners, child abusers, members of the KKK, mental patients, and others often discredited—attempt to make their lives intelligible. There are few six year olds who cannot furnish at least a rudimentary account of life in an African village, the concerns of divorcing parents, or drug-pushing in the ghetto. Hourly, our storehouse of social knowledge expands in range and sophistication.

This massive increase in knowledge of the social world lays the ground work for a second kind of learning, a *knowing how*. We learn how to place such knowledge into action, to shape it for social consumption, to act so that social life can proceed effectively. And the possibilities for placing this supply of information into effective action are constantly expanding. The Japanese businessman glimpsed on the street today, and on the television tomorrow, may well be confronted in one's office the following week. On these occasions, the rudiments of appropriate behavior are already in place. If a mate announces that he or she is thinking about divorce, the other's reaction is not likely to be dumb dismay. The drama has so often been played out on television and movie screens that one is already prepared with multiple options. If one wins a wonderful prize, suffers a humiliating loss, faces temptation to cheat, or learns of a sudden death in the family, the reactions are hardly random. One more or less knows how it goes, is more or less ready for action. Having seen it all before, one approaches a state of ennui.

In an important sense, as social saturation proceeds we become pastiches, imitative assemblages of each other. In memory, we carry others' patterns of being with us. If the conditions are favorable, we can place these patterns into action. Each of us becomes the other, a representative, or a replacement. To put it more broadly, as the century has progressed, selves become increasingly populated with the character of others....

MULTIPHRENIA

It is sunny Saturday morning, and he finishes breakfast in high spirits. It is a rare day in which he is free to do as he pleases. With relish, he contemplates his options. The back door needs fixing, which calls for a trip to the hardware store. This would allow a much-needed haircut; and, while in town, he could get a birthday card for his brother, leave off his shoes for repair, and pick up shirts at the cleaners. But, he ponders, he really should get some exercise; is there time for jogging in the afternoon? That reminds him of a championship game he wanted to see at the same time. To be taken more seriously was his ex-wife's repeated request for a luncheon talk. And shouldn't he also settle his vacation plans before all the best locations are taken? Slowly, his optimism gives way to a sense of defeat. The free day has become chaos of competing opportunities and necessities.

If such a scene is vaguely familiar, it attests only further to the pervasive effects of social saturation and the populating of the self. More important, one detects amid the hurly-burly of contemporary life a new constellation of feelings or sensibilities, a new pattern of self-consciousness. This syndrome may be termed *multiphrenia*, generally referring to the splitting of the individual into a multiplicity of self-investments. This condition is partly an outcome of self-population, but partly a result of the populated self's efforts to exploit the potentials of the technologies of relationship. In this sense, there is a cyclical spiraling toward a state of multiphrenia. As one's potentials are expanded by the technologies, so one increasingly employs the technologies for self-expression; yet, as the technologies are further utilized, so do they add to the repertoire of potentials. It would be a mistake to view this multiphrenic condition as a form of illness, for it is often suffused with a sense of expansiveness and adventure. Someday, there may indeed be nothing to distinguish multiphrenia from simply "normal living."

However, before we pass into this oceanic state, let us pause to consider some prominent features of the condition. Three of these are especially noteworthy.

VERTIGO OF THE VALUED

With the technology of social saturation, two of the major factors traditionally impeding relationships—namely time and space—are both removed. The past can be continuously renewed—via voice, video, and visits, for example—and distance poses no substantial barriers to ongoing interchange. Yet this same freedom ironically leads to a form of enslavement. For each person, passion, or potential incorporated into oneself exacts a penalty—penalty both of *being* and of *being with*. In the former case, as others are incorporated into the self, their tastes, goals, and values also insinuate themselves into one's being. Through continued interchange, one acquires, for example, a yen for Thai cooking, the desire for retirement security, or an investment in wildlife preservation. Through others, one becomes to value wholegrain breads, novels from Chile, or community politics. Yet as Buddhists have long been aware, to desire is simultaneously to become a slave of the desirable. To "want" reduces one's choice to "want not." Thus, as others are incorporated into the self, and their desires become one's own, there is an expansion of goals—of "musts," wants, and needs. Attention is necessitated, effort is exerted, frustrations are encountered. Each new desire places its demands and reduces one's liberties.

There is also the penalty of being with. As relationships develop, their participants acquire local definitions—friend, lover, teacher, supporter, and so on. To sustain the relationship requires an honoring of the definitions—both of self and other. If two persons become close friends, for example, each acquires certain rights, duties, and privileges. Most relationships of any significance carry with them a range of obligations—for communication, joint activities, preparing for the other's pleasure, rendering appropriate congratulations, and so on. Thus as relationships accumulate and expand over time, there is a steadily increasing range of phone calls to make and answer, greeting cards to address, visits or activities to arrange, meals to prepare, preparations to be made, clothes to buy, makeup to apply....

And with each new opportunity—for skiing together in the Alps, touring Australia, camping in the Adirondacks, or snorkling in the Bahamas—there are "opportunity costs." One must unearth information, buy equipment, reserve hotels, arrange travel, work long hours to clear one's desk, locate babysitters, dogsitters, homesitters .... Liberation becomes a swirling vertigo of demands.

In the professional world, this expansion of "musts" is strikingly evident. In the university of the 1950s, for example, one's departmental colleagues were often vital to one's work. One could walk but a short distance for advice, information, support, and so on. Departments were often close-knit and highly interdependent; travels to other departments or professional meetings were notable events. Today, however, the energetic academic will be linked by post, long-distance phone, fax, and electronic mail to like-minded scholars around the globe. The number of interactions possible in a day is limited only by the constraints of time. The technologies have also stimulated the development of hundreds of new organizations, international conferences, and professional meetings. A colleague recently informed me that if funds were available, he could spend his entire sabbatical traveling from one professional gathering to another. A similar condition pervades the business world. One's scope of business opportunities is no longer so limited by geography; the technologies of the age enable projects to be pursued around the world. (Colgate Tartar Control toothpaste is now sold in over forty countries.) In effect, the potential for new connection and new opportunities is practically unlimited. Daily life has become a sea of drowning demands, and there is no shore in sight.

THE EXPANSION OF INADEQUACY

It is not simply the expansion of self through relationships that hounds one with the continued sense of "ought." There is also the seeping of self-doubt into everyday consciousness, a subtle feeling of inadequacy that smothers one's activities with an uneasy sense of impending emptiness. In important respects, this sense of inadequacy is a by-product of the populating of self and the presence of social ghosts. For as we incorporate others into ourselves, so does the range of properties expand—that is, the range of what we feel a "good," "proper," or "exemplary" person should be. Many of us carry with us the "ghost of a father," reminding us of the values of honesty and hard work, or a mother challenging us to be nurturing and understanding. We may also absorb from a friend the values of maintaining a healthy body, from a lover the goal of self-sacrifice, from a teacher the ideal of worldly knowledge, and so on. Normal development leaves most people with a rich sense of personal well-being by fulfilling these goals.

But now consider the effects of social saturation. The range of one's friends and associates expands exponentially; one's past life continues to be vivid; and the mass media expose one to an enormous array of new criteria for self-evaluation. A friend from California reminds one to relax and enjoy life; in Ohio, an associate is getting ahead by working eleven hours a day. A relative from Boston stresses the importance of cultural sophistication, while a Washington colleague belittles one's lack of political savvy. A relative's return from Paris reminds one to pay more attention to personal appearance, while a ruddy companion from Colorado suggests that one grows soft.

Meanwhile, newspapers, magazines, and television provide a barrage of new criteria of self-evaluation. Is one sufficiently adventurous, clean, well-traveled, well read, low in cholesterol, slim, skilled in cooking, friendly, odor free, coiffed, frugal, burglar proof, family oriented? The list is unending. More than once, I have heard the lament of a subscriber to the Sunday *New York Times*. Each page of this weighty tome will be read by millions. Thus, each page remaining undevoured by day's end will leave one precariously disadvantaged—a potential idiot in a thousand unpredictable circumstances.

Yet the threat of inadequacy is hardly limited to the immediate confrontation with mates and media. Because many of these criteria for self-evaluation are incorporated into the self-existing within the cadre of social ghosts—they are free to speak at any moment. The problem with value is that they are sufficient unto themselves. To value justice, for example, is to say nothing of the value of love; investing in duty will blind one to the value of spontaneity. No one value in itself recognizes the importance of any alternative value. And so it is with the chorus of social ghosts. Each voice of value stands to discredit all that does not meet its standard. All the voices at odds with one's current conduct thus stand as internal critics, scolding, ridiculing, and robbing action of its potential for fulfillment. One settles in front of the television for enjoyment, and the chorus begins: "twelve year old," "couch potato," "lazy," "irresponsible".... One sits down with a good book, and again: "sedentary," "antisocial," "inefficient," "fantasist".... Join friends for a game of tennis, and "skin cancer," "shirker of household duties," "underexercised," "overly competitive" come up. Work late and it is "workaholic," "heart attack-prone," "overly ambitious," "irresponsible family member." Each moment is enveloped in the guilt born of all that was possible but now foreclosed.

RATIONALITY IN RECESSION

A third dimension of multiphrenia is closely related to the others. The focus here is on the rationality of everyday decision-making instances in which one tries to be a "reasonable person." Why, one asks, is it important for one's children to attend college? The rational reply is that a college education increases one's job opportunities, earnings, and likely sense of personal fulfillment. Why should I stop smoking? One asks, and the answer is clear that smoking causes cancer, so to smoke is simply to invite a short life. Yet these "obvious" lines of reasoning are obvious only so long as one's identity remains fixed within a particular group.

The rationality of these replies depends altogether on the sharing of opinions—of each incorporating the views of others. To achieve identity in other cultural enclaves turns these "good reasons" into "rationalizations," "false consciousness," or "ignorance." Within some subcultures, a college education is a one-way ticket to bourgeois conventionality—a white-collar job, picket fence in the suburbs, and chronic boredom. For many, smoking is an integral part of a risky lifestyle; it furnishes a sense of intensity, offbeatness, rugged individualism. In the same way, saving money for old age is "sensible" in one family, and "oblivious to the erosions of inflation" in another. For most Westerners, marrying for love is the only reasonable (if not conceivable) thing to do. But many Japanese will point to statistics demonstrating greater longevity and happiness in arranged marriages. Rationality is a vital by-product of social participation.

Yet as the range of our relationships is expanded, the validity of each localized rationality is threatened. What is rational in one relationship is questionable or absurd from the standpoint of another. The "obvious choice" while talking with a colleague lapses into absurdity when speaking with a spouse, and into irrelevance when an old friend calls that evening. Further, because each relationship increases one's capacities for discernment, one carries with oneself a multiplicity of competing expectations, values, and beliefs about "the obvious solution." Thus, if the options are carefully evaluated, every decision becomes a leap into gray vapors. Hamlet's bifurcated decision becomes all too simple, for it is no longer being or non-being that is in question, but to which of multifarious beings one can be committed.

CONCLUSION

So we find a profound sea change taking place in the character of social life during the twentieth century. Through an array of newly emerging technologies, the world of relationships becomes increasingly saturated. We engage in greater numbers of relationships, in a greater variety of forms, and with greater intensities than ever before. With the multiplication of relationships also comes a transformation in the social capacities of the individual—both in knowing how and knowing that. The relatively coherent and unified sense of self inherent in a traditional culture gives way to manifold and competing potentials. A multiphrenic condition emerges in which one swims in ever shifting, concatenating, and contentious currents of being. One bears the burden of an increasing array of oughts, of self-doubts and irrationalities. The possibility for committed romanticism or strong and single-minded modernism recedes, and the way is opened for the postmodern being ....

As belief in essential selves erodes, awareness expands of the ways in which personal identity can be created and recreated.... This consciousness of construction does not strike as a thunderbolt; rather, it eats slowly and irregularly away at the edge of consciousness. And as it increasingly colors our understanding of self and relationships, the character of this consciousness undergoes a qualitative change.... [P]ostmodern consciousness brings the erasure of the category of self. No longer can one securely determine what it is to be a specific kind of person ... or even a person at all. As the category of the individual person fades from view, consciousness of construction becomes focal. We realize increasingly that who and what we are is not so much the result of our "person essence" (real feelings, deep beliefs, and the like), but of how we are constructed in various social groups.... [T]he concept of the individual self ceases to be intelligible....

Separated at Birth: Canadian and American Culture

Michael Adams

In this selection, excerpted from his book *Fire and Ice*, Adams expands and builds upon Northrop Frye's idea that a culture is the product of a people's history and their encounter with their geography. He investigates the social values held by Americans and Canadians and addresses questions of American and Canadian culture. For example, what are the current similarities and differences in social values between the peoples of these two nation-states? How homogeneous are the social values in these two countries?

But a deeper question arose for Adams as he surveyed people to obtain answers to these attitudinal questions: "My interest ... is to find out why an initially "conservative" society like Canada has ended up producing an autonomous, inner-directed, flexible, tolerant, socially liberal, and spiritually eclectic people while an initially "liberal" society like the United States has ended up producing a people who are, relatively speaking, materialistic, outer-directed, intolerant, socially conservative, and deferential to traditional institutional authority. Why do these societies seem to prove the law of unintended consequences?" What does Adams conclude?

This book is about Canadians and Americans. It offers up the results of the pulse-taking that Environics has been performing on both sides of the border during the past decade and elaborates, through the lens of social change, the national histories that have brought Canadians to their current uneasy coexistence with their Yankee neighbours. It discusses the trajectories the two societies seem to be following—trajectories that contrary to Jeffrey Simpson's views in his book *Star-Spangled Canadians*[1] and echoed by Michael Bliss in a January 2003 series of articles in the *National Post*[2] on the evolving Canadian identity, are not ineluctably drawing together but actually diverging in subtle but important ways.

To be fair, Simpson and Bliss are certainly not alone in their belief that Canada and the United States are becoming more similar. When Ekos Research asked Canadians in May 2002 whether they thought Canada had been becoming more or less similar to the United States during the preceding ten years, a majority of respondents (58 per cent) replied that they thought Canada was becoming more American. Thirty-one per cent thought there had been no change in the two countries' similarity or difference, and a mere 9 per cent thought Canada was becoming increasingly distinct from the United States. When asked whether they wanted Canada to be more like or less like the U.S., a majority of Canadians (52 per cent) reported that they would like Canada to be less like its neighbour. Thirty-four percent wanted the two countries' identities and relationship to remain the way they are now, and only 12 per cent of Canadians desired greater convergence with the U.S.

In this book I advance the rarely heard, and even more rarely substantiated, thesis that Canadians and Americans are actually becoming increasingly different from one another....

The claims I make and the conclusions I reach are at their core based on scientific surveys of representative samples of Americans and Canadians conducted during the past decade....

In 1867, Canada's Fathers of Confederation dedicated this country to "peace, order, and good government" while the ideals set out in Thomas Jefferson's Declaration of Independence were "life, liberty, and the pursuit of happiness." Americans were the revolutionaries putting in place institutions designed to frustrate the authority of governments, while counter-revolutionary Canadians saw the authority of political institutions as central to the well-being of their country. America has long honoured the individual fighting for truth and justice; Canadians have tended to defer to elites who broker compromises between groups. The American motto is *E Pluribus Unum*, Out of Many, One. In Canada, we began as two founding European cultures, French and English, since officially expanded to a multiculturalism that includes not only more recent immigrants, but also the First Nations that were here long before Europeans arrived. The Americans separated Church and State; we entrenched state sponsorship of parochial education in our Constitution. Canada never having renounced its European political heritage (at least not as emphatically as the American revolutionaries did), the Old World ideal of noblesse oblige has survived here even into this century, informing our social assistance and public housing programs, while south of the border mass education in the service of individual achievement has been the primary public expenditure. From distinct roots, Canada and the U.S. have grown up with substantially different characters: group rights, public institutions, and deference to authority have abided north of the border, while individualism, private interests, and mistrust of authority have remained strong to the south.

But in the last quarter-century, some counterintuitive developments have occurred on both sides of the 49th parallel. Canadians have distanced themselves from traditional authority: organized religion, the patriarchal family, and political elites. Peter C. Newman has characterized recent social change in Canada as the movement from deference to defiance. Meanwhile, a greater proportion of Americans are clinging to old institutions—family, church, state, and myriad clubs, voluntary associations, even gangs—as anchors in a chaotic world. In a country where the price of untrammelled individualism is that, in an instant, illness, crime, or an injudicious investment portfolio can turn the Dream into a nightmare, many Americans are seeking refuge at church, with family, or in the gated communities that are now, according to the Community Associations Institute as quoted in the 30 August 2001 edition of *The Economist*, home to a sixth of the U.S. population. In many ways, it is Canadians who have become the true revolutionaries, at least when it comes to social life. In fact, it has become apparent to me that Canadians are at the forefront of a fascinating and important social experiment: we are coming to define a new sociological "postmodernity" characterized by multiple, flexible roles and identities while Americans, weaned for generations on ideals of freedom and independence, have in general not found adequate security and stability in their social environment to allow them to assert the personal autonomy needed to enact the kind of individual explorations—spiritual, familial, sexual—that are taking place north of the border.

. . .

Because the cultural differences between Canada and the United States tend to exist beneath the consciousness of our daily lives, it is sometimes possible to imagine that those differences do not exist. After all, on any given day, most Canadians, like most Americans, can be spotted in their natural habitats driving cars, consuming too much energy and water, spending a little less time with their nuclear families than they would like, working a little more than is healthy, watching television, and buying some things they could probably survive without. But differences— both subtle and marked—do exist, and do endure. Some are external (gun control, bilingualism, health care), but many exist only inside the minds of Canadians and Americans—in how they see the world, how they engage with it, and how they hope to shape it.

. . .

But before the big picture, I'd like to share some raw numbers. We begin our portrait of these two neighbours with a comparison of their religious convictions. Canadians are by now quite familiar with evangelists Jerry Falwell, Pat Robertson, Jimmy Swaggart, Jim and Tammy Faye Bakker (who are slowly getting back to the business), and even William

Jennings Bryan, who defended creationism in the famous Scopes Monkey trial in the 1920s. We know that Christian fundamentalism has far deeper and more enduring roots in the United States, particularly in the Bible Belt, than here in Canada. What we sometimes fail to remember is that not so long ago, Canadians were more conventionally religious than Americans. In the mid-1950s, 60 per cent of Canadians told pollsters they went to church each Sunday; the proportion in the U.S. at that time was only 50 per cent. Today, only a fifth of Canadians claim weekly church attendance (22 per cent, according to Ekos), whereas the proportion in the U.S. is 42 per cent. A 2002 Pew Research Center poll found religion to be important to 59 per cent of Americans—the highest proportion in all the developed nations surveyed—and to only 30 per cent of Canadians, a rate similar to that found in Great Britain and Italy. Nearly four in ten Canadians do not consider themselves to be members of a religious faith. In the U.S. the proportion of atheists, agnostics, or secular humanists is only 25 per cent. In less than a generation, Canadians have evolved from being much more religious than Americans to being considerably less so.

Canadians have not only rejected in large numbers the authority of religious institutions, but have brought this questioning of traditional authority closer to home. Our research shows Canadians to be far less likely than Americans to agree with the statement, "The father of the family must be master in his own home." In 1992 we found that 26 per cent of Canadians believed Father must be master (down from 42 per cent in 1983). In 1992, 42 per cent of Americans told us Dad should be on top. Since then the gap has widened: down to 20 per cent in Canada and up to 44 per cent in the U.S. in 1996, and then down even further (to 18 per cent) in Canada in 2000 and up further still (to 49 per cent) in the U.S. in that year. The widening gap between the two countries now stands at an astonishing thirty-one points, with Canadians becoming ever less deferential to patriarchal authority and Americans becoming more and more willing to Wait Till Their Father Comes Home to find out if it's okay to watch *The Simpsons.*

Paralleling this differing orientation to patriarchal authority are the two populations' attitudes toward the relative status of the sexes. In a word, Americans are more predisposed to male chauvinism than Canadians, and here again the gap is widening. In 1992, 26 per cent of Canadians told us that men are naturally superior to women, while 30 per cent of Americans felt the same way. Four years later in 1996, the proportion of Canadians believing in the innate superiority of men declined to 23 per cent while the U.S. proportion rose to 32 per cent. By 2000, the proportion in Canada stood at 24 per cent while that in the U.S. shot up to 38 per cent. It only stands to reason, many Americans seem to be telling us, that if God-fearing men are the superior beings on this planet, then they should certainly be the bosses in their own homes.

Canadians' more egalitarian views regarding the status of women and the structure of the family, plus a more skeptical view of traditional institutional authority, also seem to lead them to a more relaxed view of what constitutes a family. Over the past decade, Canadians have consistently felt that two people living together, what we used to call living common-law, in fact constitutes a family. In 2000, 71 per cent of Canadians felt a couple that shared a home were a family, up from 66 per cent in 1992. Only 54 per cent of Americans shared this view, albeit up from 49 per cent in 1992. It is almost impossible to imagine a governor of any U.S. state daring to brazenly "live in sin" with his or her "life partner" as can Ontario Premier Ernie Eves. When in 1942 the Conservatives added the adjective "Progressive" to their party name, I doubt they had common-law cohabitation in mind.

What emerges so far is a portrait of two nations evolving in unexpected directions: the once shy and deferential Canadians, who used to wait to be told by their betters what to do and how to think, have become more skeptical of traditional authority and more confident about their own personal decisions and informal arrangements. Americans, by contrast, seeking a little of the "peace and order" that Canadians hoped "good government" would provide, seem inclined to latch on to traditional institutional practices, beliefs, and norms as anchors in a national environment that is more intensely competitive, chaotic, and even violent.

Attitudes toward violence are, in fact, among the features that most markedly differentiate Canadians from Americans. In the year 2000, 50 per cent of Canadians told us they felt violence to be all around them, a high figure to be sure, but nowhere near the 76 per cent of Americans who felt the same way. Americans' responses to our questions about violence suggest that they may even be becoming inured to the violence they perceive to be ubiquitous. In 1992, 9 per cent of Canadians and 10 per cent of Americans told us that violence is a normal part of life, nothing to be concerned about. In 1996, the figure in Canada was still 9 per cent, but had grown to 18 per cent in the U.S. In 2000, 12 per cent of Canadians felt that violence in everyday life was normal, but in the same year 24 per cent of Americans felt the same way. For one American in four, representing 70 million people, violence is perceived as normal part of one's daily routine. The other three-quarters of the population, presumably, are doing all they can to avoid those 70 million, particularly if alone on the street after dark.

We found further evidence that violence is becoming more, not less, normative in America when we asked Americans to agree or disagree that when one is extremely tense or frustrated, a little violence can offer relief, and that "it's no big deal." In 1992, 14 per cent of Americans agreed with this sentiment, as did 14 per cent of Canadians we polled. In 1996, the proportion was 10 per cent in Canada but zoomed to 27 per cent in the

U.S. By 2000, the proportion in Canada was back up to 14 per cent, but had surged further to 31 per cent in America, nearly one-third of the population. Again, you might not want to confront one of these folks when they're feeling a bit on edge, particularly when you remember that many of them (including the U.S. Attorney General) believe their Constitution guarantees them the right to bear firearms.

America is and always has been a very competitive society, nurtured by the myth of the American Dream, which suggests that anyone with a little vision and a lot of hard work can achieve material success. Sociologist Seymour Martin Lipset points out that in all categories, crime rates in America are about three times higher than they are in other industrialized countries. Lipset suggests as an explanation for this phenomenon the following: the American Dream, and the concomitant imperative to achieve material success, are so strong in America that many people pursue the goals of wealth and status in reckless, sometimes even criminal, ways. The end is of such monumental importance that the means become almost irrelevant.

Our polling found some interesting results in this area. In 1992, we asked Canadians and Americans whether they would be prepared to take "great risks" in order to get what they want. That year, nearly equal proportions of Canadians (25 per cent) and Americans (26 per cent) reported that they would indeed be prepared to take great risks to get what they wanted. The same in 1996. But by 2000 still only a quarter of Canadians were prepared to take great risks while the proportion in the U.S. increased to 38 per cent—a full eleven points higher than in Canada.

Americans are prepared to put a lot more on the line than Canadians to achieve their version of the American Dream, including personal risks to life and limb. They are also, as it turns out, more willing than Canadians to risk the lives and limbs of others to achieve the same ends. In 1992, 10 per cent of Canadians and only 9 per cent of Americans told us that it is acceptable to use violence to get what you want. In 1996, 11 per cent of Canadians felt this way, but the proportion of Americans rose to 17 per cent. By 2000, 13 per cent of Canadians felt the use of violence, presumably on or off the ice, was an acceptable way of achieving one's objectives, while the proportion in the U.S. was 23 per cent, nearly one in four and almost double the figure in Canada.

Lipset's hypothesis about the possible relationship between crime and the deep-rooted imperative of the American Dream illuminates an interesting contradiction: frustrated by their inability to achieve the Dream by socially acceptable means, those who obtain the trappings of success unlawfully exercise excessive individualism precisely *in order* to conform.

The idea that America's ostensible commitment to individualism may mask a deep impulse toward conformity is borne out in our polling data.

We find that Americans are in fact more prone to conformity than their neighbours to the north, who reside in a land that not only tolerates but actually celebrates linguistic, ethnic, and regional group identities. We track three items that shed light on this intriguing question: do people mind changing their habits, do they relate to people who show originality in dress and behaviour, and do they relate to people who repress rather than show their emotions. Our findings are surprising. In 1992, 51 per cent of Canadians and 56 per cent of Americans reported that they did not like changing their habits. In 1996, 48 per cent of Canadians reported being stuck in their ways—a decline of three points—and 58 per cent of Americans said the same thing, an increase of two points. By 2000, we had a widening and quite significant gap: only 42 per cent of Canadians said they don't like changing their habits while 54 per cent of Americans reported the same, now a gap of twelve points showing Canadians to be less conservative and more flexible than Americans in their day-to-day routines.

. . .

Canada's history has been dominated by three great themes: building a nation and holding it together, providing a growing list of services to the Canadian people, and managing our relations with the United States.

At the time of the American Revolution, Canada was a collection of British colonies that remained under the protection of the British crown rather than join the republican experiment launched by the thirteen colonies to the south. Thanks to that revolution, we even inherited some American Tories who stood loyal to the British Empire and migrated north.

To put it in a social values context, the American colonists rejected the traditional authority of the British crown while the Canadian colonists deferred to it, or, in the case of Quebec, fashioned a pragmatic compromise between the authority the British won on the Plains of Abraham in 1759 and that of the Roman Catholic Church.

From the late eighteenth century until 1867, the northern colonies remained under British rule, although increasing numbers of colonists demanded that their governments be more responsible to them than to the colonial administrators in Britain and their agents here. Some firebrands even instigated rebellions—one in Upper Canada (Ontario) in 1837 and another in Lower Canada (Quebec) in 1837 and 1838. These were revolts against an elite of appointed officials, not revolutions against the British regime, and in neither case was there significant loss of life. Early Canadians valued a liberty based on order over a freedom derived from the chaos of mob rule, which they believed prevailed in the new republic to the south.

Whereas America was conceived in violent revolution, the Canadian colonists were counter-revolutionaries whose cautious leaders were unable to negotiate the compromises necessary for their reluctant Confederation until 1867, nearly a century after the American colonies broke from Britain. While the Canadian colonies were slowly and laboriously brokering a larger union, America was deadlocked over slavery, lurching unrelentingly toward—and ultimately embroiled in—a bloody civil war that took the lives of 620,000 soldiers representing 2 per cent of the population at that time, or nearly 6 million Americans in today's terms.

In his Declaration of Independence, Thomas Jefferson dedicated his country to the ideals of life, liberty, and the pursuit of happiness. Not to be outdone in the evocative slogan department, a century later Canada's Fathers of Confederation could see no higher pursuits than peace, order, and good government. Judged against these lofty objectives, one would have to concede that in each country "two out of three ain't bad" (Meat Loaf, 1977).

The early experience of the two countries also differed in a way that haunts America still. The southern colonies had developed an economy based on slavery, an institution the United States retained (with increasing reluctance in a number of quarters) until the Civil War in the 1860s. The Canadian economy had little use for slaves or indentured workers on plantations for cotton or any other crop. As a result, the gradual abolition of slavery by Upper Canada's first governor, John Graves Simcoe, after 1793 and later by the British government was a non-issue for Canada, except to make this country a refuge for American slaves who were able to escape their servitude via the Underground Railroad prior to Abraham Lincoln's Emancipation Proclamation of 1863. The American Dilemma, as Swedish sociologist Gunnar Myrdal aptly termed that country's legacy of slavery in his 1944 book of that title, continues to express itself today—often tragically for the large proportion of African-Americans who live in poverty and under threat of violence even amid the affluence of the world's richest country.

The American Constitution also infamously guaranteed the right of its citizens to bear arms. The Second Amendment was once understood to be a provision granting militias the power to overthrow illegitimate governments through the use of force, but it has recently been recast by Attorney General John Ashcroft as the codification of the God-given right of every man, woman, and toddler to pack heat. Canada's Constitution contained no such right, and the consequences for each country are palpable to this day. Americans kill themselves and each other with the use of firearms at ten times the rate Canadians do.

America's revolutionaries, many of whom were Deists, agnostics, or even atheists, separated Church from State. Their forebears, the Puritans, had departed Britain in search of freedom to practise their religion. In founding their own communities in the New World, the Puritans were not in turn overly generous to those with dissenting theologies; Tocqueville notes that the criminal codes of some early communities included long passages copied verbatim from Leviticus and Deuteronomy. Nevertheless, 150 years later the U.S. Bill of Rights enshrined the principle of religious freedom for Puritans and all others, declaring in the First Amendment that "Congress shall make no law respecting an establishment of religion, or prohibiting the free exercise thereof."

The Canadian colonies, on the other hand, inherited the British tradition of direct state involvement in religion. After the British conquest of Quebec in 1759–60, the British not only allowed Roman Catholics to practise their religion, but, with the 1774 Quebec Act (designed to keep Quebecers loyal as the American colonies threatened open revolt), ceded to the Church the responsibility for the education of Catholic children. Meanwhile in Upper Canada, Governor John Graves Simcoe attempted to implement Anglicanism as the state religion but failed in the face of religious pluralism in the colony. The British North America Act of 1867 entrenched in Canada's Constitution the Catholic Church's control over the education of Catholics in Quebec and elsewhere in the country. This provision sough to reciprocate similar rights granted to Protestants. America's constitutional separation of Church and State and its more market-driven approach to religion has contributed to much higher rates of religious belief and practice than we now see in countries like Canada and the United Kingdom.

Another difference in the founding ideologies of the two countries was the orientation to citizenship. The American revolutionaries envisioned their country as the Biblical "City upon a Hill," a shining beacon for all who shared the Enlightenment ideals of free speech, religion, and commerce as well as progress, science, and rationality. People from all nations of the world would be welcome to cast off the chains of feudalism and migrate to the home of the brave and the land of the free. Out of many, there would be one, *E Pluribus Unum*, a proud American living in one nation, and, since the 1950s when the Pledge of Allegiance was updated, "under God." Some might argue that this ideal of unity and ultimate sameness has not been honoured from the outset, beginning with the exclusion of all but property-owning Caucasian males from the voters' list in American's first presidential election in 1789, a group that formed less than 10 per cent of the population.

In spite of many gaps between the ideal and the reality that seem obvious to us today, Americans have generally honoured their self-evident truths by welcoming migrants from around the world to join their melting pot, to become unhyphenated Americans willing to join the struggle for success and to send their sons to fight and if necessary die for their new country even against their former homelands.

Canada, by contrast, had no aspiration to mould an archetypal Canadian out of its three founding nations—French, English, and Aboriginal—or subsequent waves of newcomers from every corner of the planet. Each of the founding groups found themselves in their own enclaves. In the case of the Aboriginals, relocation was often forced and to be followed by various abuses; in the case of the French in Quebec, the enclave has always enjoyed considerable sovereignty. Sociologist John Porter characterized Canada in 1965 as a Vertical Mosaic, with the descendants of the English and the Scots at the top of the socio-economic hierarchy. According to Porter, all groups lived more or less peaceably in their communities, whatever their position in the pyramid, but had little to do with one another—a place for everyone and everyone in his or her place. In 1945 novelist Hugh MacLennan characterized English- and French-speaking Canada as Two Solitudes; this even in his native Montreal, where each comprised about half the population of what was then Canada's largest metropolis. The ethnic hierarchy of Canada today bears little resemblance to the descriptions of 1945 or even 1965, and the ideology of multiculturalism has promoted more positive attitudes toward racial and ethnic minorities north of the border than the melting pot creed has in the republic to the south.

The seeds of this compartmentalized but generally peaceful society are to be found in large part in the gradual decision by the British after their defeat of the armies of France on the Plains of Abraham in 1759 to allow 60,000 French habitants to retain their language and religion rather than attempt their assimilation into what were then very small Anglo-Saxon colonies in Canada. By the mid-nineteenth century, when the English-speaking Canadian provinces were more populous, so too, thanks to the "revenge of the cradle," was Quebec's French-speaking minority, which was able to successfully resist further calls for assimilation (most famously that of Britain's Lord Durham in 1839, who saw the absorption of the French as a solution to the "two nations" that he found "warring within the bosom of a single state"). The subsequent union of Canada East (Quebec) and Canada West (Ontario) ultimately proved unworkable. But the Confederation of those two colonies, as well as New Brunswick and Nova Scotia in 1867 and subsequently six others, has proven more lasting (although certainly not without its shaky moments).

. . .

Orientation to religion, government institutions, and founding ideology. These three factors fundamentally differentiate Canada and the United States, and this has long been the case. But these foundations have expressed themselves in the latter part of the twentieth century in some unanticipated ways. First let us look at the present realities that we or a French count might have anticipated 200 years ago. The United States has become the greatest nation on earth. It is the world's dominant economic and military power, and the leading innovator in the new information and biotechnologies. It is still the only nation on earth capable of mounting a concerted effort in exploring at the same time the human genome and the solar system. Its citizens have, on average, the highest standard of living on the planet, nearly half of the world's billionaires (242 out of 538 cited by *Forbes* in 2002) even after the dot-com/telecom implosion, and 60 per cent of its millionaires, the largest elite ever known in history. (They are celebrated, with typical American understatement, as "masters of the universe.") And America is, after all, the nation that gave the world the gifts of jazz, baseball, and the giant, heady leap that was Neil Armstrong's moon walk. Not bad for thirteen former colonies that started from scratch little more than two centuries ago.

With these results, who can argue with the ideological commitment to rationality, science, technology, pragmatism, and the free forces of expression, religious commitment, and the marketplace? But how, ask Canadians, can a people so adept at making a living not figure out how to live? How can they allow so many of their fellow citizens to live in Third World squalor only blocks or a few kilometres from their fortress enclaves? How can so many believe they will win life's lottery when experience consistently shows that only one in a hundred will do so and only a handful will ever be an Oprah Winfrey, a Michael Jordan, a Tiger Woods, or a Bill Gates? And yet Americans themselves and millions and millions on this planet dream of the one big chance America advertises, a chance that almost anywhere else in the world would be impossible even to fantasize.

As at the outset, America is a more competitive society than Canada. It is more innovative. It is also more violent and more racist. Americans worship money and success more than Canadians do. Americans are more willing to take risks in the hope that they might win than to insure against disaster in the fear that they might lose. Is it the relative economic and military importance of these two countries that explain the difference? Or is it the weather? Yes on both counts, plus those founding experiences, ideologies, and institutions, and the accidents of history that shape everyday life experience much more than the vaunted contemporary forces of global commerce and technology.

Certainly the growing gap in social values between our two countries during the 1990s described in this book must be at least partly attributable to America's emerging, after the collapse of the Soviet empire in 1989, as the world's only superpower, perhaps the most powerful ever to have existed on earth. This unique new status reduces America's need to forge multilateral alliances against a powerful and threatening adversary and encourages the U.S. to revert to the more aggressive unilateralism it demonstrated in its conquest of the American West and in President Monroe's nineteenth-century doctrine proclaiming America's right to control affairs in the western hemisphere, a doctrine now implicitly writ large over the entire planet. In this century, American exceptionalism becomes the *realpolitik* of globalization, something I believe will act to further differentiate the values of the United States from those of the rest of the developed world.

History is very much with us. The violence that was America is America. The moralism—good guys, bad guys, right and wrong, you're either with us or against us, establish moral superiority, wait for provocation and then blow them away—that was America remains America. In the first decade of the twenty-first century, we have an American government that now believes it can go it alone, with former allies relegated to towing the line, some enthusiastically, others resentfully. America is also the idealistic, some would say naïve, dedication not just to a life that seems perfectly reasonable to all but martyrs with a cause; but to an individual liberty that often undermines the collective good and to a pursuit of happiness that often leads to the most trivial and narcissistic pursuits. The right of everyone to bear arms leads to the highest murder rate in the developed world. The pursuit of happiness leads to strange and saccharine theme parks, mindless television and movie fare, addiction, gambling, substance abuse, and even the often pathetic pursuit of salvation promised by huckster Christian evangelists—all diversions to compensate for a deep spiritual deficit. Absent the religion and civil society, the decline of which political scientist Robert Putnam has documented, one is left with a society hell-bent for nihilism.

Canada and the countries of Europe try to balance market forces with public policy, to reconcile the tendency for the rich to get richer and create an all but impenetrable elite with a social welfare state and policies to redistribute income from the haves to the have-nots. Such countries recognize individual rights but try to balance them with the rights of collectivities. These societies are more likely than Americans to realize that individuals can have too much freedom and that freedoms can be exercised irresponsibly by individuals to their own and others' detriment. Canadians put greater value than Americans on peace, order, and good (read activist) government. This is the aspiration of a conservative people, as opposed to the eighteenth-century liberalism that appealed to the American revolutionaries.

The diseases of an all but untrammelled individualism are, of course, not without their desirable counterpoints. America is a more dynamic society than Canada, more creative, more innovative, more exciting, and more fun. According to *The Economist* (16 November 2002, p. 52), 700 of the world's 1,200 leading scientists work in the United States; these are the people we rely on to find the cure for cancer, the antidote for AIDS, and the key to Alzheimer's, and to best the long list of diseases that afflict people in every part of the planet.

Americans are wealthier than most people and are free to purchase all the things that money can buy: material possessions, symbols of status, and extravagant experiences. But the country is also a vastly more dangerous place, not only because of its glut of guns, but because it has refused so consistently to value any common good, and has, over time, become something of a war of all against all. Its people, therefore, live more stressful lives, and grasp at extreme or exclusionary forms of order (often dressed up as "the way things used to be—in the good old days before [insert supposed plague here: lawlessness, godlessness, rock 'n' roll, immigration, television, feminism]") in an attempt to stay the chaos of social and economic life.

Some of the sites at which Canada's difference from the U.S. is most apparent are, somewhat surprisingly, our cities. In examining the size and density of the communities in which we live, we find that a counter-intuitive evolution has taken place. Canada, as any schoolchild knows, is the world's second largest country after Russia, but in terms of population contains a modest 30 million or so. The U.S. is a large country too, the world's fourth largest, but numbers roughly 280 million people.

What is astonishing is that in spite of all this vast northern space, Canadians are huddled in relatively few large urban centres, mostly a few kilometres north of the Canada–U.S. border. More than a third of Canadians live in one of three metropolitan areas: Toronto, Montreal, or Vancouver. In contrast, America's three largest metropolitan areas, New York, Los Angeles, and Chicago, represent only 16 per cent of the United States population.

Canada is a more urban country than the United States. It is also more multicultural. Whereas 11 per cent of Americans are foreign born the figure for Canada is 18 per cent. Moreover, a large proportion of America's foreign born are from Mexico; in Canada they are drawn from virtually everywhere on the planet, with very large populations being East and South Asian.

As in the United States, first- and second-generation immigrants tend to congregate in cities where entry-level jobs, now often in the service sector, are located and where they are more likely to find support from previous waves of immigrants from their homelands.

What is fascinating about Canada's cities is their cosmopolitan livability, their relatively low rates of crime and interracial and inter-ethnic conflict. Toronto is arguably the world's most multicultural city, but has a murder rate only slightly higher than fifty years ago when it was predominantly Anglo-Saxon. The homicide rate in Metro Toronto (the core area of the city) has increased slightly from 1.4 per 100,000 in the 1959–61 period, when its population was approximately 1.5 million, to 2.2 per 100,000 in the 1999–2001 period, when its population had grown to approximately 2.6 million. Compare this with murder rates in major U.S. cities: in 1999 rates per 100,000 in New York City, Chicago, and Los Angeles were 8.9, 22.7, and 11.6 respectively. In the U.S. capital of

Washington, D.C., the rate was a whopping 46.4 per 100,000, in contrast to only 0.36—three murders—in 2001 in Canada's capital of Ottawa where, thankfully in this case, nothing much ever happens. (The only U.S. city listed by the Census Bureau whose murder rate is lower than Toronto's is Honolulu.)

. . .

It is interesting indeed that these two New World nations have each won the sweepstakes in two international competitions: the Americans for the highest standard of living on the planet and the Canadians for the best quality of life. The Americans have done this by being motivated by the notion of individual achievement; the Canadians by balancing individual autonomy with a sense of collective responsibility. We are each twenty-first-century expressions of the ideas of our ancestors and the institutions they built. America honours traditionally masculine qualities; Canada honours qualities that are more traditionally feminine. America honours the lone warrior fighting for truth and justice, the father who is master of his lonely house on the prairie, and a few good men planting the Stars and Stripes on a distant planet. Canada honours compromise, harmony, and equality. Americans go where no man has gone before; Canadians follow hoping to make that new place livable.

If American historian Samuel P. Huntington is right, the twenty-first century will be an often violent clash of civilizations. In that event, we will all be grateful for American economic and military leadership. If, however, the challenges of the twenty-first century will be addressing the growing disparities between rich and poor and the degradation of the earth's ecology, then let us hope Canada and kindred nations can muster the courage to show us another path into the future.

As I have demonstrated in this book, the founding ideas and institutions of each country have given rise to unanticipated consequences. I have found Americans to be more deferential to institutions than Canadians. This is counterintuitive. I have found Canadians to be less anomic, aimless, and alienated from their society than are Americans, who are nominally a more religious people. This too is counterintuitive. And, perhaps most surprising, I have found Canadians to be a more autonomous people than Americans, less outer-directed and less conformist. This too is contrary to the stereotype of Americans as a nation of individualists.

The key to these apparent anomalies, I believe, is the consequence of America's single-minded pursuit of individual achievement in the absence of peace, order, and good government. By adolescence and often earlier in life, Americans find themselves in an intense, often dangerous struggle for survival—or a winner-take-all quest for success. In such a context, traditional authorities serve as anchors: a strong father, a strong police force, a strong military, a strong nation, the president and commander-in-chief. In such a world there is little tolerance for subtlety, nuance, or shades of grey. Life is a Manichean struggle between good and evil, winners and losers, and the only way good will prevail is by being the strongest, vanquishing the "evil empire" or the "axis of evil" or the next incarnation of the forces of evil. Bruce Springsteen, American icon and perpetual valedictorian of the school of hard knocks, summed it up in his aptly named tune "Atlantic City": "Down here it's just winners and losers—and don't get caught on the wrong side of that line." In this world, individuals must choose their side, fall into line, and follow their leader into battle. There is little room for individual autonomy in such a scenario.

Canadians, however, have found themselves throughout their history to be in an interdependent world. After the Conquest of the French by the British army on the Plains of Abraham in 1759, it was decided by the authorities not to vanquish or assimilate the Quebec colonists but to accommodate their collective aspirations to preserve their religion, language, and culture. In the nineteenth century when America suffered a bloody civil war over slavery, Canada experienced a few rebellions, but in the end negotiated compromises that eventually led to Confederation in 1867. Good never triumphed over evil in Canada. Rather, opposing forces, often more than two, fighting over geography, religion, language, or the spoils of power, eventually came to some sort of accommodation—usually with little loss of life, especially when compared with the U.S. Civil War and the near annihilation of Aboriginals as Americans settled the West.

Our founding ideas, our institutions, and then the experience of building our two nations have been very different: one by conquest, the other by compromise. This Canadian penchant for going halfway rather than fighting it out to see who's left standing expressed itself in the twentieth century with the recognition that Canada was not only bilingual but also multicultural. And now, with the establishment of the new northern jurisdiction of Nunavut (one of whose official languages is Inuktitut), Canada is formally recognizing multilingualism as well.

This penchant for compromise, I would contend, has been expressed in less dramatic but equally important ways in the everyday life of Canadians. Feminism in Canada has become much more normative and has generated much less opposition than in the United States, as our social values data have demonstrated. The equality of women is not viewed as a threat to men or to the family, to corporate hierarchy or the well-being of the nation, as it is so often in the United States.

Similarly, when homosexuals asserted their rights, they found Canadians to be far more accommodating than the people of the United States. Montreal is a gay tourism mecca, and Toronto's Gay Pride Week is among the top three in the world in terms of attendance, having drawn an estimated 1,000,000 spectators and participants in 2002—not to mention $76 million in tourist dollars (*Broadcast News*, 27 September 2002).

The point is that the "conservative" society that values "peace, order, and good government" is also the society whose people feel secure enough to acknowledge interdependence. To be interdependent means to acknowledge the essential equality of the "other." When everyone feels equal and respected, they feel they can be in control of their own destiny in a world that relies upon the enlightened self-interest rather than the subservience of others. People in such a context have the self-confidence to be autonomous. In Canada, interdependence, autonomy, and diversity work in concert.

In functional families autonomy is cultivated by treating children with respect from day one, treating them, within reason, as equals. As in the nation, the domestic realm of the family can be governed by the ideals of peace, order, and good government with an absolute minimum of violence. Along with automatic deference to patriarchal authority, Canadians have by and large abandoned the age-old shibboleths "Children should not speak until spoken to" and "Spare the rod, spoil the child." Americans are still divided on the fundamental issues of what it is to be a father, a mother, or a child. In America, a father whose son comes out of the closet and declares his homosexuality is more likely to say, "You are no longer my son." In Canada such a father is more likely to find a way to adapt.

Thus, in Canada, the culture of accommodation that has been our socio-historical tradition expresses itself today as social liberalism, multiculturalism, multilingualism, multiple faiths and spiritual paths, and sometimes even as cultural fusion or hybridization. In its most postmodern form, it can exist as an openness to flexible, multiple expressions of individual personality, the leading edges of which are the flexibility of gender, age, and cultural identities. Demography as destiny is the vestige of a bygone era.

It is fascinating to see a country evolve from such deep deference to hierarchical authority to such widespread autonomy and questioning of authority—yet in the process not descending into chaos. Canadians are no longer motivated by duty, guilt, noblesse oblige, or fear of social sanction if they do not conform to group norms. Their kinder, gentler balance of freedom and equality, and of the public and the private domains, has created a tolerant, egalitarian society that enjoys freedom from potential catastrophe, danger, and violence that many on this planet envy, including many Americans.

But I don't for a moment think that most Americans believe Canada is a model to emulate. They are too focused on their own lives in a very different culture with very different institutions; their perspective on the global picture and their apparent trajectory of development are vastly different from our own. Canadians have always been congenitally introspective, perhaps because we have always felt threatened by internal cleavages. Americans, more preoccupied by conquest, are less introspective, particularly now that their dreams of global mastery are threatened by terrorism.

NOTES

1. Simpson writes, "Canadians, whether they like or acknowledge it, have never been more like Americans, and Canadian society has never been more similar to that of the United States. If the two countries are becoming more alike, and they are, this drawing together does not arise because Americans are changing. Canadians are the ones whose habits of mind, cultural preferences, economy, and political choices are becoming more American—without being American" (p. 6).

2. In a dialogue with *National Post* features writer Brian Hutchinson, University of Toronto historian and author Michael Bliss is quoted as saying: "But what strikes me is that we are becoming more similar to the Americans in our culture and in our values" (*National Post*, 18 January 2003, p. B1)

Identity Crisis—Multiculturalism: A Twentieth-Century Dream Becomes a Twenty-First-Century Conundrum

Allan Gregg

Written in journalistic rather than academic prose, Gregg's article raises an issue being debated by many sociologists and political philosophers within more liberal, secular nations such as Canada, Australia, France, and Britain, which have promoted large-scale multi-ethnic immigration. Do such policies weaken civic nationalism, identity, and participation within such societies? Or is this a worried conservative reaction, basically uncomfortable with the newly emerging ways that such societies become organized, cohere, and function?

Gregg insists there is a crisis and that addressing it requires drawing on an idea presented a century ago by Wilfrid Laurier and re-presented more recently by political theorist Charles Taylor: A multicultural nation can be brought together if its unity is a "projective one, based on a significant common future rather than a shared past." What might such a projected common future look like for a contemporary multicultural nation-state given the intensive globalization that is occurring?

Under the cover of normalcy, on July 7, 2005, the heart of London was bombed and dozens of people were killed by young Muslim men who had grown up in the same environment as their victims. The process of acculturation—at British schools and, one presumes, local pubs or Soho restaurants—had failed, and Britons were left wondering how a cluster of radicals dedicated to terrorism and to distant ideologies could spring from the nation they all share.

In another sign that all is not well in the world's diverse cities, four months later the outskirts of Paris went mad. On the night of October 27, French police chased a group of teenagers who had ventured out of

Source: Allan Gregg, "Identity Crisis," *The Walrus*, 3 (2), March, 2006, 38–47. Reprinted with permission.

their mostly Arab and African neighbourhood into the leafy suburb of Livry-Gargan. The pursuit turned deadly when three of the youths hid in a power-generation facility and two of them were electrocuted. Within four hours of this tragic accident, the streets of Clichy-sous-Bois (and adjacent communities) erupted in violence. In scenes reminiscent of Detroit and Los Angeles during the 1960s race riots, over 9,000 cars and 200 buildings were torched. France has been on edge ever since. An orchestrated attack by a terrorist cabal had besieged London, but in France something equally ominous had occurred: entire neighbourhoods of poor and alienated immigrants had protested their sense of isolation and disenfranchisement in a binge of wanton destruction.

Six weeks after the French riots, halfway around the world, roughly 5,000 white Australians took to the beaches of Cronulla, a suburb of Sydney, to attack people of Middle Eastern origin. Organized through text messaging and the Internet, this was a planned assault by aggrieved whites demanding, essentially, a return to Australia's whites-only immigration policy. The country had abandoned this openly exclusionary approach to immigration in 1973 and today Australia, along with Canada, has the most aggressive per capita immigration targets in the world. Prior to last November's outbreak of sectarian violence, Australia also had a growing international reputation for peaceful integration. The thugs who descended on Cronulla, obviously, did not endorse this national self-image.

Canada has long considered itself immune to violence rooted in ethnic divisions. By enshrining multiculturalism in our Charter of Rights and Freedoms and by promoting policies of inclusion, the argument goes, our country has created a peaceable kingdom and a model for how to manage diversity. Will Kymlicka, a Queen's University professor of philosophy and one of Canada's foremost authorities on multiculturalism, states that while the "actual practices of accommodation in Canada are not unique, Canada is unusual in the extent to which it has built these practices into its symbols and narratives of nationhood."

Before the 2006 election campaign got under way in earnest, Joe Volpe, Canada's minister of Citizenship and Immigration, sang the praises of Canadian multiculturalism, established an immigration target of 1 percent of the total population (a level equal to Australia's and triple that of the United States), and announced a goal of attracting 340,000 immigrants per year by 2010.

With an aging workforce, declining birth rates, and concerns about retirement pensions, one might expect generalized support for increased immigration. But research conducted in 2005 by my polling and market research firm, the Strategic Counsel, suggests that Canadians are far from sanguine about the country's increasing diversity. Fewer than half of those surveyed believe that Canada is currently accepting "the right amount" of immigrants, and among the remainder the overwhelming view is that we are accepting "too many" rather than "too few." Forty percent also express the view that immigrants from some countries "make a bigger and better contribution to Canada than others." The breakdown is disturbing: almost 80 percent claim that European immigrants make a positive contribution, the number falling to 59 percent for Asians, 45 percent for East Indians, and plummeting to 33 percent for those from the Caribbean.

In his landmark investigation, *Multiculturalism: The Politics of Recognition*, philosopher Charles Taylor points out that equal treatment often requires treating people in a "difference-blind fashion"—that is, "the other" must be respected in his or her historical and cultural fullness. But, when asked what the focus of multicultural policy should be, 69 percent of Canadians say immigrants should "integrate and become part of the Canadian culture," rather than "maintain their [own] identity." To some extent, it seems that Canadians, like their brethren in Europe, Australia, and elsewhere, have had their fill of multiculturalism and hyphenated citizenship.

While visitors often marvel at the multicultural mix evident on our city streets, there is growing evidence that Canada's fabled mosaic is fracturing and that ethnic groups are self-segregating. In 1981, Statistics Canada identified six "ethnic enclaves" across the country, i.e., communities in which more than 30 percent of the local population consisted of a single visible minority group. According to a recent StatsCan report, titled "Visible minority neighbourhoods in Toronto, Montréal, and Vancouver," that number had exploded to 254 ethnic enclaves by 2001. Not all of these communities are poor—for example, Richmond, British Columbia, and Markham, Ontario, whose Asian populations top 50 percent, are middle to upper-middle class—but an alarming number of them consist of people whose incomes fall far below the Canadian average. Despite good efforts and well-intentioned policies, poverty and disenfranchisement in Canada are becoming increasingly race-based.

In Toronto, after a run of black-on-black violence and the random Boxing Day murder of fifteen-year-old Jane Creba, poverty advocates and ethnocultural groups insisted that unequal access to jobs, a lack of community-based programs, and racism were plaguing the black community, especially its young men, who, seeing no future, were lashing out. While politicians treaded gingerly around the notion of race-based violence, on the streets and in homes anxious city dwellers were saying enough was enough, demanding tough justice for anyone caught with a gun, and asking whether young black men would ever be capable of integrating into mainstream society.

When, it appears, dramatically disenfranchised groups—whether they be in East London or on the periphery of Paris or in Toronto—cease to have a stake in, or feel responsible for, their country's civic culture, they are at risk of turning to violence. Over the coming years, Canada's ability to accommodate diversity is sure to become a central issue. As is the case in England, France, and other advanced liberal democracies, national unity in Canada is threatened by the growing atomization of our society along ethnic lines.

Consider the pattern in Britain. Following World War II, the United Kingdom granted "unlimited right of entry" to former colonial subjects. Its Nationality Act allowed over 300,000 West Indians to enter Britain between 1948 and 1962, with similarly large numbers coming from India and Pakistan. While the policy was generally assimilationist, visible inequality and violent outbreaks in "coloured communities" fed concerns that the complexion of British society was changing too rapidly. This led to the passage of the Commonwealth Immigrants Act in 1962, which severely restricted the flow of new arrivals from former British colonies. But numerous ethnic communities had already put down roots, expanded, and, as the years went by, attempted to establish themselves in British society. In 1981, riots in the Brixton area of south London (followed by more race-based riots in Birmingham and other English cities) contributed to more restrictive immigration.

Clearly, the integration of visible minority groups was posing special challenges, but Britain remained reliant on immigrant labour and could not simply close the doors. In the early 1990s, it addressed the issue by shifting toward Canadian-style multiculturalism, and by promoting the virtues of ethnic identity and diversity to mainstream society. More and more, mosques, temples, and other icons of ethnicity began sprouting up in British cities as visible minorities were encouraged to retain their customs and traditions. Grumblings about ethnic neighbourhoods continued but, as international markets soared and people spoke openly of the advantages of a new cosmopolitanism, criticism was muted—until last summer. Since the London bombings, British politicians across party lines have suggested that the traditional explanations for unrest and violence—poverty, inequality, etc.—cannot explain the suicidal rage of the bombers. Many argue that, within the context of a wholesale re-evaluation of citizenship and loyalty to state, the answer must lie in the very policies designed to encourage multiculturalism and celebrate diversity.

But the French situation undermines this interpretation. France has remained staunchly assimilationist. While it has opened its doors to immigrants (and former colonials) from North Africa and the Middle East—again, largely in response to shortages of unskilled labour—the emphasis on speaking French has been resolute, and little truck has been given to the construction of ethnic shrines or the wearing of foreign cultural iconography. Often criticized for being rigidly chauvinistic, France nonetheless established a relatively firm contract with new arrivals and refused to accept notions of hyphenated citizenship. One would therefore expect that if outbreaks of violence did occur, they would not be so clearly rooted in ethnicity. And yet France—like Germany, Holland, and other European countries—is now riven by colour-line politics, and the engrained sense of alienation among ethnic groups is profound.

In England and France, it appears that the recent violence is rooted in second-generation visible-minority groups with little fealty to their adopted state (and in Australia, in what immigration policy is doing to the nation). And there is growing concern that a similar sense of alienation is developing among the same class of people in Canada.

From the beginning, and for generations, immigration to this country was based on our most fundamental need—to populate and settle the unwieldy geographic mass that was to become Canada. The nation was not born of a revolution or forced to recreate itself after an empire's passing. Rather, it was perceived as a blank slate where, owing to a harsh climate and endless land, nation-building itself became the founding mythology. Formed after the US Civil War, or the "war between the states," Canada was organized around weak provinces and a strong federal government—a source of benevolence at the centre that would knit the regions together through massive projects such as the national railway. Immigration was one of Ottawa's chief responsibilities; its policies were openly integrationist and designed for those eager to assume Canada's monumental challenge. So, early in the twentieth century, Wilfrid Laurier's Liberal government set out to populate vast territories by importing "men in sheepskin coats." Ukrainians, Norwegians, Germans, and other almost exclusively European immigrants responded to the call and began descending on Canada's ports, eager for the long trek to the West. This flood reached its peak in 1913, when 400,810 immigrants—the equivalent of 1.5 million today—arrived on our shores.

Growth through immigration continued until the combined impact of the Depression, racism, and World War II caused Canada to effectively shut its doors to outsiders. But, as was the case with Britain, the war had depleted our store of labour. With millions across Europe seeking safe haven from poverty and starvation, and Canada overdue to restart its nation-building project, by the mid-1940s the immigration taps were turned on once again. Bolstered by its reputation as a liberator, Canada attracted Italians, Portuguese, Greeks, and other Europeans to its flourishing urban centres. As is reflected in the 1952 Immigration Act, entry into Canada was deemed a privilege and individuals could be barred based on ethnic affiliation. Immigration was now clearly controlled through country-of-origin quotas, which actively restricted non-white immigrants and implicitly validated the notion that nation-building requires assimilation. While still diverse, Canada grew as a white, European, and Christian nation of immigrants grateful for the opportunity to start over in a new land. And, most crucially, the federal government retained its role as central provider, thereby encouraging immigrants to develop a strong sense of civic nationalism.

By 1961, 97 percent of all immigrants came from Europe, but Canada's openly assimilationist approach began to shift in 1967, when country-of-origin quotas were replaced by a more meritocratic points system. The impetus for this change came from many quarters, including Prime Minister John Diefenbaker's early 1960s criticism of South Africa's apartheid regime, Lester Pearson's peacekeeping initiatives, and Canada's increasing involvement in the Commonwealth. Within a few short years, the impact was dramatic. West Indian immigration to Canada, for instance, ballooned from 46,000 and 3 percent of the total (many of whom were white) in the 1960s to nearly 160,000 and 11 percent in the 1970s (almost all of whom were black). But, despite its growing diversity, to a large extent Canadian-style multiculturalism emerged less out of a sense of global citizenship than from a need to deal with a pressing domestic issue: Quebec.

Alarmed at the rise of nationalist sentiment during Quebec's Quiet Revolution, in 1963 the federal government launched the Royal Commission on Bilingualism and Biculturalism. Its thinly veiled objective was to dissipate Quebecers' sense of being a conquered nation and replace the notion of "English Canada" with a bold new pact between two founding peoples. Canada would be defined by two languages and two cultures, co-existing within a federalist framework. This approach might have tempered the flames of separatism had the process not been hijacked by swelling numbers of non-British and non-French immigrants, who failed to see themselves reflected in the new vision. As Will Kymlicka wrote in 2004, "[New Canadians] worried that government funds and civil service positions would be parcelled out between British and French, leaving [white] immigrant/ethnic groups on the margins."

Confronted by an organized ethnic lobby, the government changed the terms of reference of the commission and, in the end, declared that Canada would be a multicultural society within a bilingual framework. The commission promoted the view that immigrant groups would overcome the obstacles posed by a new home and, over time, integrate, just as they had always done. Indeed, the entire genesis of the 1971 official Multiculturalism Policy suggests some ambivalence or confusion about embarking on a new national concept and a certain naïveté in the assumption that settlement would proceed largely as it had historically.

Our Centennial celebration, Expo 67, drew the world's attention to Canada, a progressive, modern state that promised universal health care, low university tuition fees, and jobs. This, combined with suggestions of a cultural mosaic, attracted large numbers of immigrants throughout the 1970s. The recession of the early 1980s stemmed the tide, but the notion of Canada as a cosmopolitan, caring, and multicultural society became even more concrete in 1988, when Brian Mulroney's Conservative government passed the Canadian Multiculturalism Act. Aggressive immigration targets and multiculturalism gained non-partisan support and became politically unassailable.

In 1984, Canada admitted only 88,239 immigrants, but the years following saw increased numbers, and by 2001, some 5.4 million Canadians aged fifteen or older were foreign-born—18.4 percent of the population. This represented the highest rate of diversity in seventy years; in Ontario and British Columbia, the figure reached nearly 35 percent.

The most significant change over the past two decades has been the increase in visible minority immigration. In 2004, only 20 percent came from Europe, while nearly 50 percent came from China, India, Pakistan, the Philippines, Korea, or Iran. For the moment, non-white Canadians represent approximately 16 percent of the population. But with more inflow, and with first-generation immigrants raising families, the figure will increase significantly in the coming years.

Recognizing that visible-minority groups faced unique obstacles to integration, the Heritage Department conducted a formal review of multi-cultural programs in 1996. The result was a more assertive mandate: "to foster an inclusive society in which people of all backgrounds, whose identities are respected and recognized as vital to an evolving Canadian identity, feel a sense of belonging and an attachment to this country, and participate fully in Canadian society." The new thrust was directed at non-immigrant society, at getting it to respect and encourage diversity. In fact, through the 1990s, the government directed funding to ethnic organizations and insisted that public institutions such as the civil service and the CRTC reflect the ethnic diversity of the country through their hiring practices. Whereas the goal of past initiatives was clearly integration, Canada had evolved into a state that promoted hyphenated citizenship.

The changes were controversial. Over and above critiques that hiring quotas inevitably lead to reverse discrimination, there were questions about whether encouraging the retention of ethnic identity would drive visible-minority groups away from mainstream society. Examining the United States, American historian Arthur Schlesinger Jr. wrote of a "cult of ethnicity" that "exaggerates differences, intensifies resentments and antagonism, drives even deeper the awful wedges between races and nationalities. The endgame is self-pity and self-ghettoization." Schlesinger's critique resonates in Canada. Recent settlement trends suggest that so-called ethnic box settlements are becoming prevalent.

In the Canadian context, as Ottawa continued devolving powers to the provinces, the sense of nationhood receded in significance. When the issues directly affecting people's lives—health care, education, cities—are overwhelmingly controlled by the provinces and there is an absence of large-scale, nation-defining projects (to follow the historic examples of the railway, the Canadian Broadcasting Corporation, medicare), the creation of a coherent national vision becomes difficult in the extreme.

Twenty years ago, roughly half of the immigrant population gravitated to Toronto, Montreal, or Vancouver. Today, nearly 80 percent does—and this is 80 percent of a much larger total. Within these growing urban centres, immigrant groups are clustering in tightly knit, ethnically homogeneous neighbourhoods partly because, according to the government's own studies, many ethnic groups feel out of place in Canada. Their first loyalty is to their group, and, against a history of the children of immigrants "moving out," today there is an increasing concentration of visible-minority groups "staying home," staying alien to host cultures and having little sense of civic nationalism.

How can this situation change? For multiculturalism to work, the native-born must accept immigrants as equals and new arrivals must demonstrate a willingness to join mainstream society by adopting the fundamental mores and values of the prevailing culture. There must also be cross-fertilization between ethnic groups and civic nationalism has to be clearly defined. According to University of Toronto sociology professor Jeffrey Reitz, recent evidence casts serious doubt that this is occurring in modern Canada. Reitz has spent his career studying the Canadian immigrant experience and, considering data on both income levels and attitudes, he believes that "multicultural policies are simply not working as well for visible minorities." Despite targeted programs to ease adjustment, Reitz's research shows that, unlike post–World War II immigrants, Canada's newest arrivals are not only failing to catch up financially, but the gap between them and non-immigrant groups is widening. Social disparities are most pronounced among visible-minority groups, and Reitz's data indicates that while "satisfaction with life" increases from the first to the second generation for white immigrants, it actually decreases among non-white immigrants.

Voting behaviour is one of many indices researched by Reitz to gauge rates of societal participation and involvement. A scant 20 percent of first-generation immigrants (that is, the foreign-born), regardless of colour, exercise their franchise. By the second generation, however, white immigrant participation rates almost quadruple, while among visible-minority groups it only doubles. Surprisingly, it appears that first-generation non-white immigrants actually enter Canada with a greater sense of belonging than white immigrants, but within a generation that feeling diminishes among visible-minority groups, while white immigrants report a growing sense of belonging and involvement.

Because immigration is most often push driven—that is, homeland conditions motivate emigration—in the main, immigrants are satisfied with their adopted country. Early on, their sense of belonging derives principally from involvement within their own ethnic communities, which Reitz reports is much higher among non-white minorities. But by the second generation, that involvement diminishes, as cultural ties loosen and expectations of their adopted home increase. According to Reitz, it is with the second generation, the same demographic responsible for the London bombings and the riots outside Paris, that ethnic tensions and alienation most clearly reveal themselves.

Unlike Britain and France, however, which began accepting visible-minority immigrants after World War II, Canada did not do so in any real numbers, until the 1970s. Consequently, second-generation immigrants represent only 14 percent of Canada's current visible-minority population. But today, two-thirds of all native-born visible minorities in Canada are under sixteen years old. Their numbers are destined to swell, and given current settlement trends and growing income disparities, Canada may indeed face the kinds of ethnic conflicts that have beset England and France. Instead of having more effective multicultural policies or greater societal tolerance, Canada has avoided these problems to date largely because it got into the visible-minority immigration game a generation later.

Political theorist Charles Taylor has analyzed the issue of achieving common objectives in a multicultural society that places a primacy on respecting difference. He concluded that Canadians can "be brought together by common purposes [but] our unity must be a projective one, based on a significant common future rather than a shared past." Some have suggested promoting diversity itself as a rallying call for all Canadians, but, again, drawing attention to difference can undermine attempts to forge an over-arching national identity. The situation has been further complicated by the emergence of intensive globalization and the necessarily diminished role of nation-states (and hence of national mythologies) that globalization has ushered in.

Canada, Britain, France, and Australia share a common dilemma. All are stable constitutional democracies that are based on the primacy of individual rights and all share secular-humanist leanings. Each recognizes the need for immigration and is coping with growing visible-minority populations, and each is struggling in a post-nation-state world where well-defined national purposes are less certain. Without grand designs or defining national projects, new immigrants run the risk of arriving and going about their business with little sense of the roles they can play in their adopted homeland. With no national mythology to adhere to, they naturally retreat to the familiar, seeking out their own communities.

Throughout Europe, nations known for their liberalism are now engaged in vigorous debate around one central question: what is more important to our national direction, inclusion under the umbrella of a unifying nationalism or the celebration of uniqueness and difference? Defenders of multiculturalism argue that these two options are not mutually exclusive and that both can be achieved by open, tolerant, and just societies. But in Britain, the decision to encourage uniqueness drove certain second-generation groups away from the mainstream and its values; in France, assimilationist policies had led to feelings of intense isolation.

In Canada, we may live in a multicultural society, but the evidence suggests that fewer and fewer of us are living in multicultural neighbourhoods. Furthermore, the tradition of immigrants clustering in a community for one generation before the next generation moves on and "melts" into mainstream culture seems to be breaking down. Large districts are evolving into areas dominated by individual ethnic groups that have chosen to live apart from those who do not share their ancestry. Meanwhile, most white Canadians would confess that the vast majority of their friends look a lot like they do and that they tend to stay within their own communities, rarely venturing into the ethnic enclaves that are burgeoning, especially in suburban Canada.

This growing sense of separateness can have troubling consequences for national identity. Just as the landmark 1954 US Supreme Court decision *Brown v. Board of Education* demonstrated that separate can never be equal, the history of segregation teaches us that the notion of citizenship cannot survive in modern liberal societies that become atomized. The absence of interaction between groups of different backgrounds invariably perpetuates cultural divisions, breeds ignorance, and leads to stereotyping and prejudice.

It is true, the attacks on the World Trade Center and the Pentagon brought grievances from distant lands to our doorstep. But they were perpetrated by foreigners, and it was their very "foreignness" that made their motivations impenetrable to the Western mind. When we learned, however, that the London bombers could have grown up playing soccer on the same pitches as the people they murdered and that entire neighbourhoods were burning only a bus ride from Paris's tonier cafés, a new face of ethnic conflict came knocking on the Western door.

The events of the last year have presented the West with a conundrum: can liberal democracies, lacking a unifying ethos, satisfy the needs of societies that are increasingly heterogeneous? Bernard Ostry, one of the principal architects of multiculturalism under Trudeau, has voiced anxiety that the experiment has gone wrong and must be reviewed. Mindful, one suspects, of the example of Australia, which also opened its arms to non-white immigrants in the past few decades, Ostry is demanding a travelling royal commission. At the beginning of the last century, Wilfrid Laurier answered the question of immigration and identity by telling new arrivals, "Let them look to the past, but let them also look to the future; let them look to the land of their ancestors, but let them look also to the land of their children." The events in Sydney, as well as those in London and Paris, suggest just how imperative it is to heed his words, as inspiring today as they were 100 years ago, when a young nation trembled before the twentieth century and wondered how it would find its way.

NOTE

References to Reitz's studies came from an interview conducted by Allan Gregg with Jeffrey Reitz at the University of Toronto.

References

Canadian Heritage, "C. 1995-96 Performance Report," in *Canadian Heritage 1997-98 Estimates*, Government of Canada. (www.canadianheritage.gc.ca).

Kymlicka, Will, "The Canadian Model of Diversity in Comparative Perspective," Eighth Standard Life Visiting Lecture, University of Edinburgh, April 29, 2004.

Laurier, Sir Wilfrid, as quoted in the *Toronto Globe*, September 2, 1905, page 1.

Schlesinger, Arthur M, Jr., *The Disuniting of America: Reflections on a Multicultural Society*, NY: W. W. Norton, 1998 (first published 1992).

Statistics Canada (author HouFeng), "Recent Immigration and the Formation of Visible Minority Neighbourhoods in Canada's Large Cities," Ministry of Industry, July, 2004 [catalogue # 11F0019MIE2004221].

Strategic Counsel, "August Survey for the *Globe and Mail*; and CTV: Immigration, Terrorism and National Security," August 7, 2005 (www.thestrategiccounsel.com).

Taylor, Charles, *Multiculturalism and "The Politics of Recognition,"* Princeton, NJ: Princeton University Press, 1992.

Teaching the Cultural with Disney

Henry Giroux

Most of us think that large media corporations like Disney have an influence on children's culture. But what is that influence? The mass media and the entertainment industry have become an educational force not only in children's culture but also in the wider civic culture. Far from being innocuous entertainment, Giroux argues, viewers of all ages "use Disney's cultural texts to make sense of their lives." Even more chilling, he declares that Disney presents a real threat to a vibrant democracy. Is Giroux being overly alarmist?

What, after all, makes Disney so appealing and successful? Central to its success and influence, he contends, is a "commodified view of innocence." This corporate packaging of fun, innocence, and purity has a political dimension to it as well. One consequence is that it infantilizes both children and adults, leaving them with overly simplified, sanitized views of history and politics. If it's important, as Giroux exhorts, to examine and question the educational and political influence of the mass media, when and how do you get children to reflect on and be critical of the cultural products they joyfully consume?

That corporate culture is rewriting the nature of children's culture becomes clear as the boundaries once maintained between spheres of formal education and entertainment collapse into each other. To be convinced of this, one only has to consider a few telling events from the growing corporate interest in schools as profit-making ventures, the production of curricular materials by toy companies, or the increasing use of school space for the advertising of consumer goods.

The organization and regulation of culture by large media corporations such as Disney profoundly influence children's culture and increasingly dominate public discourse. The concentration of control over the means of producing, circulating, and exchanging information has been matched by the emergence of new technologies that have transformed culture, especially popular culture, into the primary educational site in which youth learn about themselves, their relationship to others, and the larger world. The Hollywood film industry, television, satellite broadcasting technologies, the Internet, posters, magazines, billboards, newspapers, videos, and other media forms and technologies have transformed culture into a pivotal force for, as Stuart Hill sees it, "shaping human meaning and behavior and regulat[ing] our social practices at every turn."[1] Although the endlessly proliferating sites of media production promise unlimited access to vast stores of information, such sites are increasingly controlled by a handful of multinational corporations. The Disney corporation's share of the communication industry represents a case in point. Disney's numerous holdings include a controlling interesting in twenty television stations that reach 25 percent of all U.S. households, and ownership of over twenty-one radio stations and the largest radio network in the United States, serving 3,400 stations and covering 24 percent of all households in the country. In addition, Disney owns three music studios, the ABC network, and five motion picture studios. Other holdings include, but are not limited to, television and cable channels, book publishing companies, sports teams (the Atlanta Braves, Atlanta Hawks, and World Championship Wrestling), theme parks, insurance companies, magazines, and multimedia productions.[2]

Mass-produced images fill our daily lives and condition our most intimate perceptions and desires. At issue for parents, educators, and others is how culture, particularly media culture, has become a substantive, if not the primary, educational force in regulating the meanings, values, and tastes that set the norms and conventions that offer up and legitimate particular subject positions. In other words, media culture influences what it means to claim an identity as male, female, white, black, citizen, or noncitizen as well as defining the meaning of childhood, the national past, beauty, truth, and social agency.[3] The scope and impact of new electronic technologies as teaching machines can be seen in some rather astounding statistics. It is estimated that "the average American spends more than four hours a day watching television. Four hours a day, 28 hours a week, 1456 hours a year."[4] Don Hazen and Julie Winokur report, citing American Medical Association statistics, that the "number of hours spent in front of a television or video screen is the single biggest chunk of time in the waking life of an American child."[5]

Such statistics warrant grave concern given that the pedagogical messages often provided through such programming are shaped largely by a $130-billion-a-year advertising industry, which not only sells its products but also values, images, and identities that are largely aimed at teaching young people to be consumers. Alarmed by the growing influence of the media on young children, the American Academy of Pediatrics released a report in 1999 claiming that the influence of television viewing among the young constituted a public health issue. The report urged parents not to allow children under two years old to watch television, and recommended that older children not be allowed televisions in their bedrooms.[6] It would be reductionistic not to recognize that there is also some excellent programming that is provided to audiences, but by and large much of what is produced on television and in the big Hollywood studios panders to the lowest common denominator, defines freedom as consumer choice, and debases public discourse by reducing it to a spectacle.[7]

. . .

No identity, desire, or need appears to escape the advertiser's grip. For instance, *Disney Magazine* recently ran an ad for the Baby Mickey porcelain doll. The ad features a baby wearing a cap with the logo Disney Babies on it. The baby's pajama top also has a Mickey Mouse logo on it, and just in case the reader has missed the point, the baby is holding a Baby Mickey doll, with Baby Mickey emblazoned on its bib. The caption for the ad reads, "He drifts off to dreamland with Baby Mickey to cuddle nearby!"[8] The doll is part of the Ashton-Drake Galleries collection offered to adults who can rewrite their own memories of being a child in Disney's terms while simultaneously indulging a commodified view of innocence that they can use to introduce their own infants to Disney's version of childhood. The ad appeals to innocence as it appropriates it at one of its most seductive and vulnerable moments.

In this case, innocence is emptied of any substantive content so that it can be commodified and exploited. Disney disingenuously presents innocence as that untouched psychic space in which the brutal world is forgotten, and, as Ernest Larsen notes, "the fullness of fantasy reliably, if pathetically compensates for the emptiness of reality."[9] Larsen argues that one of Walt Disney's greatest insights was that he "knew that we all believe ourselves to be like unto children. And he knew how to exploit the pathos and comedy—but especially the pathos—of that universal delusion of innocence."[10]

The Disney commercial fantasy machine uses innocence as a representational image to infantilize the very adults at whom it is aimed. The appeal to fantasy in Disney's perfectly scripted world functions to disable rather than liberate the imagination. Nothing is out of place in Disney's landscape, and under the rubric of *community* Disney purposely "confuses the personal and the corporate."[11] Within this context critical agency is replaced by corporate planning, allowing Disney to edit out conflict, politics, and contradictions, thus relieving adults and children of having to make choices beyond those that allow them to indulge in the fantasy of unfettered consumerism. Of course, the larger issue is that the commercialization of the media and the culture, in general, limits the choices that children and adults can make in extending their sense of agency beyond a commercial culture that enshrines an intensely myopic, narcissistic, and conservative sense of self and society . . . .

Beyond this notion of fantasy and entertainment that supports a loss of faith in public institutions and participatory democratic politics is a pedagogical model that suggests that those engaged with Disney culture become "quiet" citizens just as cast members in Disney's theme-park labor force are required to become utterly compliant and obsequious. Jane Kuenz captures this sentiment in the response of one of the Disney's "cast members" at Disney World, who explains, "You've got to keep your mouth shut. You can't tell them your opinion. You have to do everything they say. The Disney Way. Never say anything negative. Everything's positive. There's never a no. You never say I don't know. If you don't know something you find out fast, even on your own after work."[12] Within this type of model of cultural and moral regulation, Disney's image of innocence is completely nullified next to the power it exercises in dominating public discourse and undermining the social and political capacities necessary for individuals to sustain even the most basic institutions of democracy....

Consider the enormous control that a handful of transnational corporations have over the diverse properties that shape popular and media culture. Joshua Karliner has noted that "51 of the largest 100 economies in the world are corporations."[13] Moreover, the United States media is dominated by fewer than ten conglomerates, whose annual sales range from $10 to $27 billion. These include major corporations such as Time Warner, General Electric, Disney, Viacom, TCI, and Westinghouse. Not only are these firms major producers of much of the entertainment, news, culture, and information that permeates our daily lives, they also produce media software and have networks for distribution that include television and cable channels as well as retail stores.[14] Similarly, the heads of these corporations have amassed enormous amounts of personal wealth, further contributing to the shameless disparity of income between company CEOs and factory workers, on the one hand, and the new global elite and the rest of the world's population on the other. For instance, in 1997 Disney CEO Michael Eisner was paid more than $575 million. Russell Mokhiber and Rubert Weissman explain that, in addition to his $750,000 salary, Eisner claimed a $9.9 million bonus and cashed in on $565 million in stock options.[15] CEOs like Michael Eisner, Bill Gates, Warren Buffett, and others are part of an exclusive club of 358 global billionaires whose collective income, explains Zygmunt Bauman, "equals the combined incomes of [the] 2.3 billion poorest people (45 percent of the world's population)."[16] Much of what young people learn today is brought to them by a handful of corporate elite, including Eisner, Rupert Murdoch, and others, who have little regard for what kids learn beyond what it means for them to become consumers.

The recognition that this global elite and the corporations they own and control are involved in every aspect of cultural production—ranging from the production of identities, representations, and texts to control over the production, circulation, and distribution of cultural goods—is not a new insight to theorists of cultural studies. What has been missing from such work is an analysis of the educational role that such corporations play in promoting a public pedagogy that uses, as Raymond Williams has pointed out, the educational force of all of its institutions, resources, and relationships to actively and profoundly teach an utterly privatized notion of citizenship.[17] All too often within the parameters of such a public pedagogy, consumption is the only form of citizenship being offered to children, and democracy is privatized through an emphasis on egoistic individualism, low levels of participation in political life, and a diminishing of the importance of noncommodified public spheres. In what follows, I want to point to some theoretical and political implications for focusing on corporations as "teaching machines" engaged in a particular form of public pedagogy. In doing so, I will focus specifically on the Disney corporation and what I will call its discourse of innocence as a defining principle in structuring its public pedagogy of commercialism within children's culture.

DISNEY AND THE POLITICS OF INNOCENCE

Within the last decade, the rise of corporate power and its expansions into all aspects of everyday life has grown by leaps and bounds.[18] One of the most visible examples of such growth can be seen in the expanding role that the Walt Disney Company plays in shaping popular culture and daily life in the United States and abroad. The Disney Company is an exemplary model of the new face of corporate power at the beginning of the twenty-first century. Like many other megacorporations, its focus is on popular culture, and it continually expands its reach to include not only theme parks but television networks, motion pictures, cruise lines, baseball and hockey teams, Broadway theater, and a children's radio program. What is unique about Disney is that, unlike Time Warner or Westinghouse, its brand name is synonymous with the notion of childhood innocence. As an icon of American culture and middle-class family values, Disney actively appeals to both parental concerns and children's fantasies as it works hard to transform every child into a lifetime consumer of Disney products and ideas. In this scenario, a contradiction emerges between the company's cutthroat commercial ethos and a Disney culture that presents itself as the paragon of virtue and childlike innocence.

Disney has built its reputation on both profitability and wholesome entertainment. Largely removing itself from issues of power, politics, and ideology, it embraces a pristine self-image associated with the magic of pixie dust and Main Street USA. Yet this is merely the calculating rhetoric of a corporate giant whose annual revenues in 1997 and 1998 exceeded

$22 billion as a result of its ability to manufacture, sell, and distribute culture on a global scale, making it the world's most powerful leisure icon.[19] Michael Ovitz, a former Disney executive, touches on the enormous power Disney wields, claiming, "Disney isn't a company as much as it is a nation-state with its own ideas and attitudes, and you have to adjust to them."[20]

. . .

Disney has given new meaning to the politics of innocence as a narrative for shaping public memory and for producing a "general body of identifications" that promote a packaged and sanitized version of American history.[21] Innocence also serves as a rhetorical device that cleanses the Disney image of the messy influence of commerce, ideology, and power. In other words, Disney's strategic association with childhood, a world cleansed of contradictions and free of politics, represents not just the basic appeal of its theme parks and movies, but also provides a model for defining corporate culture separate from the influence of corporate power. . . . [B]ehind the rhetoric of innocence is the reality of a company that, Charles Kernaghan of the National Labor Committee argues, uses subcontractors to produce Disney clothing and toys in countries not only connected to military dictatorships but also actively engaged in child labor. The myth of innocence and fun seems all the more insidious given the fact, according to Russell Mokhiber and Robert Weissman, that in recent years Disney has outsourced production of its "clothing and toys to sweatshops in Haiti, Burma, Vietnam, China and elsewhere."[22] For example, one Disney subcontractor pays approximately one thousand factory workers in Vietnam "six to eight cents an hour, far below the subsistence wage estimated at 32 cents an hour."[23] In Haiti, workers produce Mickey Mouse pajamas; Pocahontas, Donald Duck, and Lion King T-shirts; and Hunchback of Notre Dame sweatshirts while being paid an hourly wage of 38 cents, or $3.30 a day.[24]

. . .

Eisner's toy-store promotional image is sullied by more than the company's use of sweatshops, child labor, and the commercial carpet bombing of children. It is also tarnished by Disney's often hostile and demeaning labor practices endured by its American workers—work relations scripted from beginning to end and supervised, in part, by hired spies and informants (known as shoppers) who contribute to an enforced "culture of mutually generating suspicion and dependence."[25] Eisner also willfully refuses to acknowledge responsibility for the role that Disney plays in harnessing children's identities and desires to an ever-expanding sphere of consumption; for editing public memory to reconstruct an American past in its own image; and for setting limits on democratic public life by virtue of its controlling influence on the media and its increasing presence in the schools. Education is never innocent, because it always presupposed a particular view of citizenship, culture, and society, and yet it is this very appeal to innocence, bleached of any semblance of politics and commerce that has become a defining feature of Disney culture and pedagogy.

The Walt Disney Company's attachment to the appeal of innocence provides a rationale for Disney to both reaffirm its commitment to children's pleasure and to downplay any critical assessments of the role Disney plays as a benevolent corporate power in sentimentalizing childhood innocence as it simultaneously commodifies it. Stripped of the historical and social constructions that give it meaning, innocence in the Disney universe becomes an atemporal, ahistorical, apolitical, and atheoretical space where children share a common bond free of the problems and conflicts of adult society. Disney both markets this ideal and presents itself as a corporate parent who safeguards this protective space for children by magically supplying the fantasies that nourish it and keep it alive....

MAKING THE POLITICAL MORE PEDAGOGICAL

Although my focus is on Disney because of its particular attempt to mystify its corporate agenda with appeals to fun, innocence, and purity, the seriousness of the threat to a vibrant democracy that Disney and other corporations present through their ownership of the media and their control over information cannot underestimated. I don't mean to suggest that Disney is engaged in a conspiracy to undermine American youth or democracy around the world. Nor do I want to suggest that Disney is part of an evil empire incapable of providing joy and pleasure to the millions of kids and adults who visit its theme parks, watch its videos and movies, or buy products from its toy stores. On the contrary, the main issue here is that such entertainment now takes place under conditions which, Toby Miller points out, the media "becomes a critical site for the articulation of a major intellectual shift in the ground of public discourse ... in which pricing systems are now brought to bear on any problem, anytime, anywhere."[26] In other words, media conglomerates such as Disney are not merely producing harmless entertainment, disinterested news stories, or unlimited access to the information age; nor are they removed from the realm of power, politics, and ideology. But recognition of the pleasure that Disney provides should not blind us to the realization that it is about more than the production of entertainment and enjoyment.

I also want to avoid suggesting that the effect of Disney films, radio stations, theme parks, magazines, and other products is the same for all those who are exposed to them. Disney is not a self-contained system of unchanging formal conventions. Disney culture, like all cultural formations, is riddled with contradictions; rather than viewing the Disney empire as monolithic, it is important to emphasize that within Disney culture there are also potentially subversive moments and pleasures.

In fact, any approach to studying Disney must address the issue of why so many kids and adults love Disney, and experience its theme parks, plays, and travel opportunities as a counterlogic that allows them to venture beyond the present while laying claim to unrealized dreams and hopes. For adults, Disney's theme parks offer an invitation to adventure, a respite from the drudgery of work, and an opportunity to escape from the alienation and boredom of daily life. As Susan Willis points out, Disney invites adults to construct a new sense of agency founded on joy and happiness and to do so by actively participating in their own pleasures, whether it be a wedding ceremony, a cruise ship adventure, or a weekend at the Disney Institute. Disney's appeal to pleasure and the "child in all of us" is also rooted in a history that encompasses the lives of many baby boomers. These adults have grown up with Disney culture and often "discover some nostalgic connection to [their] childhood" when they enter into the Disney cultural apparatus. In this sense, Willis notes, Disney can be thought of as an "immense nostalgia machine whose staging and specific attractions are generationally coded to strike a chord with the various age categories of its guests."[27] Similarly, Disney's power lies, in part, in its ability to tap into the lost hopes, abortive dreams, and utopian potential of popular culture.

Disney's appeal to the relationship between fantasy and dreams becomes all the more powerful against a broader American landscape in which cynicism has become a permanent fixture. Disney's invitation to a world where "the fun always shines" does more than invoke the utopian longing and promise of the sun-drenched vacation; it also offers an acute sense of the extraordinary in the ordinary, a powerful antidote to even the most radical forms of pessimism. But at the same time Disney's utopia points beyond the given while remaining firmly within it. As Ernst Bloch has pointed out, genuine wishes are felt here at the start but these are often siphoned off within constructions of nostalgia, fun, and childhood innocence that undercut the utopian dream of "something else"—that which extends beyond what the market and a commodity-crazed society can offer.[28] ...

Although it is true that people mediate what they see, buy, wear, and consume, and bring different meanings to the texts and products that companies like Disney produce, it is crucial that any attempt to deal with the relationship between culture and politics not stop with such a recognition but investigate both its limits and its strengths, particularly in dealing with the three- to eight-year-old crowd.[29] ...

How viewers use Disney's cultural texts to make sense of their lives, or how such texts mobilize those pleasures, identifications, and fantasies that connect audiences to the broader issues that make up their everyday experiences are crucial issues that need to be addressed in order to understand how the media do their pedagogical work without reducing respondents to passive dupes.[30] However, the ways in which such messages, products, and conventions "work" on audiences is one that must be left open to the investigation of particular ethnographic interventions and/or pedagogical practices. There is no virtue, ideologically or politically, in simply pronouncing what Disney means—as if that is all there is to do. I am suggesting a very different approach to Disney, one that highlights the pedagogical and the contextual by raising questions about Disney—such as what role it plays in shaping childhood identity, public memory, national identity, gender roles, or in suggesting who qualifies as an American or what the role of consumerism is in American life—that expand the scope of inquiry in order to allow people to enter into such a discussion in a way that they ordinarily might not have. Disney needs to be engaged as a public discourse, and doing so means offering an analysis that prompts civic discourse and popular culture to rub up against each other. Such an engagement represents both a pedagogical intervention and a way of recognizing the changing contexts in which any text must be understood and engaged.

Questioning what Disney teaches is part of a much broader inquiry regarding what it is that parents, children, educators and other cultural workers need to know in order to critique and challenge, when necessary, those institutional and cultural forces that have a direct impact on public life. Such inquiry is all the more important at a time when corporations hold such an inordinate amount of power in shaping children's culture into a largely commercial endeavour, using their various cultural technologies as teaching machines to relentlessly commodify and homogenize all aspects of everyday life, and in this sense posing a potential threat to the real freedoms associated with a substantive democracy. Yet questioning what megacorporations such as Disney teach also means appropriating the most resistant and potentially subversive ideas, practices, and images at work in their various cultural productions.

NOTES

I have developed many of the ideas presented in this chapter in Henry A. Giroux, *The Mouse That Roared: Disney and the End of Innocence* (Lanham, Md.: Rowman and Littlefield, 1999).

1. Stuart Hall, "The Centrality of Culture: Notes on the Cultural Revolutions of Our Time," in Kenneth Thompson, ed., *Media and Cultural Regulation* (Thousand Oaks, Calif.: Sage, 1997), p. 232.

2. A list of Disney's holdings, cited in "The National Entertainment State Media Map," *The Nation*, June 3, 1996, pp. 23–26. See also, Robert W. McChesney, "Oligopoly:

The Media Game Has Fewer and Fewer Players," *The Progressive*, November 1999, p. 22.

3. The concentrated power of the media market by corporations is evident in the following figures. According to Robert W. McChesney, "In cable, Time Warner and TCI [now AT & T] control 47.4 percent of all subscribers; in radio, Westinghouse, in addition to owning the CBS Television network, now owns 82 radio stations; in books, Barnes & Noble and Borders sell 45 percent of all books in the United States.... In newspapers, only 24 cities compared to 400 fifty years ago support two or more daily newspapers." Robert W. McChesney, "Global Media for the Global Economy," in Don Hazen and Julie Winokur, eds., *We the Media* (New York: The New Press, 1997), p. 27.

4. Don Hazen and Julie Winokur, "Children Watching TV," Hazen and Winokur, eds., *We the Media* (New York: The New Press, 1997), p. 64.

5. Ibid.

6. See Lawrie Mifflin, "Pediatricians Urge Limiting TV Watching," *New York Times*, August 4, 1999, pp. A1, A4.

7. Pierre Bourdieu's analysis of the systemic corruption of television in France is equally informative when applied to the United States. See his *On Television*, trans. Priscilla Parkhurst Ferguson (New York: The New Press, 1998).

8. *Disney Magazine,* Fall 1998, p. 61.

9. Ernest Larsen, "Compulsory Play," *The Nation*, March 16, 1998, p. 31.

10. Ibid., p. 32.

11. Karen Klugman, et al., *Inside the Mouse: Work and Play at Disney World* (Durham, N.C.: Duke University Press, 1995), p. 119.

12. Klugman, et al., *Inside the Mouse: Work and Play at Disney World*, p. 119.

13. Joshua Karliner , "Earth Predators," *Dollars and Sense*, July/August 1998, p. 7. For an extensive analysis of media concentration and its effects on democracy, see Robert W. McChesney, *Rich Media, Poor Democracy* (Urbana, Ill.: University of Illinois Press, 1999).

14. Robert W. McChesney, *Corporate Media and the Threat to Democracy* (New York: Seven Stories Press, 1997), p. 18. There is an enormous amount of detailed information about the new global conglomerates and their effects on matters of democracy, censorship, free speech, social policy, national identity, and foreign policy. For example, see such classics as Herbert I. Schiller, *Culture Inc.: The Corporate Takeover of Public Expression* (New York: Oxford University Press, 1989); Edward S. Herman and Noam Chomsky, *Manufacturing Consent* (New York: Pantheon, 1988); Ben H. Bagdikian, *The Media Monopoly*, 4th ed. (Boston: Beacon Press, 1992); George Gerbner and Herbert I. Schiller, eds., *Triumph of the Image* (Boulder, Colo.: Westview Press, 1992); Douglas Kellner, *Television and the Crisis of Democracy* (Boulder, Colo.: Westview Press, 1990); Philip Schlesinger, *Media, State and Nation* (London: Sage, 1991); John Fiske, *Media Matters* (Minneapolis: University of Minnesota Press, 1994); Jeff Cohen and Norman Solomon, *Through the Media Looking Glass* (Monroe, Maine: Common Courage Press, 1995); and Erik Barnouw, *Conglomerates and the Media* (New York: The New Press, 1997).

15. Russell Mokhiber and Robert Weissman, *Corporate Predators: The Hunt for Mega-Profits and the Attack on Democracy* (Munroe, Maine: Common Courage Press, 1999), p. 167.

16. Zygmunt Bauman, *Globalization: The Human Consequences* (New York: Columbia University Press, 1998), p. 70.

17. Raymond Williams, *The Politics of Modernism* (London: Verso Press, 1989).

18. Two recent critical commentaries can be found in Robert W. McChesney, *Corporate Media and the Threat to Democracy* (New York: Seven Stories Press, 1997); and Erik Barnouw, et al., eds., *Conglomerates and the Media* (New York: The New Press, 1997).

19. Figures taken from Michael D. Eisner , "Letter to Shareholders," *The Walt Disney Company 1997 Annual Report* (Burbank, Calif.: The Walt Disney Company, 1997), p. 2. For the 1998 reference, see Lauren R. Rublin, "Cutting Back the Magic," *Barron's*, July 26, 1999, p. 28.

20. Michael Ovitz, cited in Peter Bart, "Disney's Ovitz Problem Raises Issues for Showbiz Giants," *Daily Variety*, December 16, 1996, p. 1.

21. Kenneth Burke, *A Rhetoric of Motives* (Berkeley and Los Angeles: University of California Press, 1962), p. 26.

22. Mokhiber and Weissman, *Corporate Predators*, p. 168.

23. Ibid.

24. Ray Sanches, " 'Misery' for Haitian Workers," *New York Newsday*, June 16, 1996, pp. A4, A30.

25. Karen Klugman, et al., *Inside the Mouse*, p. 125. For an excellent chapter on Disney's relations with its workers at Disney World, see Jane Kuenz , "Working at the Rat," also in Klugman, et al., *Inside the Mouse*, pp. 110–61.

26. Toby Miller, *Technologies of Truth* (Minneapolis: University of Minnesota Press, 1998), p. 90.

27. Susan Willis, "Problem with Pleasure," in Klugman, et al., eds., *Inside the Mouse: Work and Play at Disney World,* p. 5.

28. See Ernst Bloch, *The Utopian Function of Art and Literature,* trans. Jack Zipes and Frank Mecklenburg (Cambridge, Mass.: MIT Press, 1988). Ernst Bloch, *The Principle of Hope*, vol. 1, trans. Neville Plaice, Stephen Plaice, and Paul Knight (Cambridge, Mass.: MIT Press, 1986).

29. I invoke here Meaghan Morris's argument in which she identifies the chief error of cultural studies to be the narcissistic identity that is made "between the knowing subject of cultural studies, and a collective subject, the 'people.' " The people in this discourse "have no necessary defining characteristics—except an indomitable capacity to 'negotiate' readings, generate new interpretations, and remake the materials of culture.... So against the hegemonic force of the dominant classes, 'the people' in fact represent the most creative energies and functions of critical reading. In the end they are not simply the cultural student's object of study, [but] his native informants. The people are also the textually delegated, allegorical emblem of the critic's own activity." See Meaghan Morris, "Banality in Cultural Studies," *Discourse* 10:2 (Spring/Summer 1988), p. 17.

30. This issue is raised in an interesting way in Roger Silverstone, "So Who Are These People," *Sight and Sound* 9:5 (May 1999), pp. 28–29.

PART FOUR

Social Institutions: Ordering Collective Life

Another fundamental sociological problem concerns the very existence of social order itself. We don't usually wake up in the morning consciously reminding ourselves that we have to help create a group that will stick together today. So when, where, and how does it happen? How are we capable of uniting to form groups and societies in the first place? How is organization and order accomplished and sustained? Why doesn't society break down into a chaotic struggle between conflicting interest groups, or collapse into a war of all against all? This set of questions refers to what is usually termed the social order problem and is linked to any discussion of how human (or, more pointedly, civil) society is possible.

Some sociologists have stressed the importance of members obeying a ruling authority. Others have highlighted the importance of group consensus concerning shared values, beliefs, norms, and ideas. Still others have emphasized the power of the ruling class and the role of ideology in establishing and maintaining social order. However, all of these explanations point to the significant role that social institutions perform in ordering collective life and enabling everyday social interaction to occur as it does.

Social institutions are themselves part of a group's social structure, and social structure is a formal way of describing and thinking about how group life is organized. Analogies to a building's architecture or biological organisms have been made to help make concrete the idea of social structure. For example, how would you describe the structure of the building you are in? How many floors does it have? How many rooms and doors and sets of stairs? For what purpose is it organized in the way it is? And, of course, where are you located in this building? Now think of a group or organization you are part of. How would you describe the way it is structured, the various statuses that members have, and the roles that they play? Where are you located in this structure?

While there are different ways of depicting a group's structure, the consequences of social structure or a group's form of organization are very real. A group's social structure establishes the opportunities and constraints experienced by individuals as they live out their lives with others within a society. Together the social positions, group divisions, ranking systems, social relationships, and social institutions regulate how individuals act and interact in particular social situations. How you act at home with your family is probably significantly different from the way you act in the classroom or at your workplace. It is likely, though, that there are particular and recurrent patterns of interaction that are associated with each of these settings. What you are experiencing in each of these situations is the orderly character of social structure.

But these places—the home, the classroom, and the workplace—are not just physical locations. They are physical sites that point to, and often support, the social institutions that order collective life. The concept of social institution itself refers to the systems which emerge to manage essential problems common to all societies. While your college or university may be a specific educational institution, the formal educational system as a whole is a social institution. Similarly with family. Your family is a specific institution, but the social institution we call family preceded your own family and will certainly outlast it as well.

Social institutions can be described and analyzed in both cultural and social structural terms. Culturally, the focus is on the system of beliefs, norms, and values that define and regulate how a group addresses problems common to all societies. Structurally, the focus is on the social relationships and organizing principles that underlie the formal framework of relationships set up to address the problem.

The articles in this section concern some of the major social institutions found in modern society and examine how they order collective life on a micro and/or macro level. In particular they take up the social institutions of the family, gender, work, sports, media, politics, and the global political economic system. We have selected articles which identify strains and contradictions inherent in these social institutions, and which pose challenging questions about them. We are familiar with the coercive and persuasive power of these institutions. But what would a world without them look like?

The Family

Claude Levi-Strauss

In this article Levi-Strauss explores the universality and origins of the family institution. Drawing on anthropological studies of marriage and family arrangements around the world, he argues that the origins of marriage and the family are not essentially natural, as we generally assume, but social. Levi-Strauss finds that a recognizable structure underlies the diverse marriage and family patterns, and that it involves a division of labour among the sexes and an incest taboo among members of the family grouping. The morality associated with both of these indicates that, like the family itself, they are not based in nature though we customarily explain them in naturalistic terms. Levi-Strauss concludes that they both function to establish conditions of reciprocity in which groups are forced to establish relationships of interdependency and exchange with one another, thereby making human society possible.

What evidence can be provided to support the author's claim that the family and the gendered division of labour are not natural arrangements? What are the implications of his inquiry for current debates about the changing family?

The word *family* is so plain, the kind of reality to which it refers is so close to daily experience that one may expect to be confronted in this chapter with a simple situation. Anthropologists, however, are a strange breed; they like to make even the "familiar" look mysterious and complicated. As a matter of fact, the comparative study of the family among many different peoples has given rise to some of the most bitter arguments in the whole history of anthropological thought and probably to its more spectacular reversal.

During the second half of the nineteenth century and the beginning of the twentieth, anthropologists were working under the influence of biological evolutionism. They were trying to organize their data so that the institutions of the simpler people would correspond to an early state of the evolution of mankind, while our own institutions were related to the more advanced or developed forms. And since, among ourselves, the

Source: Claude Levi-Strauss, "The Family," from *Man, Culture and Society* (2005), edited by Harry Shapiro, pp. 142–170. By permission of Oxford University Press, Inc.

family founded on monogamic marriage was considered as the most praiseworthy and cherished institution, it was immediately inferred that savage societies—equated for the purpose with the societies of man at the beginning of its existence—could only have something of a different type. Therefore, facts were distorted and misinterpreted; even more, fanciful "early" stages of evolution were invented, such as "group marriage" and "promiscuity" to account for the period when man was still so barbarous that he could not possibly conceive of the niceties of the social life it is the privilege of civilized man to enjoy. Every custom different from our own was carefully selected as a vestige of an older type of social organization.

This way of approaching the problem became obsolete when the accumulation of data made obvious the following fact: the kind of family featured in modern civilization by monogamous marriage, independent establishment of the young couple, warm relationship between parents and offspring, etc., while not always easy to recognize behind the complicated network of strange customs and institutions of savage peoples, is at least conspicuous among those which seem to have remained on—or returned to—the simplest cultural level.

There are two ways of interpreting this preeminence of the family at both ends of the scale of development of human societies. Some writers have claimed that the simpler peoples may be considered as a remnant of what can be looked at as a "golden age," prior to the submission of mankind to the hardships and perversities of civilization; thus, man would have known in that early stage the bliss of monogamic family only to forgo it late until its more recent Christian rediscovery. The general trend, however, ... has been that more and more anthropologists have become convinced that familial life is present practically everywhere in human societies, even in those with sexual and educational customs very remote from our own. Thus, after they had claimed for about fifty years that the family, as modern societies knew it, could only be a recent development and the outcome of a slow and long-lasting evolution, anthropologists now lean toward the opposite conviction, i.e., that the family, consisting of a more or less durable union, socially approved, of a man, a woman, and their children, is a universal phenomenon, present in each and every type of society.

These extreme positions, however, suffer equally from oversimplification. It is well known that, in very rare cases, family bonds cannot be claimed to exist. A telling example comes from the Nayar, a very large group living on the Malabar coast of India. In former times, the warlike type of life of the Nayar men did not allow them to found a family. Marriage was a purely symbolical ceremony which did not result in a permanent tie between a man and a woman. As a matter of fact, married women were permitted to have as many lovers as they wished. Children belonged exclusively to the mother line, and familial as well as land authority was exercised not by the ephemeral husband but by the wife's brothers. Since land was cultivated; by an inferior caste, subservient to the Nayar, a woman's brothers were as completely free as their sister's husband or lovers to devote themselves to military activities.

There are a large number of human societies which, although they did not go quite as far as the Nayar in denying recognition to the family as a social unit, have nevertheless limited this recognition by their simultaneous admission of patterns of a different type. For instance, the Masai and the Chagga, both of them African tribes, did recognize the family as a social unit. However, and for the same reason as the Nayar, this was not true for the younger class of adult men who were dedicated to warlike activities and consequently were not allowed to marry and found a family. They used to live in regimental organizations and were permitted, during that period, to have promiscuous relations with the younger class of adult girls. Thus, among these peoples, the family did exist side by side with a promiscuous, non-familial type of relations between the sexes.

During recent years anthropologists have taken great pains to show that, even among people who practice wife-lending, either periodically in religious ceremonies or on a statutory basis (as where men are permitted to enter into a kind of institutional friendship entailing wife-lending among members), these customs should not be interpreted as survivals of "group marriage" since they exist side by side with, and even imply, recognition of the family. It is true enough that, in order to be allowed to lend one's wife, one should first get one. However, if we consider the case of some Australian tribes as the Wunambal of the northwestern part of the continent, a man who would not lend his wife to her other potential husbands during ceremonies would be considered as "very greedy," i.e., trying to keep for himself a privilege intended by the social group to be shared between numerous persons equally entitled to it. And since that attitude toward sexual access to a woman existed along with the official dogma that men have no part in physiological procreation (therefore doubly denying any kind of bond between the husband and his wife's children), the family becomes an economic grouping where man brings the products of his hunt and the woman those of her collecting and gathering. Anthropologists, who claim that this economic unit built upon a "give and take" principle is a proof of the existence of the family even among the lowest savages, are certainly on no sounder basis than those who maintain that such a kind of family has little else in common than the word used to designate it with the family as it has been observed elsewhere.

The same relativistic approach is advisable in respect to the polygamous family. The word polygamy, it should be recalled, refers to polygyny, that is, a system where a man is entitled to several wives, as well as to polyandry, which is the complementary system where several husbands share one wife.

Now it is true that in many observed cases, polygamous families are nothing else than a combination of several monogamous families, although the same person plays the part of several spouses. For instance, in some tribes of Bantu Africa, each wife lives in a separate hut with her children, and the only difference with the monogamous family results from the fact that the same man plays the part of husband to all his wives. There are other instances, however, where the situation is not so clear. Among the Tupi-Kawahib of central Brazil, a chief may marry several women who may be sisters, or even a mother and her daughters by former marriage; the children are raised together by the women, who do not seem to mind very much whether they nurse their own children or not; also, the chief willingly lends his wives to his younger brothers, his court officers, or to visitors. Here we have not only a combination of polygyny and polyandry, but, the mix-up is increased even more by the fact that the co-wives may be united by close consanguineous ties prior to their marrying the same man.

As to polyandry proper, it may sometimes take extreme forms, as among the Toda where several men, usually brothers, share one wife, the legitimate father of the children being the one who has performed a special ceremony and who remains legal father of all the children to be born until another husband decides to assume the right of fathership by the same process. In Tibet and Nepal, polyandry seems to be explained by occupational factors of the same type as those already stated for the Nayar: for men living a semi-nomadic existence as guides and bearers, polyandry provides a good chance that there will be, at all times, at least one husband at hand to take care of the homestead.

Therefore, it becomes apparent why the problem of the family should not be approached in a dogmatic way. As a matter of fact, this is one of the more elusive questions in the whole field of social organization. Of the type of organization which prevailed in the early stages of mankind, we know very little, since the remnants of man during the Upper Palaeolithic Period of about 50,000 years ago consist principally of skeletal fragments and stone implements which provide only a minimum of information on social customs and laws. On the other hand, when we consider the wide diversity of human societies which have been observed since, let us say, Herodotus' time until present days, the only thing which can be said is as follows: monogamic, conjugal family is fairly frequent. Wherever it seems to be superseded by different types of organization, this generally happens in very specialized and sophisticated societies and not, as was previously expected, in the crudest and simplest types. Moreover, the few instances of non-conjugal family (even in its polygamous form) establish beyond doubt that the high frequency of the conjugal type of social grouping does not derive from a universal necessity. It is at least conceivable that a perfectly stable and durable society could exist without it. Hence, the difficult problem: if there is no natural law making the family universal, how can we explain why it is found practically everywhere?

In order to try to solve the problem, let us try first to define the family, not by integrating the numerous factual observations made in different societies nor even by limiting ourselves to the prevailing situation among us, but by building up an ideal model of what we have in mind when we use the word *family*. It would then seem that this word serves to designate a social group offering at least three characteristics: (1) it finds its origin in marriage; (2) it consists in husband, wife, and children born out of their wedlock, though it can be conceived that other relatives may find their place close to that nuclear group; and (3) the family members are united together by (a) legal bonds, (b) economic, religious, and other kinds of rights and obligations, (c) a precise network of sexual rights and prohibitions, and a varying and diversified amount of psychological feelings such as love, affection, respect, awe, etc. We will now proceed to a close examination of several aspects in the light of the available data.

As we have already noticed, marriage may be monogamous or polygamous. It should be pointed out immediately that the first kind is not only more frequently found than the second, but even much more than a cursory inventory of human societies would lead to believe. Among the so-called polygamous societies, there are undoubtedly a substantial number which are authentically so; but many others make a strong difference between the "first" wife who is the only true one, endowed with the full right attached to the marital status, while the other ones are sometimes little more than official concubines. Besides, in all polygamous societies, the privilege of having several wives is actually enjoyed by a small minority only. This is easily understandable, since the number of men and women in any random grouping is approximately the same with a normal balance of about 110 to 100 to the advantage of either sex.

Therefore, it is not necessary to wonder a great deal about the predominance of monogamic marriage in human societies. That monogamy is not inscribed in the nature of man is sufficiently evidenced by the fact that polygamy exists in widely different forms and in many types of societies; on the other hand, the prevalence of monogamy results from the fact that, unless special conditions are voluntarily or involuntarily brought about, there is, normally, about just one woman available for each man. In modern societies, moral, religious, and economic reasons have officialized monogamous marriage (a rule which is in actual practice breached by such different means as premarital freedom, prostitution, and adultery). But in societies which are on a much lower cultural level and where there is no prejudice against polygamy, and even where polygamy may be actually permitted or desired, the same result can be brought about by the lack of social or economic differentiation, so that each man has neither the means, nor the power, to obtain more than one wife and, consequently, everybody is obliged to make a virtue of necessity.

If there are many different types of marriage to be observed in human societies—whether monogamous or polygamous, and in the last case, polygynous, polyandrous, or both; and whether by exchange, purchase, free choice or imposed by the family, etc.—the striking fact is that everywhere a distinction exists between marriage, i.e., a legal, group-sanctioned bond between a man and a woman, and the type of permanent or temporary union resulting either from violence or consent alone.

In the first place, nearly all societies grant a very high rating to the married status. Wherever age-grades exist, either in an institutional way or as non-crystallized forms of grouping, some connection is established between the younger adolescent group and bachelorhood, less young and married without children, and adulthood with full rights, the latter going usually on par with the birth of the first child.

What is even more striking is the true feeling of repulsion which most societies have toward bachelorhood. Generally speaking it can be said that, among the so-called primitive tribes, there are no bachelors, simply for the reason that they could not survive.

. . .

This is true of the bachelor and also, to a lesser extent, of a couple without children. Indeed they can make a living, but there are many societies where a childless man (or woman) never reaches full status within the group, or else, beyond the group, in this all-important society which is made up of dead relatives, and where one can only expect recognition as ancestor through the cult, rendered to him or her by one's descendants. Conversely, an orphan finds himself in the same dejected condition as a bachelor. As a matter of fact, both terms provide sometimes the strongest insults existing in the native vocabulary. Bachelors and orphans can even be merged together with cripples and witches, as if their conditions were the outcome of some kind of supernatural malediction.

The interest shown by the group in the marriage of its members can be directly expressed, as it is the case among us where prospective spouses, if they are of marriageable age, have first to get a licence and then to secure the services of an acknowledged representative of the group to celebrate their union. Although this direct relationship between the individuals, on the one hand, and the group as a whole, on the other, is known at least sporadically in other societies, it is by no means a frequent case. It is almost a universal feature of marriage that it is originated, not by the individuals but by the groups concerned (families, lineage, clans, etc.), and that it binds the groups before and above the individuals. Two kinds of reasons bring about this result: on the one hand, the paramount importance of being married tends to make parents, even in very simple societies, start early to worry about obtaining a suitable mate for their offspring and this, accordingly, may lead to children being promised to each other from infancy. But above all, we are confronted here with that strange paradox to which we shall have to return later on, namely, that although marriage gives birth to the family, it is the family, or rather families, which produce marriage as the main legal device at their disposal to establish an alliance between themselves. As a New Guinea native put it, the real purpose of getting married is not so much to obtain a wife but to secure brothers-in-law. If marriage takes place between groups rather than individuals, a large number of strange customs become immediately clearer. For instance, we understand why in some parts of Africa, where descent follows the father's line, marriage becomes final only when the woman has given birth to a male child, thus fulfilling its function of maintaining her husband's lineage.... But whatever the way in which the collectivity expresses its interest in the marriage of its members, whether through the authority vested in strong consanguineous groups, or more directly through the intervention of the State, it remains true that marriage is not, is never, and cannot be a private business.

We have to look for cases as extreme as the Nayar, already described, to find societies where there is not, at least, a temporary de facto union of the husband, wife, and their children. But we should be careful to note that, while such a group among us constitutes the family and is given legal recognition, this is by no means the case in a large number of human societies. Indeed, there is a maternal instinct which compels the mother to care for her children and makes her find a deep satisfaction in exercising those activities, and there are also psychological drives which explain that a man may feel warmly toward the offspring of a woman with whom he is living, and the development of which he witnesses step by step, even if he does not believe (as is the case among the tribes who are said to disclaim physiological paternity) that he had any actual part in their procreation.

The great majority of societies, however, do not show a very active interest in a kind of grouping which, to some of them at least (including our own), appears so important. Here, too, it is the groups which are important, not the temporary aggregate of the individual representatives of the group. For instance, many societies are interested in clearly establishing the relations of the offspring with the father's group on the one hand, and with the mother's group on the other, but they do it by differentiating strongly the two kinds of relationships. Territorial rights may be inherited through one line, and religious privileges and obligations through the other. Or else, status from one side, magical techniques from the other.

In most of contemporary India and in many parts of western and eastern Europe, sometimes as late as the nineteenth century, the basic social unit was constituted by a type of family which should be described as *domestic* rather than *conjugal*: ownership of the land and of the homestead, parental authority and economic leadership were vested in the eldest living ascendant, or in the community of brothers issued from the same ascendant. In the Russian *bratsvo*, the south-Slavic *zadruga*, the French *maisnie*, the family actually consisted of the elder or the surviving brothers, together with their wives, married sons with their wives and unmarried daughters, and so on down to the great-grandchildren. Such large groups, which could sometimes include several dozen persons living and working under a common authority, have been designated as *joint families* or *extended families*. Both terms are useful but misleading, since they imply that these large units are made up of small conjugal families. As we have already seen, while it is true that the conjugal family limited to mother and children is practically universal, since it is based on the physiological and psychological dependency which exists between them at least for a certain time, and that the conjugal family consisting of husband, wife, and children is almost as frequent for psychological and economic reasons which should be added to those previously mentioned, the historical process which has led among ourselves to the legal recognition of the conjugal family is a very complex one: it has been brought about only in part through an increasing awareness of a natural situation. But there is little doubt that, to a very large extent, it has resulted from the narrowing down to a group, as small as can be, the legal standing of which, in the past of our institutions, was vested for centuries in very large groups. In the last instance, one would not be wrong in disallowing the terms joint family and extended family. Indeed, it is rather the   conjugal family which deserves the name of *restricted family*.

We have just seen that when the family is given a small functional value, it tends to disappear even below the level of the conjugal type. On the contrary, when the family has a great functional value, it becomes actualized much above that level. Our would-be universal conjugal family, then, corresponds more to an unstable equilibrium between extremes than to a permanent and everlasting need coming from the deepest requirements of human nature.

To complete the picture, we have finally to consider cases where the conjugal family differs from our own, not so much on account of a different amount of functional value, but rather because its functional value is conceived in a way qualitatively different from our own conceptions.

As will be seen later on, there are many peoples for whom the kind of spouse one should marry is much more important than the kind of match they will make together. These people are ready to accept unions which to us would seem not only unbelievable, but in direct contradiction with the aims and purposes of setting up a family. For instance, the Siberian

Chukchee were not in the least abhorrent to the marriage of a mature girl of let us say about twenty, with a baby-husband two or three years old. Then, the young woman, herself a mother by an authorized lover, would nurse together her own child and her little husband. Like the North American Mohave, who had the opposite custom of a man marrying a baby girl and caring for her until she became old enough to fulfil her conjugal duties, such marriages were thought of as very strong ones, since the natural feelings between husband and wife would be reinforced by the recollection of the parental care bestowed by one of the spouses on the other. These are by no means exceptional cases to be explained by extraordinary mental abnormalities. Examples could be brought together from other parts of the world: South America, both highland and tropical, New Guinea, etc.

As a matter of fact, the examples just given still respect, to some extent, the duality of sexes which we feel is a requirement of marriage and raising a family. But in several parts of Africa, women of high rank were allowed to marry other women and have them bear children through the services of unacknowledged male lovers, the noble woman being then entitled to become the "father" of her children and to transmit to them, according to the prevalent father's right, her own name, status, and wealth. Finally, there are the cases, certainly less striking, where the conjugal family was considered necessary to procreate the children but not to raise them, since each family did endeavour to retain somebody else's children (if possible of a higher status) to raise them while their own children were similarly retained (sometimes before they were born) by another family. This happened in some parts of Polynesia, while "fosterage," i.e., the custom whereby a son was sent to be raised by his mother's brother, was a common practice on the Northwest Coast of America as well as in European feudal society.

During the course of centuries we have become accustomed to Christian morality, which considers marriage and setting up a family as the only way to prevent sexual gratification from being sinful. That connection has been shown to exist elsewhere in a few scattered instances; but it is by no means frequent. Among most people, marriage has very little to do with the satisfaction of the sexual urge, since the social setup provides for many opportunities which can be not only external to marriage, but even contradictory to it. For instance, among the Muria of Bastar, in central India, when puberty comes, boys and girls are sent to live together in communal huts where they enjoy a great deal of sexual freedom, but after a few years of such leeway they get married according to the rule that no former adolescent lovers should be permitted to unite. Then, in a rather small village, each man is married to a wife whom he has known during his younger years as his present neighbour's (or neighbours') lover.

On the other hand, if sexual considerations are not paramount for marriage purposes, economic necessities are found everywhere in the first place. We have already shown that what makes marriage a fundamental need in tribal societies is the division of labour between the sexes.

Like the form of the family, the division of labour stems more from social and cultural considerations than from natural ones. Truly, in every human group, women give birth to children and take care of them, and men rather have as their specialty hunting and warlike activities. Even there, though, we have ambiguous cases: of course men never give birth to babies, but in many societies, as we have seen with the couvade, they are made to act as if they did. And there is a great deal of difference between the Nambikwara father nursing his baby and cleaning it when it soils itself, and the European nobleman of not long ago to whom his children were formally presented from time to time, being otherwise confined to the women's quarters until the boys were old enough to be taught riding and fencing.

When we turn to activities less basic than child-rearing and war-making, it becomes still more difficult to discern rules governing the division of labour between the sexes. The Bororo women till the soil while among the Zuni this is man's work; according to tribe, hut-building, pot-making, weaving, may be incumbent upon either sex. Therefore, we should be careful to distinguish the fact of the division of labour between the sexes which is practically universal, from the way according to which different tasks are attributed to one or the other sex, where we should recognize the same paramount influence of cultural factors, let us say the same artificiality which presides over the organization of the family itself.

Here, again, we are confronted with the same question we have already met with: if the natural reasons which could explain the division of labour between the sexes do not seem to play a decisive part, as soon as we leave the solid ground of women's biological specialization in the production of children, why does it exist at all? The very fact that it varies endlessly according to the society selected for consideration shows that, as for the family itself, it is the mere fact of its existence which is mysteriously required, the form under which it comes to exist being utterly irrelevant, at least from the point of view of any natural necessity. However, after having considered the different aspects of the problem, we are now in a position to perceive some common features which may bring us nearer to an answer than we were at the beginning of this chapter. Since family appears to us as a positive social reality, perhaps the only positive social reality, we are prone to define it exclusively by its positive characteristics. Now it should be pointed out that whenever we have tried to show what the family is, at the same time we were implying what it is not, and the negative aspects may be as important as the others. To return to the division of labour we were just discussing, when it is stated that one sex must perform certain tasks, this also means that the other sex is forbidden to do them. In that light, the sexual division of labour is nothing else than a device to institute a reciprocal state of dependency between the sexes.

The same thing may be said of the sexual side of the family life. Even if it is true, as we have shown, that the family can be explained on sexual grounds, since for many tribes, sexual life and the family are by no means as closely connected as our moral norms would make them, there is a negative aspect which is much more important: the structure of the family, always and everywhere, makes certain types of sexual connections impossible, or at least wrong.

Indeed, the limitations may vary to a great extent according to the culture under consideration. In ancient Russia, there was a custom known as *snokatchestvo* whereby a father was entitled to a sexual privilege over his son's young wife; a symmetrical custom has been mentioned in some parts of southeastern Asia where the persons implied are the sister's son and his mother's brother's wife. We ourselves do not object to a man marrying his wife's sister, a practice which English law still considered incestuous in the mid-nineteenth century. What remains true is that every known society, past or present, proclaims that if the husband-wife relationship, to which, as just seen, some others may eventually be added, implies sexual rights, there are other relationships equally derived from the familial structure, which make sexual connections inconceivable, sinful, or legally punishable. The universal prohibition of incest specifies, as a general rule, that people considered as parents and children, or brother or sister, even if only by name, cannot have sexual relations and even less marry each other. In some recorded instances—such as ancient Egypt, pre-Columbian Peru, also some African, southeast Asian, and Polynesian kingdoms—incest was defined far less strictly than elsewhere. Even there, however, the rule existed, since incest was limited to a minority group, the ruling class....

The space at our disposal is too short to demonstrate that, in this case as previously, there is no natural ground for the custom. Geneticists have shown that while consanguineous marriages are likely to bring ill effects in a society which has consistently avoided them in the past, the danger would be much smaller if the prohibition had never existed, since this would have given ample opportunity for the harmful hereditary characteristics to become apparent and be automatically eliminated through selection: as a matter of fact, this is the way breeders improve the quality of their subjects. Therefore, the dangers of consanguineous marriages are the outcome of the incest prohibition rather than actually explaining it. Furthermore, since very many primitive peoples do not share our belief in biological harm resulting from consanguineous marriages, but have entirely different theories, the reason should be sought elsewhere, in a way more consistent with the opinions generally held by mankind as a whole.

The true explanation should be looked for in a completely opposite direction, and what has been said concerning the sexual division of labour may help us to grasp it. This has been explained as a device to make the sexes mutually dependent on social and economic grounds, thus establishing clearly that marriage is better than celibacy. Now, exactly in the same way that the principle of sexual division of labour establishes a mutual dependency between the sexes, compelling them thereby to perpetuate themselves and to found a family, the prohibition of incest establishes a mutual dependency between families, compelling them, in order to perpetuate themselves, to give rise to new families. It is through a strange oversight that the similarity of the two processes is generally over-looked on account of the use of terms as dissimilar as *division,* on the one hand, and *prohibition* on the other. We could easily have emphasized only the negative aspect of the division of labour by calling it a prohibition of tasks; and conversely, outlined the positive aspects of incest-prohibition by calling it the principle of division of marriageable rights between families. For incest-prohibition simply states that families (however they should be defined) can marry between each other and that they cannot marry inside themselves.

We now understand why it is so wrong to try to explain the family on the purely natural grounds of procreation, motherly instinct, and psychological feelings between man and woman and between father and children. None of these would be sufficient to give rise to a family, and for a reason simple enough: for the whole of mankind, the absolute requirement for the creation of a family is the previous existence of two other families, one ready to provide a man, the other one a woman, who will through their marriage start a third one, and so on indefinitely. To put it in other words: what makes man really different from the animal is that, in mankind, a family could not exist if there were no society; i.e., a plurality of families ready to acknowledge that there are other links than consanguineous ones, and that the natural process of filiation can only be carried on through the social process of affinity.

How this interdependency of families has become recognized is another problem which we are in no position to solve because there is no reason to believe that man, since he emerged from his animal state, has not enjoyed a basic form of social organization, which, as regards the fundamental principles, could not be essentially different from our own. Indeed, it will never be sufficiently emphasized that, if social organization had a beginning, this could only have consisted in the incest prohibition since, as we have just shown, the incest prohibition is, in fact, a kind of remodelling of the biological conditions of mating and procreation (which know no rule, as can be seen from observing animal life), compelling them to become perpetuated only in an artificial framework of taboos and obligations. It is there, and only there, that we find a passage from nature to culture, from animal to human life, and that we are in a position to understand the very essence of their articulation.

As Taylor has shown almost a century ago, the ultimate explanation is probably that mankind has understood very early that, in order to free itself from a wild struggle for existence, it was confronted with the very simple choice of "either marrying-out or being killed-out." The alternative was between biological families living in juxtaposition and endeavouring to remain closed, self-perpetuating units, overridden by their fears, hatreds, and ignorances, and the systematic establishment, through the incest prohibition, of links of intermarriage between them, thus succeeding to build, out of the artificial bonds of affinity, a true human society, despite, and even in contradiction with, the isolating influence of consanguinity.

In order to ensure that families will not become closed and that they will not constitute progressively as many self-sufficient units, we satisfy ourselves with forbidding marriage between near relatives. The number of social contacts which any given individual is likely to maintain outside his or her own family is great enough to afford a good probability that, on the average, the hundreds of thousands of families constituting at any given moment a modern society will not be permitted to "freeze," if one may say so. On the contrary, the greatest possible freedom for the choice of a mate (submitted to the only condition that the choice has to be made outside the restricted family) ensures that these families will be kept in a continuous flow and that a satisfactory process of continuous "mix-up" through intermarriage will prevail among them, thus making for a homogeneous and well-blended social fabric.

Conditions are quite different in the so-called primitive societies: there, the global figure of the population is a small one, although it may vary from a few dozen up to several thousands. Besides, social fluidity is low and it is not likely that many people will have a chance to get acquainted with others, during their lifetime, except within the limits of the village, hunting territory, etc., though it is true that many tribes have tried to organize occasions for wider contacts, for instance, during feasts, tribal ceremonies, etc. Even in such cases, however, the chances are limited to the tribal group, since most primitive peoples consider that the tribe is a kind of wide family, and that the frontiers of mankind stop together with the tribal bonds themselves.

Given such conditions, it is still possible to ensure the blending of families into a well-united society by using procedures similar to our own, i.e., a mere prohibition of marriage between relatives without any kind of positive prescriptions as to where and whom one should correctly marry. Experience shows, however, that this is only possible in small societies under the condition that the diminutive size of the group and the lack of social mobility be compensated by widening to a considerable extent the range of prohibited degrees. It is not only one's sister or daughter that, under such circumstances, one should not marry, but any woman with whom blood relationship may be traced, even in the remotest possible way. Very small groups with a low cultural level and a loose political and social organization, such as some desert tribes of North and South America, provide us with examples of that solution.

However, the great majority of primitive peoples have devised another method to solve the problem. Instead of confining themselves to a statistical process, relying on the probability that certain interdictions being set up, a satisfactory equilibrium of exchanges between the biological families will spontaneously result, they have preferred to invent rules which every individual and family should follow carefully, and from which a given form of blending, experimentally conceived of as satisfactory, is bound to arise.

Whenever this takes place, the entire field of kinship becomes a kind of complicated game, the kinship terminology being used to distribute all the members of the group into different categories, the rule being that the category of the parents defines either directly or indirectly the category of the children, and that, according to the categories in which they are placed, rules of kinship and marriage have provided modern anthropology with one of its more difficult and complicated chapters. Apparently ignorant and savage peoples have been able to devise fantastically clever codes which sometimes require, in order to understand their workings and effects, some of the best logical and even mathematical minds available in modern civilization. Therefore, we will limit ourselves to explaining the crudest principles which are the more frequently met with.

One of these is, undoubtedly, the so-called rule of cross-cousin marriage, which has been taken up by innumerable tribes all over the world. This is a complex system according to which collateral relatives are divided into two basic categories: "parallel" collaterals, when the relationship can be traced through two siblings of the same sex, and "cross" collaterals, when the relationship is traced through two siblings of opposite sex. For instance, my paternal uncle is a parallel relative and so is my maternal aunt; while the maternal uncle on the one hand, the paternal aunt on the other, are cross-relatives.

Now, the startling fact about this distinction is that practically all the tribes which make it claim that parallel relatives are the same thing as the closest ones on the same generation level: my father's brother is a "father," my mother's sister a "mother," my parallel-cousins are like brothers and sisters to me, and my parallel-nephews like children. Marriage with any of these would be incestuous and is consequently forbidden. On the other hand, cross-relatives are designated by special terms of their own, and it is among them that one should preferably find a mate.

All these distinctions (to which others could be added) are fantastic at first sight because they cannot be explained on biological or psychological grounds. But, if we keep in mind what has been explained in the preceding section, i.e., that all the marriage prohibitions have as their only purpose to establish a mutual dependency between the biological families, or, to put it in stronger terms, that marriage rules express the refusal, on the part of society, to admit the exclusive existence of the biological family, then everything becomes clear. For all these complicated sets of rules and distinctions are nothing but the outcome of the processes according to which, in a given society, families are set up against each other for the purpose of playing the game of matrimony.

The female reader, who may be shocked to see womankind treated as a commodity submitted to transactions between male operators, can easily find comfort in the assurance that the rules of the game would remain unchanged should it be decided to consider the men as being exchanged by women's groups. As a matter of fact, some very few societies, of a highly developed matrilineal type, have to a limited extent attempted to express things that way. And both sexes can be comforted from a still different (but in that case slightly more complicated) formulation of the game whereby it would be said that consanguineous groups consisting of both men and women are engaged in exchanging together bonds of relationships.

The important conclusion to be kept in mind is that the restricted family can neither be said to be the element of the social group, nor can it be claimed to result from it. Rather, the social group can only become established in contradistinction, and to some extent in compliance, with the family.

Society belongs to the realm of culture, while the family is the emanation, on the social level, of those natural requirements without which there could be no society and indeed no mankind. As a philosopher of the sixteenth century has said, man can only overcome nature by complying with its laws. Therefore, society has to give the family some amount of recognition. And it is not so surprising that, as geographers have also noticed with respect to the use of natural land resources, the greatest amount of compliance with the natural laws is likely to be found at both extremes of the cultural scale: among the simpler peoples as well as among the more highly civilized. Indeed, the first ones are not in a position to afford paying the price of too great a departure, while the second have already suffered from enough mistakes to understand that compliance is the best policy. This explains why, as we have already noticed, the small, relatively stable, monogamic restricted family seems to be given greater recognition, both among the more primitive peoples and in modern societies, than in what may be called (for the sake of the argument) the intermediate levels.

Men as Success Objects and Women as Sex Objects: A Study of Personal Advertisements

Simon Davis

Given the current concern with achieving gender equality, what do women and men look for in a mate? In this study the author conducts a content analysis of personal advertisements in newspapers as a way of examining sex stereotyping. Although he finds some surprising results, given the many cultural changes that have occurred in gender relations, what remains unanalyzed is how and why stereotypes of what constitutes a desirable partner persist.

Previous research has indicated that, to a large extent, selection of opposite-sex partners is dictated by traditional sex stereotypes (Urberg 1979). More specifically, it has been found that men tend to emphasize sexuality and physical attractiveness in a mate to a greater extent than women (e.g., Deaux and Hanna 1984; Harrison and Saeed 1977; Nevid 1984); this distinction has been found across cultures, as in the study by Stiles and colleagues (1987) of American and Icelandic adolescents.

The relatively greater preoccupation with casual sexual encounters demonstrated by men (Hite 1987, 184) may be accounted for by the greater emotional investment that women place in sex; Basow (1986, 80) suggests that the "gender differences in this area (different meaning attached to sex) may turn out to be the strongest of all gender differences."

Women, conversely, may tend to emphasize psychological and personality characteristics (Curry and Hock 1981; Deaux and Hanna 1984), and to seek longevity and commitment in a relationship to a greater extent (Basow 1986, 213).

Sources: Simon Davis, "Men as Success Objects and Women as Sex Objects: A Study of Personal Advertisements," *Sex Roles* Vol. 23, Nos. 1/2. Copyright © 1990. Reprinted with kind permission from Springer Science and Business Media.

Lorne Tepperman, Postscript, "Erotic Property," from *Everyday Life: A Reader,* McGraw-Hill/Primis 1996. Reprinted with permission from the author.

Women may also seek financial security more so than men (Harrison and Saeed 1977). Regarding this last point, Farrell (1986, 25) suggests that the tendency to treat men as success objects is reflected in the media, particularly in advertisements in women's magazines. On the other hand, men themselves may reinforce this stereotype in that a number of men still apparently prefer the traditional marriage with working husband and unemployed wife (Basow 1986, 210).

Men have traditionally been more dominant in intellectual matters, and this may be reinforced in the courting process: Braito (1981) found in his study that female coeds feigned intellectual inferiority with their dates on a number of occasions. In the same vein, Hite, in her 1981 survey, found that men were less likely to seek intellectual prowess in their mate (108).

The mate selection process has been characterized in at least two ways. Harrison and Saeed (1977) found evidence for a matching process, where individuals seeking particular characteristics in a partner were more likely to offer those characteristics in themselves. This is consistent with the observation that "like attracts like" and that husbands and wives tend to resemble one another in various ways (Thiessen and Gregg 1980). Additionally, an exchange process may be in operation, wherein a trade-off is made with women offering "domestic work and sex for financial support" (Basow 1986, 213).

With respect to sex stereotypes and mate selection, the trend has been for "both sexes to believe that the other sex expects them to live up to the gender stereotype" (Basow 1986, 209).

Theoretical explanations of sex stereotypes in mate selection range from the sociobiological (Symons 1987) to radical political views (Smith, 1973). Of interest in recent years has been demographic influences, that is, the lesser availability of men because of population shifts and marital patterns (Shaevitz 1987, 40). Age may differentially affect women, particularly when children are desired; this, combined with women's generally lower economic status [particularly when unmarried (Halas 1981, 124)], may mean that the need to "settle down" into a secure, committed relationship becomes relatively more crucial for women.

The present study looks at differential mate selection by men and women as reflected in newspaper companion ads. Using such a forum for the exploration of sex stereotypes is not new; for instance, in the study by Harrison and Saeed (1977) cited earlier, the authors found that in such ads women were more likely to seek financial security and men to seek attractiveness; a later study by Deaux and Hanna (1984) had similar results, along with the finding that women were more likely to seek psychological characteristics, specific personality traits, and to emphasize the quality and longevity of the relationship. The present study may be seen as a follow-up of this earlier research, although on this occasion using a Canadian setting. Of particular interest was the following: Were traditional stereotypes still in operation, that is, women being viewed as sex objects and men as success objects (the latter defined as financial and intellectual accomplishments)?

METHOD

Personal advertisements were taken from the *Vancouver Sun*, which is the major daily newspaper serving Vancouver, British Columbia. The *Sun* is generally perceived as a conservative, respectable journal—hence it was assumed that people advertising in it represented the "mainstream." It should be noted that people placing the ads must do so in person. For the sake of this study, gay ads were not included. A typical ad would run about 50 words, and included a brief description of the person placing it and a list of the attributes desired by the other party. Only the parts pertaining to the attributes desired in the partner were included for analysis. Attributes that pertained to hobbies or recreations were not included for the purpose of this study.

The ads were sampled as follows: Only Saturday ads were used, since in the *Sun* the convention was for Saturday to be the main day for personal ads, with 40 to 60 ads per edition—compared to only 2 to 4 ads per edition on weekdays. Within any one edition *all* the ads were included for analysis. Six editions were randomly sampled, covering the period of September 30, 1988, to September 30, 1989. The attempt to sample through the calendar year was made in an effort to avoid an unspecified seasonal effect. The size of the sample (six editions) was large enough to meet goodness-of-fit requirements for statistical tests.

The attributes listed in the ads were coded as follows:

1. *Attractive:* specified that a partner should be, for example, "pretty" or "handsome."
2. *Physique:* similar to 1; however, this focused not on the face but rather on whether the partner was "fit and trim," "muscular," or had "a good figure." If it was not clear if body or face was being emphasized, this fell into variable (1) by default.
3. *Sex:* specified that the partner should have, for instance, "high sex drive," or should be "sensuous" or "erotic," or if there was a clear message that this was an arrangement for sexual purposes ("lunch-time liaisons—discretion required").
4. *Picture:* specified that the partner should include a photo in his/her reply.
5. *Profession:* specified that the partner should be a professional.
6. *Employed:* specified that the partner should be employed, e.g., "must hold steady job" or "must have steady income."
7. *Financial:* specified that the partner should be, for instance, "financially secure" or "financially independent."

8. *Education:* specified that the partner should be, for instance, "well educated" or "well read," or should be a "college grad."
9. *Intelligence:* specified that the partner should be "intelligent," "intellectual," or "bright."
10. *Honest:* specified, for instance, that the partner should be "honest" or have "integrity."
11. *Humor:* specified "sense of humor" or "cheerfulness."
12. *Commitment:* specified that the relationship was to be "long term" or "lead to marriage," or some other indication of stability and longevity.
13. *Emotion:* specified that the partner should be "warm," "romantic," "emotionally supportive," "emotionally expressive," "sensitive," "loving," "responsive," or similar terms indicating an opposition to being cold and aloof.

In addition to the 13 attribute variables, two other pieces of information were collected: The length of the ad (in lines) and the age of the person placing the ad. Only if age was exactly specified was it included; if age was vague (e.g., "late 40s") this was not counted.

Variables were measured in the following way: Any ad requesting one of the 13 attributes was scored once for that attribute. If not explicitly mentioned, it was not scored. The scoring was thus "all or nothing," e.g., no matter how many times a person in a particular ad stressed that looks were important it was only counted as a single score in the "attractive" column; thus, each single score represented one person. Conceivably, an individual ad could mention all, some, or none of the variables. Comparisons were then made between the sexes on the basis of the variables, using percentages and chi-squares. Chi-square values were derived by cross-tabulating gender (male/female) with attribute (asked for/not asked for). Degrees of freedom in all cases equaled one. Finally, several of the individual variables were collapsed to get an overall sense of the relative importance of (a) physical factors, (b) employment factors, and (c) intellectual factors.

RESULTS

A total of 329 personal ads were contained in the six newspaper editions studied. One ad was discarded in that it specified a gay relationship, leaving a total of 328. Of this number, 215 of the ads were placed by men (65.5%) and 113 by women (34.5%).

The mean age of people placing ads was 40.4. One hundred and twenty-seven cases (38.7%) counted as missing data in that the age was not specified or was vague. The mean age for the two sexes was similar: 39.4 for women (with 50.4% of cases missing) and 40.7 for men (with 32.6% of cases missing).

TABLE 1 Gender Comparison for Attributes Desired in Partner

Variable	Gender		Chi-square
	Desired by Men **(n = 215)**	**Desired by Women** **(n = 113)**	
1. Attractive	76 (35.3%)	20 (17.7%)	11.13^a
2. Physique	81 (37.7%)	27 (23.9%)	6.37^a
3. Sex	25 (11.6%)	4 (3.5%)	6.03^a
4. Picture	74 (34.4%)	24 (21.2%)	6.18^a
5. Profession	6 (2.8%)	19 (16.8%)	20.74^a
6. Employed	8 (3.7%)	12 (10.6%)	6.12^a
7. Financial	7 (3.2%)	22 (19.5%)	24.26^a
8. Education	8 (3.7%)	8 (7.1%)	1.79 (ns)
9. Intelligence	22 (10.2%)	24 (21.2%)	7.46^a
10. Honest	20 (9.3%)	17 (15.0%)	2.44 (ns)
11. Humor	36 (16.7%)	26 (23.0%)	1.89 (ns)
12. Commitment	38 (17.6%)	31 (27.4%)	4.25^a
13. Emotion	44 (20.5%)	35 (31.0%)	4.36^a

^{a} Significant at the .05 level.

Sex differences in desired companion attributes are summarized in Table 1. It will be seen that for 10 of the 13 variables a statistically significant difference was detected. The three largest differences were found for attractiveness, professional and financial status. To summarize the table: in the case of attractiveness, physique, sex, and picture (physical attributes) the men were more likely than the women to seek these. In the case of professional status, employment status, financial status, intelligence, commitment, and emotion (nonphysical attributes) the women were more likely to seek these. The women were also more likely to specify education, honesty and humor, however not at a statistically significant level.

The data were explored further by collapsing several of the categories: the first 4 variables were collapsed into a "physical" category, variables 5-7 were collapsed into an "employment" category, and variables 8 and 9 were collapsed into an "intellectual" category. The assumption was that the collapsed categories were sufficiently similar (within the three new categories) to make the new larger categories conceptually meaningful; conversely, it was felt the remaining variables (10-13) could not be meaningfully collapsed any further.

Sex differences for the three collapsed categories are summarized in Table 2. Note that the Table 2 figures were not derived simply by adding the numbers in the Table 1 categories: recall that for variables 1-4 a subject could specify all, one, or none; hence simply adding the Table 1 figures would be biased by those individuals who were more effusive in specifying

TABLE 2 Gender Comparison for Physical, Employment, and Intellectual Attributes Desired in Partner

Variable	Gender		
	Desired by Men (n = 215)	Desired by Women (n = 113)	Chi-square
Physical	143	50	15.13[a]
(collapsing variables 1-4)	(66.5%)	(44.2%)	
Employment	17	47	51.36[a]
(collapsing variables 5-7)	(7.9%)	(41.6%)	
Intellectual	29	31	9.65[a]
(collapsing 8 and 9)	(13.5%)	(27.4%)	

[a] Significant at the .05 level.

various physical traits. Instead, the Table 2 categories are (like Table 1) all or nothing: whether a subject specified one or all four of the physical attributes it would only count once. Thus, each score represented one person.

In brief, Table 2 gives similar, although more exaggerated, results to Table 1. (The exaggeration is the result of only one item of several being needed to score within a collapsed category.) The men were more likely than the women to specify some physical attribute. The women were considerably more likely to specify that the companion be employed, or have a profession, or be in good financial shape. And the women were more likely to emphasize the intellectual abilities of their mate....

DISCUSSION

Sex Differences

This study found that the attitudes of the subjects, in terms of desired companion attributes, were consistent with traditional sex role stereotypes. The men were more likely to emphasize stereotypically desirable feminine traits (appearance) and deemphasize the non-feminine traits (financial, employment, and intellectual status). One inconsistency was that emotional expressiveness is a feminine trait but was emphasized relatively less by the men. Women, on the other hand, were more likely to emphasize masculine traits such as financial, employment, and intellectual status, and valued commitment in a relationship more highly. One inconsistency detected for the women concerned the fact that although emotional expressiveness is not a masculine trait, the women in this sample asked for it, relatively more than the men, anyway. Regarding this last point, it may be relevant to refer to Basow's (1986, 210) conclusion that "women prefer relatively androgynous men, but men, especially traditional ones, prefer relatively sex-typed women."

These findings are similar to results from earlier studies, e.g., Deaux and Hanna (1984), and indicate that at this point in time and in this setting sex role stereotyping is still in operation....

Methodological Issues

Content analysis of newspaper ads has its strengths and weaknesses. By virtue of being an unobtrusive study of variables with face validity, it was felt some reliable measure of gender-related attitudes was being achieved. That the mean age of the men and women placing the ads was similar was taken as support for the assumption that the two sexes in this sample were demographically similar. Further, sex differences in desired companion attributes could not be attributed to differential verbal ability in that it was found that length of ad was similar for both sexes.

On the other hand, there were some limitations. It could be argued that people placing personal ads are not representative of the public in general. For instance, with respect to this study, it was found that the subjects were a somewhat older group—mean age of 40—than might be found in other courting situations. This raises the possibility of age being a confounding variable. Older singles may emphasize certain aspects of a relationship, regardless of sex. On the other hand, there is the possibility that age differentially affects women in the mate selection process, particularly when children are desired. The strategy of controlling for age in the analysis was felt problematic in that the numbers for analysis were fairly small, especially given the missing data, and further, that one cannot assume the missing cases were not systematically different (i.e., older) from those present.

References

Basow, S. 1986. *Gender Stereotypes: Traditions and Alternatives.* Pacific Grove, CA: Brooks/Cole.

Curry, T., and R. Hock. 1981. "Sex Differences in Sex Role Ideals in Early Adolescence." *Adolescence* 16: 779–789.

Deaux, K., and R. Hanna. 1984. "Courtship in the Personals Column: The Influence of Gender and Sexual Orientation." *Sex Roles* 11: 363–375.

Farrell, W. 1986. *Why Men Are the Way They Are.* New York: Berkeley Books.

Halas, C. 1981. *Why Can't a Woman Be More Like a Man?* New York: Macmillan.

Harrison, A., and L. Saeed. 1977. "Let's Make a Deal: An Analysis of Revelations and Stipulations in Lonely Hearts Advertisements." *Journal of Personality and Social Psychology* 35: 257–264.

Hite, S. 1981. *The Hite Report on Male Sexuality.* New York: Knopf.

___. 1987. *Women and Love: A Cultural Revolution in Progress.* New York: Knopf.

Nevid, J. 1984. "Sex Differences in Factors of Romantic Attraction." *Sex Roles* 11: 401–411.

Shaevitz, M., 1987. *Sexual Static.* Boston: Little, Brown.

Stiles, D., J. Gibbon, S. Hardardottir, and J. Schnellmann. 1987. "The Ideal Man or Woman as Described by Young Adolescents in Iceland and the United States." *Sex Roles* 17: 313–320.

Symons, D. 1987. "An Evolutionary Approach." In J. Geer and W. O'Donohue (eds.), *Theories of Human Sexuality.* New York: Plenum Press.

Thiessen, D., and B. Gregg. 1980. "Human Assortive Mating and Genetic Equilibrium: An Evolutionary Perspective." *Ethology and Sociobiology* 1: 111–140.

Urberg, K. 1979. "Sex Role Conceptualization in Adolescents and Adults." *Developmental Psychology* 15: 90–92.

POSTSCRIPT

Generally, when people mate they look for qualities that complement their own and compensate for their weaknesses. What sociologists call "heterogamy" offers a large range of possibilities. It allows a person to trade off one quality or characteristic for another: for instance, youth and beauty for wealth and status. However, all such exchanges occur within a context of acquiring "erotic (or sexual) property."

Sociologist Randall Collins believes that sexual property is the key to family structure. According to him, when people mate they take possession of erotic property, that is, exclusive sexual rights to someone's body. Even in our society, marriage is a system of property relations, and men have far more property rights over women than women have over men.

Consider our society's ideas about infidelity. Adultery is treated as a serious infraction because (Collins claims) it violates the right to exclusive sexual access. The same reasoning explains why most societies have placed a heavy emphasis on the virginity of the bride—but not the groom—at marriage. In almost every case, the property system underlying marriage is one in which men own women's bodies.

Today, of course, women enjoy an unparalleled degree of independence in our society. Yet an examination of the newspaper advertisements that people use to attract mates still reveals men looking for sex objects—erotic property—and women looking for steady, successful partners.

Is There a Family?

*Jane Collier, Michelle Z. Rosaldo,
and Sylvia Yanagisako*

> *"When I use a word,"* Humpty Dumpty said in a rather scornful tone, *"it means just what I choose it to mean—neither more nor less."*
>
> *"The question is,"* said Alice, *"whether you can make words mean so many different things."*
>
> *"The question is,"* said Humpty Dumpty, *"which is to be master ... that's all."*
>
> (from Lewis Carroll, "Through the Looking Glass," in *Alice in Wonderland*, W. W. Norton and Co., 1971: 163.)

One of the most problematic terms in the social sciences is *family*. Part of the problem is trying to determine common patterns among the remarkable diversity of family forms and kinship systems that anthropologists have found. Another part is the equally remarkable gap that exists in modern societies, between the family as a normative social institutional form ("The Family") and the concrete household arrangements and family groupings in which we actually live. A third part concerns the politics involved in defining family, an issue which feminist scholarship has brought to the forefront of the debate.

In this essay the authors examine what social scientists have traditionally meant by the family. They consider what historical biases led anthropologist Bronislaw Malinowski and his successors to the mistaken view that the family was a universal human institution. They draw on other social theorists and researchers to argue for viewing "The Family" not as a concrete institution designed to fulfill universal human needs, but as an ideological construct associated with the modern state.

Source: Jane Collier, Michelle Z. Rosaldo and Sylvia Yanagisako, "Is there a family?" from Barrie Thorne and Marilyn Yalom *Rethinking the Family: Some Feminist Questions*. Northeastern University Press, 1992. Reprinted with permission from the authors.

Despite the diversity of families across North America, and the many criticisms of the ideal as an ideological construct, the cultural idea of the family persists as an external, constraining social fact. Why?

The authors leave the reader with provocative questions. Are we demanding too much of the family? Are we being naïve in expecting the family to provide us with a "love" that money cannot buy? What, exactly, do we want our families to do?

This essay poses a rhetorical question in order to argue that most of our talk about families is clouded by unexplored notions of what families "really" are like. It is probably the case, universally, that people expect to have special connections with their genealogically closest relations. But a knowledge of genealogy does not in itself promote understanding of what these special ties are about. The real importance of The Family in contemporary social life and belief has blinded us to its dynamics. Confusing ideal with reality, we fail to appreciate the deep significance of what are, cross-culturally, various ideologies of intimate relationship, and at the same time we fail to reckon with the complex human bonds and experiences all too comfortably sheltered by a faith in the "natural" source of a "nurture" we think is found in the home.

. . .

MALINOWSKI'S CONCEPT OF THE FAMILY

In 1913 Bronislaw Malinowski published a book called *The Family Among the Australian Aborigines*[1] in which he laid to rest earlier debates about whether all human societies had families. During the nineteenth century, proponents of social evolution argued that primitives were sexually promiscuous and therefore incapable of having families because children would not recognize their fathers.[2] Malinowski refuted this notion by showing that Australian aborigines, who were widely believed to practice "primitive promiscuity," not only had rules regulating who might have intercourse with whom during sexual orgies but also differentiated between legal marriages and casual unions. Malinowski thus "proved" that Australian aborigines had marriage, and so proved that aboriginal children had fathers, because each child's mother had but a single recognized husband.

Malinowski's book did not simply add data to one side of an ongoing debate. It ended the debate altogether, for by distinguishing coitus from conjugal relationships, Malinowski separated questions of sexual behavior from questions of the family's universal existence. Evidence of sexual promiscuity was henceforth irrelevant for deciding whether families existed. Moreover, Malinowski argued that the conjugal relationship, and therefore The Family, had to be universal because it fulfilled a universal human need. As he wrote in a posthumously published book:

> The human infant needs parental protection for a much longer period than does the young of even the highest anthropoid apes. Hence, no culture could endure in which the act of reproduction, that is, mating, pregnancy, and childbirth, was not linked up with the fact of legally-founded parenthood, that is, a relationship in which the father and mother have to look after the children for a long period, and, in turn, derive certain benefits from the care and trouble taken.[3]

In proving the existence of families among Australian aborigines, Malinowski described three features of families that he believed flowed from The Family's universal function of nurturing children. First, he argued that families had to have clear boundaries, for if families were to perform the vital function of nurturing young children, insiders had to be distinguishable from outsiders so that everyone could know which adults were responsible for the care of which children. . . .

Second, Malinowski argued that families had to have a place where family members could be together and where the daily tasks associated with child rearing could be performed. . . .

Finally, Malinowski argued that family members felt affection for one another—that parents who invested long years in caring for children were rewarded by their own and their children's affections for one another. Malinowski felt that long and intimate association among family members fostered close emotional ties, particularly between parents and children, but also between spouses. . . .

Malinowski's book on Australian aborigines thus gave social scientists a concept of The Family that consisted of a universal function, the nurturance of young children, mapped onto (1) a bounded set of people who recognized one another and who were distinguishable from other like groups; (2) a definite physical space, a hearth and home; and (3) a particular set of emotions, family love. This concept of The Family as an institution for nurturing young children has been enduring, probably because nurturing children is thought to be the primary function of families in modern industrial societies. The flaw in Malinowski's argument is the flaw common to all functionalist arguments: Because a social institution is observed to perform a necessary function does not mean either that the function would not be performed if the institution did not exist or that the function is responsible for the existence of the institution.

Later anthropologists have challenged Malinowski's idea that families always include fathers, but, ironically, they have kept all the other aspects of his definition. For example, later anthropologists have argued that the basic social unit is not the nuclear family including father but the unit composed of a mother and her children: "Whether or not a mate becomes attached to the mother on some more or less permanent basis is a variable matter."[4] In removing father from the family, however, later anthropologists have nevertheless retained Malinowski's concept of The Family as a functional unit, and so have retained all the features Malinowski took such pains to demonstrate....

Modern anthropologists may have removed father from The Family, but they did not modify the basic social science concept of The Family in which the function of child rearing is mapped onto a bounded set of people who share a place and who "love" one another. Yet it is exactly this concept of The Family that we, as feminist anthropologists, have found so difficult to apply. Although the biological facts of reproduction, when combined with a sufficiently elastic definition of marriage, make it possible for us, as social scientists, to find both mother-child units and Malinowski's conjugal-pairs-plus-children units in every human society, it is not at all clear that such Families necessarily exhibit the associated features Malinowski "proved" and modern anthropologists echo.

An outside observer, for example, may be able to delimit family boundaries in any and all societies by identifying the children of one woman and that woman's associated mate, but natives may not be interested in making such distinctions. In other words, natives may not be concerned to distinguish family members from outsiders, as Malinowski imagined natives should be when he argued that units of parents and children have to have clear boundaries in order for child-rearing responsibilities to be assigned efficiently. Many languages, for example, have no word to identify the unit of parents and children that English speakers call a "family." Among the Zinacantecos of southern Mexico, the basic social unit is identified as a "house," which may include from one to twenty people.[5] ... In Zinacanteco society, the boundary between "houses" is linguistically marked, while the boundary between "family" units is not.

Just as some languages lack words for identifying units of parents and children, so some "families" lack places. Immature children in every society have to be fed and cared for, but parents and children do not necessarily eat and sleep together as a family in one place. Among the Mundurucu of tropical South America, for example, the men of a village traditionally lived in a men's house together with all the village boys over the age of thirteen; women lived with other women and young children in two or three houses grouped around the men's house.[6] In Mundurucu society, men and women ate and slept apart....

Finally, people around the world do not necessarily expect family members to "love" one another. People may expect husbands, wives, parents, and children to have strong feelings about one another, but they do not necessarily expect prolonged and intimate contact to breed the loving sentiments Malinowski imagined as universally rewarding parents for the care they invested in children. The mother-daughter relationship, for example, is not always pictured as warm and loving. In modern Zambia, girls are not expected to discuss personal problems with, or seek advice from, their mothers. Rather, Zambian girls are expected to seek out some older female relative to serve as confidante.[7] ...

Of course, anthropologists have recognized that people everywhere do not share our deep faith in the loving, self-sacrificing mother, but in matters of family and motherhood, anthropologists, like all social scientists, have relied more on faith than evidence in constructing theoretical accounts. Because we *believe* mothers to be loving, anthropologists have proposed, for example, that a general explanation of the fact that men marry mother's brothers' daughters more frequently than they marry father's sisters' daughters is that men naturally seek affection (i.e., wives) where they have found affection in the past (i.e., from mothers and their kin).[8]

LOOKING BACKWARD

The Malinowskian view of The Family as a universal institution ... corresponds, as we have seen, to that assumed by most contemporary writers on the subject. But a consideration of available ethnographic evidence suggests that the received view is a good deal more problematic than a naive observer might think. If Families in Malinowski's sense are *not* universal, then we must begin to ask about the biases that, in the past, have led us to misconstrue the ethnographic record. The issues here are too complex for thorough explication in this essay, but if we are to better understand the nature of "the family" in the present, it seems worthwhile to explore the question, first, of why so many social thinkers continue to believe in Capital-Letter Families as universal institutions, and second, whether anthropological tradition offers any alternatives to a "necessary and natural" view of what our families are. Only then will we be in a position to suggest "new anthropological perspectives" on the family today.

Our positive critique begins by moving backward. In the next few pages, we suggest that tentative answers to both questions posed above lie in the nineteenth-century intellectual trends that thinkers like Malinowski were at pains to reject. During the second half of the nineteenth century, a number of social and intellectual developments—among them, the evolutionary researches of Charles Darwin; the rise of "urban problems" in fast-growing cities; and the accumulation of data on non-Western peoples by missionaries and agents of the colonial states—contributed to what most of us would now recognize as the beginnings of modern social science.... At base, a sense of "progress" gave direction to their thought, whether, like Spencer,[9] they believed "man" had advanced from the love of violence to a more civilized love of peace or, like Engels,[10] that humanity had moved from primitive promiscuity and incest toward monogamy and "individual sex love." Proud of their position in the modern world, some of these writers claimed that rules of force had been transcended by new rules of law,[11] while others thought that feminine "mysticism" in the past had been supplanted by a higher male "morality."[12]

At the same time, and whatever else they thought of capitalist social life (some of them criticized, but none wholly abhorred it), these writers also shared a sense of moral emptiness and a fear of instability and loss. Experience argued forcefully to them that moral order in their time did not rest on the unshakable hierarchy—from God to King to Father in the home—enjoyed by Europeans in the past.[13] Thus, whereas Malinowski's functionalism led him to stress the underlying continuities in all human social forms, his nineteenth-century predecessors were concerned to understand the facts and forces that set their experiential world apart. They were interested in comparative and, more narrowly, evolutionary accounts because their lives were torn between celebration and fear of change. For them, the family was important not because it had at all times been the same but because it was at once the moral precondition for, the triumph of, and the victim of developing capitalist society....

Given this purpose and the limited data with which they had to work, it is hardly surprising that the vast majority of what these nineteenth-century writers said is easily dismissed today. They argued that in simpler days such things as incest were the norm; they thought that women ruled in "matriarchal" and peace-loving states or, alternatively, that brute force determined the primitive right and wrong. None of these visions of a more natural, more feminine, more sexy, or more violent primitive world squares with contemporary evidence about what, in technological and organizational terms, might be reckoned relatively "primitive" or "simple" social forms. We would suggest, however, that whatever their mistakes, these nineteenth-century thinkers *can* help us rethink the family today, at least in part because we are (unfortunately) their heirs, in the area of prejudice, and partly because their concern to characterize difference and change gave rise to insights much more promising than their functionalist critics may have thought.

To begin, although nineteenth-century evolutionary theorists did not believe The Family to be universal, the roots of modern assumptions can be seen in their belief that women are, and have at all times been, defined by nurturant, connective, and reproductive roles that *do not change* through time. Most nineteenth-century thinkers imaged social development as a process of differentiation from a relatively confused (and thus incestuous) and indiscriminate female-oriented state to one in which men fight, destroy their "natural" social bonds, and then forge public and political ties to create a human "order." For some, it seemed reasonable to assume that women dominated, as matriarchs, in the undifferentiated early state, but even these theorists believed that women everywhere were "mothers" first, defined by "nurturant" concerns and thus excluded from the business competition, cooperation, social ordering, and social change propelled and dominated by their male counterparts. And so, while nineteenth-century writers differed in their evaluations of such things as "women's status," they all believed that female reproductive roles made women different from and complementary to men and guaranteed both the relative passivity of women in human history and the relative continuity of "feminine" domains and functions in human societies. Social change consisted in the acts of men, who left their mothers behind in shrinking homes....

For nineteenth-century evolutionists, women were associated, in short, with an unchanging biological role and a romanticized community of the past, while men were imaged as the agents of all social process. And though contemporary thinkers have been ready to dismiss manifold aspects of their now-dated school of thought, on this point we remain, perhaps unwittingly, their heirs. Victorian assumptions about gender and the relationship between competitive male markets and peace-loving female homes were not abandoned in later functionalist schools of thought at least in part because pervasive sexist biases make it easy to forget that women, like men, are important actors in *all* social worlds....

If most modern social scientists have inherited Victorian biases that tend ultimately to support a view uniting women and The Family to an apparently unchanging set of biologically given needs, we have at the same time failed to reckon with the one small area in which Victorian evolutionists were right. They understood, as we do not today, that families—like religions, economies, governments, or courts of law—are *not* unchanging but the product of various social forms, that the relationships of spouses and parents to their young are apt to be different things in different social orders. More particularly, although nineteenth-century writers had primitive society all wrong, they were correct in insisting that *family* in the modern sense—a unit bounded, biologically as well as legally defined, associated with property, self-sufficiency, with affect and a space "inside" the home—is something that emerges not in Stone Age caves but in complex state-governed social forms. Tribal peoples may speak readily of lineages, households, and clans, but—as we have seen—they rarely have a word denoting Family as a particular and limited group of kin; they rarely worry about differences between legitimate and illegitimate heirs or find themselves concerned (as we so often are today) that what children and/or parents do reflects on their family's public image and self-esteem. Political influence in tribal groups in fact consists in adding children to one's home and, far from distinguishing Smith from Jones, encouraging one's neighbours to join one's household as if kin. By contrast, modern bounded Families try to keep their neighbors out....

In short, what the Victorians recognized and we have tended to forget is, first, that human social life has varied in its "moral"—we might say its "cultural" or "ideological"—forms, and so it takes more than making babies to make Families. And having seen The Family as something more than a response to omnipresent, biologically given needs, they realized too that Families do not everywhere exist; rather, The Family (thought to be universal by most social scientists today) is a moral and ideological unit that appears, not universally, but in particular social orders. The Family as we know it is not a "natural" group created by the claims of "blood" but a sphere of human relationships shaped by a state that recognizes Families as units that hold property, provide for care and welfare, and attend particularly to the young—a sphere conceptualized as a realm of love and intimacy *in opposition* to the more "impersonal" norms that dominate modern economies and politics.... Stated otherwise, because our notions of The Family are rooted in a contrast between "public" and "private" spheres, we will not find that Families like ours exist in a society where public and political life is radically different from our own.

. . .

TOWARD A RETHINKING

Our perspective on families therefore compels us to listen carefully to what the natives in other societies say about their relationships with genealogically close kin. The same is true of the natives in our own society. Our understanding of families in contemporary American society can be only as rich as our understanding of what The Family represents symbolically to Americans. A complete cultural analysis of The Family as an American ideological construct, of course, is beyond the scope of this essay. But we can indicate some of the directions such an analysis would take and how it would deepen our knowledge of American families.

One of the central notions in the modern American construct of The Family is that of nurturance. When antifeminists attack the Equal Rights Amendment, for example, much of their rhetoric plays on the anticipated loss of the nurturant, intimate bonds we associate with The Family. Likewise, when pro-life forces decry abortion, they cast it as the ultimate denial of nurturance. In a sense, these arguments are variations of a functionalist view that weds families to specific functions. The logic of the argument is that because people need nurturance, and people get nurtured in The Family, then people need The Family. Yet if we adopt the perspective that The Family is an ideological unit rather than merely a functional unit, we are encouraged to subject this syllogism to closer scrutiny. We can ask, first, What do people mean by nurturance? Obviously, they mean more than mere nourishment—that is, the provision of food, clothing, and shelter required for biological survival. What is evoked by the word nurturance is a certain kind of relationship: a relationship that entails affection and love, that is based on cooperation as opposed to competition, that is enduring rather than temporary, that is noncontingent rather than contingent upon performance, and that is governed by feeling and morality instead of law and contract.

The reason we have stated these attributes of The Family in terms of oppositions is because in a symbolic system the meanings of concepts are often best illuminated by explicating their opposites. Hence, to understand our American construct of The Family, we first have to map the larger system of constructs of which it is only a part. When we undertake such an analysis of The Family in our society, we discover that what gives shape to much of our conception of The Family is its symbolic opposition to work and business, in other words, to the market relations of capitalism. For it is in the market, where we sell our labor and negotiate contract relations of business, that we associate with competitive, temporary, contingent relations that must be buttressed by law and legal sanctions.

The symbolic opposition between The Family and market relations renders our strong attachment to The Family understandable, but it also discloses the particularity of our construct of The Family. We can hardly be speaking of a universal notion of The Family shared by people everywhere and for all time because people everywhere and for all time have not participated in market relations out of which they have constructed a contrastive notion of the family.

The realization that our idea of The Family is part of a set of symbolic oppositions through which we interpret our experience in a particular society compels us to ask to what extent this set of oppositions reflects real relations between people and to what extent it also shapes them.... [W]e are interested in understanding how people come to summarize their experience in folk constructs that gloss over the diversity, complexity, and contradictions in their relationships. If, for example, we consider the second premise of the aforementioned syllogism—the idea that people get "nurtured" in families—we can ask how people reconcile this premise with the fact that relationships in families are not always this simple or altruistic. We need not resort to the evidence offered by social historians (e.g., Philippe Aries[14] and Lawrence Stone[15]) of the harsh treatment and neglect of children and spouses in the history of the Western family, for we need only read our local newspaper to learn of similar abuses among contemporary families....

The point is not that our ancestors or our contemporaries have been uniformly mean and nonnurturant to family members but that we have all been both nice and mean, both generous and ungenerous, to them. In like manner, our actions toward family members are not always motivated by selfless altruism but are also motivated by instrumental self-interest. What is significant is that, despite the fact that our complex relationships are the result of complex motivations, we ideologize relations within The Family as nurturant while casting relationships outside The Family—particularly in the sphere of work and business—as just the opposite.

We must be wary of oversimplifying matters by explaining away those disparities between our notion of the nurturant Family and our real actions toward family members as the predictable failing of imperfect beings. For there is more here than mere disjunction of the ideal and the real. The American construct of The Family, after all, is complex enough to comprise some key contradictions. The Family is seen as representing not only the antithesis of the market relations of capitalism; it is also sacralized in our minds as the last stronghold against The State, as the symbolic refuge from the intrusions of a public domain that constantly threatens our sense of privacy and self-determination. Consequently, we can hardly be surprised to find that the punishments imposed on people who commit physical violence are lighter when their victims are their own family members.[16] Indeed, the American sense of the privacy of the things that go on inside families is so strong that a smaller percentage of homicides involving family members are prosecuted than those involving strangers.[17] We are faced with the irony that in our society the place where nurturance and noncontingent affection are supposed to be located is simultaneously the place where violence is most tolerated.

There are other dilemmas about The Family that an examination of its ideological nature can help us better understand. For example, the hypothesis that in England and the United States marriages among lower-income ("working-class") groups are characterized by a greater degree of "conjugal role segregation" than are marriages among middle-income groups has generated considerable confusion. Since Bott observed that working-class couples in her study of London families exhibited more "segregated" conjugal roles than "middle-class" couples, who tended toward more "joint" conjugal roles,[18] researchers have come forth with a range of diverse and confusing findings.... Other researchers, however, have raised critical methodological questions about how one goes about defining a joint activity and hence measuring the degree of "jointness" in a conjugal relationship.[19] Platt's finding that couples who reported "jointness" in one activity were not particularly likely to report "jointness" in another activity is significant because it demonstrates that "jointness" is not a general characteristic of a relationship that manifests itself uniformly over a range of domains. Couples carry out some activities and tasks together or do them separately but equally; they also have other activities in which they do not both participate. The measurement of the "jointness" of conjugal relationships becomes even more problematic when we recognize that what one individual or couple may label a "joint activity," another individual or couple may consider a "separate activity." ...

In other words, the distinction Bott drew among "joint," "differentiated," and "autonomic" (independent) relationships summarized the way people thought and felt about their activities rather than what they were observed to actually do. Again, it is not simply that there is a disjunction between what people say they do and what they in fact do. The more cogent point is that the meaning people attach to action, whether they view it as coordinated and therefore shared or in some other way, is an integral component of that action and cannot be divorced from it in our analysis. ...

Finally, the awareness that The Family is not a concrete "thing" that fulfills concrete "needs" but an ideological construct with moral implications can lead to a more refined analysis of historical change in the American or Western family than has devolved upon us from our functionalist ancestors. The functionalist view of industrialization, urbanization, and family change depicts The Family as responding to alterations in economic and social conditions in rather mechanistic ways. As production gets removed from the family's domain, there is less need for strict rules and clear authority structures in the family to accomplish productive work. At the same time, individuals who now must work for wages in impersonal settings need a haven where they can obtain emotional support and gratification. Hence, The Family becomes more concerned with "expressive" functions, and what emerges is the modern "companionate family." In short, in the functionalist narrative The Family and its constituent members "adapt" to fulfill functional requirements created for it by the industrialization of production. Once we begin to view The Family as an ideological unit and pay due respect to it as a moral statement, however, we can begin to unravel the more complex, dialectical process through which family relationships and The Family as a construct were mutually transformed. We can examine, for one, the ways in which people and state institutions acted, rather than merely reacted, to assign certain functions to groupings of kin by making them legally responsible for these functions .... We can begin to understand the consequences of social reforms and wage policies for the age and sex inequalities in families. And we can elucidate the interplay between these social changes and the cultural transformations that assigned new meanings and modified old ones to make The Family what we think it to be today.

Ultimately, this sort of rethinking will lead to a questioning of the somewhat contradictory modern views that families are things we need (the more "impersonal" the public world, the more we need them) and at the same time that loving families are disappearing. In a variety of ways, individuals today *do* look to families for a "love" that money cannot buy and find; our contemporary world makes "love" more fragile than most of us hope and "nurturance" more self-interested than we believe.[20] But what we fail to recognize is that familial nurturance and the social forces that turn our ideal families into mere fleeting dreams are *equally* creations of the world we know *today*. Rather than think of the ideal family as a world we lost (or, like the Victorians, as a world just recently achieved), it is important for us to recognize that while families symbolize deep and salient modern themes, contemporary families are unlikely to fulfill our equally modern nurturant needs.

We probably have no cause to fear (or hope) that The Family will dissolve. What we can begin to ask is what we *want* our families to do. Then, distinguishing our hopes from what we have, we can begin to analyze the social forces that enhance or undermine the realization of the kinds of human bonds we need.

NOTES

1. Bronislaw Malinowski, *The Family Among the Australian Aborigines* (London: University of London Press, 1913).
2. Lewis Henry Morgan, *Ancient Society* (New York: Holt, 1877).
3. Bronislaw Malinowski, *A Scientific Theory of Culture* (Chapel Hill; University of North Carolina Press, 1944), p. 99.
4. Robin Fox, *Kinship and Marriage* (London: Penguin, 1967), p. 39.
5. Evon Z. Vogt, *Zinacantan: A Maya Community in the Highlands of Chiapas* (Cambridge, Mass.: Harvard University Press, 1969).
6. Yolanda and Robert Murphy, *Women of the Forest* (New York: Columbia University Press, 1974).
7. Ilsa Schuster, *New Women of Lusaka* (Palo Alto: Mayfield, 1979).
8. George C. Homans and David M. Schneider, *Marriage, Authority, and Final Causes* (Glencoe, Ill.: Free Press, 1955).
9. Herbert Spencer, *The Principles of Sociology*, vol. 1, *Domestic Institutions* (New York: Appleton, 1973).
10. Frederick Engels, *The Origin of Family, Private Property and the State*, in *Karl Marx and Frederick Engels: Selected Works*, vol. 2 (Moscow: Foreign Language Publishing House, 1955).
11. John Stuart Mill, *The Subjection of Women* (London: Longmans, Green, Reader and Dyer, 1869).
12. J. J. Bachofen, *Das Mutterrecht* (Stuttgart, 1861).
13. Elizabeth Fee, "The Sexual Politics of Victorian Social Anthropology," in *Clio's Banner Raised*, ed. M. Hartman and L. Banner (New York: Harper & Row, 1974).
14. Philippe Aries, *Centuries of Childhood*, trans. Robert Baldick (New York: Vintage, 1962).
15. Lawrence Stone, *The Family, Sex, and Marriage in England 1500–1800* (London: Weidenfeld and Nicholson, 1977).

16. Henry P. Lundsgaarde, *Murder in Space City: A Cultural Analysis of Houston Homicide Patterns* (New York: Oxford University Press, 1977).

17. Ibid.

18. Elizabeth Bott, *Family and Social Network: Roles, Norms, and External Relationships in Ordinary Urban Families* (London: Tavistock, 1957).

19. John Platt, "Some Problems in Measuring the Jointness of Conjugal Role-Relationships," *Sociology* 3 (1969): 287–97; Christopher Turner "Conjugal Roles and Social Networks: A Re-examination of an Hypothesis," *Human Relations* 20 (1967): 121–30; and Morris Zelditch, Jr., "Family, Marriage and Kinship," in *A Handbook of Modern Sociology,* ed. R. E. L. Faris (Chicago: Rand McNally, 1964), pp. 680–707.

20. Rayna Rapp, "Family and Class in Contemporary America: Notes Toward an Understanding of Ideology," *Science and Society* 42 (Fall 1978): 278–300.

The Time Bind: When Work Becomes Home and Home Becomes Work

Arlie Russell Hochschild

Hochschild begins her research guided by conventional assumptions regarding work and home. These two realms were thought to be unmistakably divided, the purpose and nature of each clearly and significantly different than the other. What was revealed through her research was a new model of family and work life. How would you characterize this model? What did she discover about how the meaning of work has changed for women working outside the home? How does gender influence one's perception of home and work? Why is it that mothers reported more positive states at work, and women increasingly felt the centre of their social worlds was no longer the home? What are the implications of this for the future of the family?

If working parents are "deciding" to work full time and longer, what experiences at home and work might be influencing them to do so? When I first began this research, I assumed that home was "home" and work was "work"—that each was a stationary rock beneath the moving feet of working parents. I assumed as well that each stood in distinct opposition to the other. In a family, love and commitment loom large as ends in themselves and are not means to any further end. As an Amerco parent put it, "I work to live; I don't live to work." However difficult family life may be at times, we usually feel family ties offer an irreplaceable connection to generations past and future. Family is our personal embrace with history.

Jobs, on the other hand, earn money that, to most of us, serves as the means to other ends. To be sure, jobs can also allow us to develop skills or friendships, and to be part of a larger work community. But we seldom envision the workplace as somewhere workers would freely choose to spend their time. If in the American imagination the family has a touch of the sacred, the realm of work seems profane.

Source: Excerpt from "Family Values and Reversed Worlds," from *The Time Bind: When Work Becomes Home and Home Becomes Work* by Arlie Russell Hoschschild, © 1997 by Arlie Russell Hochschild. Reprinted by permission of Henry Holt and Company, LLC.

In addition, I assumed, as many of us do, that compared to the workplace, home is a more pleasant place to be. This is after all one reason why employers pay workers to work and don't pay them to stay home. The very word "work" suggests to most of us something unpleasant, involuntary, even coerced.

If the purpose and nature of family and work differ so drastically in our minds, it seemed reasonable to assume that people's emotional experiences of the two spheres would differ profoundly, too. In *Haven in a Heartless World*, the social historian Christopher Lasch drew a picture of family as a "haven" where workers sought refuge from the cruel world of work. Painting in broad strokes, we might imagine a picture like this: At the end of a long day, a weary worker opens his front door and calls out, "Hi, Honey! I'm home!" He takes off his uniform, puts on a bathrobe, opens a beer, picks up the paper, and exhales. Whatever its strains, home is where he's relaxed, most himself. At home, he feels that people know him, understand him, appreciate him for who he really is. At home, he is safe.

At work, our worker is "on call," ready to report at a moment's notice, working flat out to get back to the customer right away. He feels "like a number." If he doesn't watch out, he can take the fall for somebody else's mistakes. This, then, is Lasch's "heartless world," an image best captured long ago in Charlie Chaplin's satirical *Modern Times*. In that film, Charlie acts the part of a hapless factory hand on an automated assembly line moving so fast that when he takes a moment to scratch his nose, he falls desperately behind. Dwarfed by the inhuman scale of the workplace, pressured by the line's relentless pace, Charlie quickly loses his humanity, goes mad, climbs into the giant machine that runs the conveyor belt, and becomes a machine part himself.

It was just such images of home and work that were challenged in one of my first interviews at Amerco. Linda Avery, a friendly thirty-eight-year-old mother of two daughters, is a shift supervisor at the Demco Plant, ten miles down the valley from Amerco headquarters. Her husband, Bill, is a technician in the same plant. Linda and Bill share the care of her sixteen-year-old daughter from a previous marriage and their two-year-old by working opposite shifts, as a full fifth of American working parents do. "Bill works the 7 A.M. to 3 P.M. shift while I watch the baby," Linda explained. "Then I work the 3 P.M. to 11 P.M. shift and he watches the baby. My older daughter works at Walgreens after school."

When we first met in the factory's breakroom over a couple of Cokes, Linda was in blue jeans and a pink jersey, her hair pulled back in a long blond ponytail. She wore no makeup, and her manner was purposeful and direct. She was working overtime, and so I began by asking whether Amerco required the overtime, or whether she volunteered for it. "Oh, I put in for it," she replied with a low chuckle. But, I wondered aloud, wouldn't she and her husband like to have more time at home to⸌ finances and company policy permitting. Linda took off her s⸜ glasses, rubbed her whole face, folded her arms, resting her elbows on table, and approached the question by describing her life at home:

> I walk in the door and the minute I turn the key in the lock my older daughter is there. Granted, she needs somebody to talk to about her day.... The baby is still up. She should have been in bed two hours ago and that upsets me. The dishes are piled in the sink. My daughter comes right up to the door and complains about anything her stepfather said or did, and she wants to talk about her job. My husband is in the other room hollering to my daughter, "Tracy, I don't *ever* get any time to talk to your mother, because you're always monopolizing her time before I even get a chance!" They all come at me at once.

To Linda, her home was not a place to relax. It was another workplace. Her description of the urgency of demands and the unarbitrated quarrels that awaited her homecoming contrasted with her account of arriving at her job as a shift supervisor:

> I usually come to work early just to get away from the house. I get there at 2:30 P.M., and people are there waiting. We sit. We talk. We joke. I let them know what's going on, who has to be where, what changes I've made for the shift that day. We sit there and chit-chat for five or ten minutes. There's laughing, joking, fun. My coworkers aren't putting me down for any reason. Everything is done with humor and fun from beginning to end, though it can get stressful when a machine malfunctions.

For Linda, home had become work and work had become home. Somehow, the two worlds had been reversed. Indeed, Linda felt she could only get relief from the "work" of being at home by going to the "home" of work. As she explained,

> My husband's a great help watching our baby. But as far as doing housework or even taking the baby when I'm at home, no. He figures he works five days a week; *he's* not going to come home and clean. But he doesn't stop to think that I work *seven* days a week. Why should I have to come home and do the housework without help from anybody else? My husband and I have been through this over and over again. Even if he would just pick up from the kitchen table and stack the dishes for me, that would make a big difference. He does nothing. On his weekends off, I have to provide a sitter for the baby so he can go fishing. When I have a day off, I have the

baby all day long without a break. He'll help out if I'm not here, but the minute I am, all the work at home is mine.

With a light laugh, she continued, "So I take a lot of overtime. The more I get out of the house, the better I am. It's a terrible thing to say, but that's the way I feel." Linda said this not in the manner of a new discovery, a reluctant confession, or collusion between two working mothers—"Don't you just want to get away sometimes?"—but in a matter-of-fact way. This was the way life was.

Bill, who was fifty-six when I first met him, had three grown children from a contentious first marriage. He told me he felt he had already "put in his time" to raise them and now was at a stage of life in which he wanted to enjoy himself. Yet when he came home afternoons he had to "babysit for Linda."

In a previous era, men regularly escaped the house for the bar, the fishing hole, the golf course, the pool hall, or, often enough, the sweet joy of work. Today, as one of the women who make up 45 percent of the American workforce, Linda Avery, overloaded and feeling unfairly treated at home, was escaping to work, too. Nowadays, men and women both may leave unwashed dishes, unresolved quarrels, crying tots, testy teenagers, and unresponsive mates behind to arrive at work early and call out, "Hi, fellas, I'm here!"

Linda would have loved a warm welcome from her family when she returned from work, a reward for her day of labors at the plant. At a minimum, she would have liked to relax, at least for a little while. But that was hard to do because Bill, on *his* second shift at home, would nap and watch television instead of engaging the children. The more Bill slacked off on his shift at home, the more Linda felt robbed of rest when she was there. The more anxious the children were, or the messier the house was when she walked in the door, the more Linda felt she was simply returning to the task of making up for being gone.

For his part, Bill recalled that Linda had wanted a new baby more than he had. So now that they were the parents of a small child, Bill reasoned, looking after the baby should also be more Linda's responsibility. Caring for a two-year-old after working a regular job was hard enough. Incredibly, Linda wanted him to do more. That was her problem though, not his. He had "earned his stripes" with his first set of children.

Early Saturday mornings, while Linda and the kids were rustling about the house, Bill would get up, put his fishing gear and a six-pack of beer into his old Ford truck, and climb into the driver's seat. "Man, I slam that truck door shut, frraaammm!, and I'm ready to go! I figure I *earned* that time."

Both Linda and Bill felt the need for time off, to relax, to have fun, to feel free, but they had not agreed that it was Bill who needed a break more than Linda. Bill simply climbed in his truck and took his free time. This irritated Linda because she felt he *took* it at her expense. Largely in response to her resentment, Linda grabbed what she also called "free time"—at work.

Neither Linda nor Bill Avery wanted more time at home, not as things were arranged. Whatever images they may have carried in their heads about what family and work should be like, the Averys did not feel their actual home was a haven or that work was a heartless world.

Where did Linda feel most relaxed? She laughed more, joked more, listened to more interesting stories while on break at the factory than at home. Working the 3 P.M. to 11 P.M. shift, her hours off didn't coincide with those of her mother or older sister who worked in town, nor with those of her close friends and neighbors. But even if they had, she would have felt that the true center of her social world was her plant, not her neighborhood. The social life that once might have surrounded her at home she now found at work. The sense of being part of a lively, larger, ongoing community—that too, was at work. In an emergency, Linda told me, she would sacrifice everything for her family. But in the meantime, the everyday "emergencies" she most wanted to attend to, that challenged rather than exhausted her, were those she encountered at the factory. Frankly, life there was more fun.

How do Linda and Bill Avery fit into the broader picture of American family and work life? Psychologist Reed Larson and his colleagues studied the daily emotional experiences of mothers and fathers in fifty-five two-parent Chicago families with children in the fifth to eighth grades. Some of the mothers cared for children at home, some worked part time, others full time, while all the fathers in the study worked full time. Each participant wore a pager for a week, and whenever they were beeped by the research team, each wrote down how he or she felt: "happy, unhappy, cheerful-irritable, friendly-angry." The researchers found that men and women reported a similar range of emotional states across the week. But fathers reported more "positive emotional states" at home; mothers, more positive emotional states at work. This held true for every social class. Fathers like Bill Avery relaxed more at home; while mothers like Linda Avery did more housework there. Larson suggests that "because women are constantly on call to the needs of other family members, they are less able to relax at home in the way men do." Wives were typically in better moods than their husbands at home only when they were eating or engaging in "family transport." They were in worse moods when they were doing "child-related activities" or "socializing" there. Men and women each felt most at ease when involved in tasks they felt less obliged to do, Larson reports. For women, this meant first shift work; for men, second.

A recent study of working mothers made another significant discovery. Problems at home tend to upset women more deeply than problems at work. The study found that women were most deeply affected by family stress—and were more likely to be made depressed or physically ill by it—even when stress at the workplace was greater. For women, current research on stress does not support the common view of home as a sanctuary and work as a "jungle." However hectic their lives, women who do paid work, researchers have consistently found, feel less depressed, think better of themselves, and are more satisfied with life than women who don't do paid work. One study reported that, paradoxically, women who work feel more valued at home than women who stay home.

In sum, then, women who work outside the home have better physical and mental health than those who do not, and not simply because healthier women go to work. Paid work, the psychologist Grace Baruch argues, "offers such benefits as challenge, control, structure, positive feedback, self-esteem ... and social ties." Reed Larson's study found, for example, that women were no more likely than men to see coworkers as friendly, but when women made friendly contact it was far more likely to lift their spirits.

As a woman quoted by Baruch put it, "A job is to a woman as a wife is to a man."

For Linda Avery self-satisfaction, well-being, high spirits, and work were inextricably linked. It was mainly at work, she commented, that she felt really good about herself. As a supervisor, she saw her job as helping people, and those she helped appreciated her. She mused,

> I'm a good mom at home, but I'm a better mom at work. At home, I get into fights with Tracy when she comes home late. I want her to apply to a junior college; but she's not interested, and I get frustrated with her, because I want so much for her. At work, I think I'm better at seeing the other person's point of view. People come to me a lot, because I'm good at helping them.

Often relations at work seemed more manageable. The "children" Linda Avery helped at work were older and better able to articulate their problems than her own children. The plant where she worked was clean and pleasant. She knew everyone on the line she supervised. Indeed, all the workers knew each other, and some were even related by blood, marriage, or, odd as it may sound, by divorce. One coworker complained bitterly that a friend of her husband's ex-wife was keeping track of how much overtime she worked in order to help this ex-wife make a case for increasing the amount of his child support. Workers sometimes carried such hostilities generated at home into the workplace. Yet despite the common assumption that relations at work are emotionally limited, meaningful friendship often blossom. When Linda Avery joined coworkers for a mug of beer at a nearby bar after work to gossip about the "spy" who was tracking the deadbeat dad's new wife's overtime, she was among real friends: Research shows that work friends can be as important as family members in helping both men and women cope with the blows of life. The gerontologist Andrew Sharlach studied how middle-aged people in Los Angeles dealt with the death of a parent. He found that 73 percent of the women in the sample, and 64 percent of the men, responded that work was a "helpful resource" in copying with a mother's death.

Amerco regularly reinforced the family-like ties of coworkers by holding recognition ceremonies honoring particular workers or entire self-managed production teams. The company would decorate a section of the factory and serve food and drink. The production teams, too, had regular get-togethers. The halls of Amerco were hung with plaques praising workers for recent accomplishments. Such recognition luncheons, department gatherings, and, particularly in the ranks of clerical and factory workers, exchange of birthday gifts were fairly common workday events.

At its white-collar offices, Amerco was even more involved in shaping the emotional culture of the workplace and fostering an environment of trust and cooperation in order to bring out everyone's best. At the middle and top levels of the company, employees were invited to periodic "career development seminars" on personal relations at work. The centerpiece of Amerco's personal-relations culture was a "vision" speech that the CEO had given called "Valuing the Individual," a message repeated in speeches, memorialized in company brochures, and discussed with great seriousness throughout the upper reaches of the company. In essence, the message was a parental reminder to respect others. Similarly, in a new-age recasting of an old business slogan ("The customer is always right"), Amerco proposed that its workers "Value the internal customer." This meant: Be as polite and considerate to your coworkers as you would be to Amerco customers. "Value the internal customer" extended to coworkers the slogan, "Delight the customer." Don't just work with your coworkers, delight them.

"Employee empowerment," "valuing diversity," and "work-family balance"—these catchphrases, too, spoke to a moral aspect of work life. Though ultimately tied to financial fain, such exhortations—and the policies that followed from them—made workers feel the company was concerned with people, not just money. In many ways, the workplace appeared to be a site of benign social engineering where workers came to feel appreciated, honored, and liked. On the other hand, how many recognition ceremonies for competent performance were going on at home? Who was valuing the internal customer there?

After thirty years with Amerco, Bill Avery felt, if anything, over-qualified for his job, and he had a recognition plaque from the company to prove it. But when his toddler got into his fishing gear and he blew up at her and she started yelling, he felt impotent in the face of her rageful screams—and nobody was there to back him up. When his teenage stepdaughter reminded him that she saw him, not as an honorable patriarch, but as an infantile competitor for her mother's attention, he felt humiliated. At such moments, he says, he had to resist the impulse to reach for the whiskey he had given up five years earlier.

Other fathers with whom I talked were less open and self-critical about such feelings, but in one way or another many said that they felt more confident they could "get the job done" at work than at home. As one human resource specialist at Amerco reflected,

> We used to joke about the old "Mother of the Year Award." That doesn't exist anymore. Now, we don't know a meaningful way to reward a parent. At work, we get paid and promoted for doing well. At home, when you're doing the right thing, chances are your kids are giving you hell for it.

If a family gives its members anything, we assume it is surely a sense of belonging to an ongoing community. In its engineered corporate cultures, capitalism has rediscovered communal ties and is using them to build its new version of capitalism. Many Amerco employees spoke warmly, happily, and seriously of "belonging to the Amerco family," and everywhere there were visible symbols of this belonging. While some married people have dispensed with their weddings rings, people proudly wore their "Total Quality" pins or "High Performance Team" tee-shirts, symbols of their loyalty to the company and of its loyalty to them. In my interviews, I heard little about festive reunions of extended families, while throughout the year, employees flocked to the many company-sponsored ritual gatherings.

In this new model of family and work life, a tired parent flees a world of unresolved quarrels and unwashed laundry for the reliable orderliness, harmony, and managed cheer of work. The emotional magnets beneath home and workplace are in the process of being reversed. In truth, there are many versions of this reversal going on, some more far-reaching than others. Some people find in work a respite from the emotional tangles at home. Others virtually marry their work, investing it with an emotional significance once reserved for family, while hesitating to trust loved ones at home. If Linda and Bill Avery were not yet at that point, their situation was troubling enough, and by no means restricted to a small group. Overall, this "reversal" was a predominant pattern in about a fifth of Amerco families, and an important theme in over half of them.

We may be seeing here a trend in modern life destined to affect us all. To be sure, few people feel totally secure either at work or at home. In the last fifteen years, massive waves of downsizing have reduced the security workers feel even in the most apparently stable workplaces. At the same time, a rising divorce rate has reduced the security they feel at home. Although both Linda and Bill felt their marriage was strong, over the course of their lives, each had changed relationships more often than they had changed jobs. Bill had worked steadily for Amerco for thirty years, but he had been married twice; and in the years between marriages, he had lived with two women and dated several more. Nationwide, half the people who marry eventually divorce, most within the first seven years of marriage. Three-quarters of divorced men and two-thirds of divorced women remarry, but remarried couples are more likely than those in first marriages to divorce. Couples who only live together are even more likely to break up than couples who marry. Increasing numbers of people are getting their "pink slips" at home. Work may become their rock.

The Natural Religion

Michael Novak

Sports and religion? In this article Novak uses examples from the United States to examine how contemporary sports generates a "civic religion" by drawing on the same natural impulses as established religions and by organizing and dramatizing sports events in a religious way. Whether it's football in the U.S., hockey in Canada, or soccer in Europe and other parts of the world, Novak argues that sports have become an expression of "natural religion."

But what are the deep natural impulses which are religious and which sports draws upon? Can such seemingly secular activities, exciting and meaningful as they may be to competitors and fans, really be considered an expression of religiosity?

A sport is not a religion in the same way that Methodism, Presbyterianism, or Catholicism is a religion. But these are not the only kinds of religion. There are secular religions, civil religions. The United States of America has sacred documents to guide and to inspire it: The Constitution, the Declaration of Independence, Washington's Farewell Address, Lincoln's Gettysburg Address, and other solemn presidential documents. The President of the United States is spoken to with respect, is expected to exert "moral leadership"; and when he walks among crowds, hands reach out to touch his garments. Citizens are expected to die for the nation, and our flag symbolizes vivid memories, from Fort Sumter to Iwo Jima, from the Indian Wars to Normandy: memories that moved hardhats in New York to break up a march that was "desecrating" the flag. Citizens regard the American way of life as though it were somehow chosen by God, special, uniquely important to the history of the human race. "Love it or leave it," the guardians of orthodoxy say. Those on the left, who do not like the old-time patriotism, have a new kind: they evince unusual outrage when this nation is less than fully just, free, compassionate, or good—in short, when it is like all the other nations of human history. America should be *better.* Why?

Source: From Michael Novak, *The Joy of Sports: Endzones, Bases, Baskets, Balls and the Consecration of the American Spirit*, Basic Books, 1976. Reprinted by permission of SLL/Sterling Lord Literistic, Inc. Copyright 1976 by Michael Novak.

The institutions of the state generate a civil religion; so do the institutions of sport. The ancient Olympic games used to be both festivals in honor of the gods and festivals in honor of the state—and that has been the classical position of sports ever since. The ceremonies of sports overlap those of the state on one side, and those of the churches on the other. At the Super Bowl in 1970, clouds of military jets flew in formation, American flags and patriotic bunting flapped in the wind, ceremonies honored prisoners of war, clergymen solemnly prayed, thousands sang the national anthem. Going to a stadium is half like going to a political rally, half like going to church. Even today, the Olympics are constructed around high ceremonies, rituals, and symbols. The Olympics are not barebones athletic events, but religion and politics as well.

Most men and women don't separate the sections of their mind. They honor their country, go to church, and also enjoy sports. All parts of their lives meld together.

Nor am I indulging in metaphor when I say that nearly every writer about sports lapses into watery religious metaphor. So do writers on politics and sex. Larry Merchant says television treated the Super Bowl "as though it were a solemn high mass." Words like *sacred, devotion, faith, ritual, immortality*, and *love* figure often in the language of sports. Cries like "You gotta believe!" and "life and death" and "sacrifice" are frequently heard.

But that is not what I mean. I am arguing a considerably stronger point. I am saying that sports flow outward into action from a deep natural impulse that is radically religious: an impulse of freedom, respect for ritual limits, a zest for symbolic meaning, and a longing for perfection. The athlete may of course be pagan, but sports are, as it were, natural religions. There are many ways to express this radical impulse: by the asceticism and dedication of preparation; by a sense of respect for the mysteries of one's own body and soul, and for powers not in one's own control; by a sense of awe for the place and time of competition; by a sense of fate; by a felt sense of comradeship and destiny; by a sense of participation in the rhythms and tides of nature itself.

Sports, in the second place, are organized and dramatized in a religious way. Not only do the origins of sports, like the origins of drama, lie in religious celebrations; not only are the rituals, vestments, and tremor of anticipation involved in sports events like those of religions. Even in our own secular age and for quite sophisticated and agnostic persons, the rituals of sports really work. They do serve a religious function: they feed a deep human hunger, place humans in touch with certain dimly perceived features of human life within this cosmos, and provide an experience of at least a pagan sense of godliness.

Among the godward signs in contemporary life, sports may be the single most powerful manifestation. I don't mean that participation in sports, as athlete or fan, makes one a believer in "God," under whatever concept, image, experience, or drive to which one attaches the name. Rather, sports drive one in some dark and generic sense "godward." In the language of Paul Tillich, sports are manifestations of concern, of will and intellect and passion. In fidelity to that concern, one submits oneself to great bodily dangers, even to the danger of death. Symbolically, too, to lose is a kind of death.

Sports are not the highest form of religion. They do not exclude other forms. Jews, Christians, and others will want to put sports in second place, within a scheme of greater ultimacy. ...

For some, it may require a kind of conversion to grasp the religiousness at the heart of sports. Our society has become secular, and personal advancement obliges us to become pragmatic, glib, superficial, and cynical. Our spirits often wither. Eyes cannot see; ears cannot hear. The soil of our culture is not always fertile for religious life. Americans must read religious messages in foreign languages. And so many will, at first, be tempted to read what I am saying as mere familiar metaphor. A change of perspective, and of heart, may be necessary.

Sports are religious in the sense that they are organized institutions, disciplines, and liturgies; and also in the sense that they teach religious qualities of heart and soul. In particular, they recreate symbols of cosmic struggle, in which human survival and moral courage are not assured. To this extent, they are not mere games, diversions, pastimes. Their power to exhilarate or depress is far greater than that. To say "It was only a game" is the psyche's best defense against the cosmic symbolic meaning of sports events. And it is partly true. For a game is a symbol; it is not precisely identified with what it symbolizes. To lose symbolizes death, and it certainly feels like dying; but it is not death. The same is true of religious symbols like Baptism or the Eucharist; in both, the communicants experience death, symbolically, and are reborn, symbolically. If you give your heart to the ritual, its effects upon your inner life can be far-reaching. Of course, in all religions many merely go through the motions. Yet even they, unaware, are surprised by grace. A Hunter pursues us everywhere, in churches and stadia alike, in the pews and bleachers, and occasionally in the pulpit and the press box.

Something has gone wrong in sports today. It went wrong in medieval Christendom, too. A proverb in Chaucer expresses it: *Radix malorum cupiditas* (The root of all evils is greed). True in the fourteenth century, it is as modern as television: Money corrupts. Nothing much changes down the centuries, only the props and the circumstances. With every day that passes, the "new" world recreates the "old." The ancient sources of corruption in Athens, Constantinople, Alexandria, and Rome are as vigorous in New York, Boston, and Washington as the preparations for Olympic games. Then as now, the hunger for excellence, for perfection-in-act, for form and beauty, is expressed in the straining muscles and fiercely determined wills of heroes of the spirit: of athletes, artists, and even, sometimes, political giants like a Pericles or Cicero. In the corruption of a slave state, a fleshpot, Homer wrote of deeds of beauty. Through his writing, pieces from the flames were salvaged. So it is in every age. Rise and fall are as steady as the seasons of our sports.

But Homer seems to be nodding nowadays. Larry Merchant of the New York *Post*, no Homer he, called his column "Fun and Games" and has been called the pioneer "of modern skepticism and irreverence toward sports." Merchant modestly replies to praise: "I must state for posterity that I was merely part of a broad-based movement. ... We were irreverent, debunking heroes and myths that didn't stand up to scrutiny." Shucks, folks. Sports-writing "has changed conceptually." His own self-image isn't bad: "We were humanistic. ... We saw ballparks as funhouses, not temples. We ... [dug] for the hows and whys and whos." The intellectually fearless skeptics. Sports isn't religion. It's entertainment. "A baseball game," writes Robert Lipsyte, formerly a New York *Times* sportswriter and the author of *SportsWorld*, "is a staged entertainment, and baseball players are paid performers."

Jewish and Protestant writers draw on different intellectual traditions from mine, of course, but from my point of view, Catholic that I was born, any religion worthy of the name thrives on irreverence, skepticism, and high anticlericalism. A religion without skeptics is like a bosom never noticed (which isn't entirely farfetched, since at least one writer has said that covering sports for the New York *Times* is like being Racquel Welch's elbow). When Catholicism goes sour, as periodically down the centuries it does, almost always the reason is a dearth of critics or, worse, the death of heretics. A nonprophet church decays. When things go well, it is because critics condemn what is going ill. A decent religion needs irreverence as meat needs salt.

Temples do not require whispering. Jesus knocked temple tables over, jangling metal coins on the stones. The root of the religious sense is not the stifling of questions. It lies in asking so many questions that the true dimensions of reality begin to work their own mysterious awe. No one is less religious than the pleasantly contented pragmatist after lunch. Nothing is more religious than fidelity to the drive to understand. For that drive is endless, and satisfied by nothing on earth. It is the clearest sign in our natures that our home is not here; that we are out of place; and that to be restless, and seeking, is to be what we most are.

Sports are not merely fun and games, not merely diversions, not merely entertainment. A ballpark is not a temple, but it isn't a fun house either. A baseball game is not an entertainment, and a ballplayer is considerably more than a paid performer. No one can explain the passion, commitment, discipline, and dedication involved in sports by evasions like these....

Those who think sports are merely entertainment have been bemused by an entertainment culture. Television did not make sports possible—not even great, highly organized sports. College football and major league baseball thrived for decades without benefit of television. Sports made television commercially successful. No other motive is so frequently cited as a reason for shelling out money for a set. (Non-sports fans, it appears, are the least likely Americans to have sets.)

In order to be entertained, I watch television: primetime shows. They slide effortlessly by. I am amused, or distracted, or engrossed. Good or bad, they help to pass the time pleasantly enough. Watching football on television is totally different. I don't watch football to pass the time. The outcome of the games affects me. I care. Afterward, the emotion I have lived through continues to affect me. Football is not entertainment. It is far more important than that. If you observe the passivity of television viewers being entertained, and the animation of fans watching a game on television, the difference between entertainment and involvement of spirit becomes transparent. Sports are more like religion than like entertainment. Indeed, at a contest in the stadium, the "entertainment"— the bands, singers, comedians, balloons, floats, fireworks, jets screaming overhead—pales before the impact of the contest itself, like lemonade served to ladies while the men are drinking whiskey.

Television is peculiarly suited to football, and vice versa; the case is special. . . .

On Monday nights, when television carries football games, police officers around the nation know that crime rates will fall to low levels otherwise reached only on Mother's Day and Christmas.

We are, as I said, too close to sports to appreciate their power. Besides, our education is rigorously pragmatic and factual: those things are real which can be counted. It teaches us nothing about play, or myth, or spirit. So we are totally unprepared to speak about the things we love the most. Our novelists write poorly of women, of love, of tragedy. Our religious sensibilities, which in some are warm with fervor, in our major publicists are chill. Grown men among us are virtually inarticulate about anything that touches our souls. Grunts, groans, and silence. Being cool. . . . We have no truly popular operas, or suitably complex literature, or plays in which our entire population shares. The streets of America, unlike the streets of Europe, do not involve us in stories and anecdotes rich with a thousand years of human struggle. Sports are our chief civilizing agent. Sports are our most universal art form. Sports tutor us in the basic lived experiences of the humanist tradition.

The hunger for perfection in sports cleaves closely to the driving core of the human spirit. It is the experience of this driving force that has perennially led human beings to break forth in religious language. This force is in us, it is ours. Yet we did not will its existence, nor do we command it, nor is it under our power. It is there unbidden. It is greater than we, driving us beyond our present selves. "Be ye perfect," Jesus said, "as your heavenly Father is perfect." The root of human dissatisfaction and restlessness goes as deep into the spirit as any human drive—deeper than any other drive. It *is* the human spirit. Nothing stills it. Nothing fulfills it. It is not a need like a hunger, a thirst, or an itch, for such needs are easily satisfied. It is a need even greater than sex; orgasmic satisfaction does not quiet it. "Desire" is the word by which coaches call it. A drivenness. Distorted, the drive for perfection can propel an ugly and considerably less than perfect human development. True, straight, and well targeted, it soars like an arrow toward the proper beauty of humanity. Sports nourish this drive as well as any other institution in our society. If this drive is often distorted, as it is, even its distortions testify to its power, as liars mark out the boundaries of truth.

Sport, in a word, are a form of godliness. That is why the corruptions of sports in our day, by corporations and television and glib journalism and cheap public relations, are so hateful. If sports were entertainment, why should we care? They are far more than that. So when we see them abused, our natural response is the rise of vomit in the throat.

It may be useful to list some of the elements of religions, to see how they are imitated in the world of sports.

. . .

[S]uppose you are an anthropologist from Mars. You come suddenly upon some wild, adolescent tribes living in territories called the "United States of America." You try to understand their way of life, but their society does not make sense to you. Flying over the land in a rocket, you notice great ovals near every city. You descend and observe. You learn that an oval is called a "stadium." It is used, roughly, once a week in certain seasons. Weekly, regularly, millions of citizens stream into these concrete doughnuts, pay handsomely, are alternately hushed and awed and outraged and screaming mad. (They demand from time to time that certain sacrificial personages be "killed.") You see that the figures in the rituals have trained themselves superbly for their performances. The combatants are dedicated. So are the dancers and musicians in tribal dress who occupy the arena before, during, and after the combat. You note that, in millions of homes, at corner shrines in every household's sacred room, other citizens are bound by invisible attraction to the same events. At critical moments, the most intense worshipers demand of the less attentive silence. Virtually an entire nation is united in a central public rite. Afterward, you note exultation or depression among hundreds of thousands, and animation almost everywhere.

Some of the elements of a religion may be enumerated. A religion, first of all, is organized and structured. Culture is built on cult. Accordingly, a religion begins with ceremonies. At these ceremonies, a few surrogates perform for all. They need not even believe what they are doing. As professionals, they may perform so often that they have lost all religious instinct; they may have less faith than any of the participants. In the official ceremonies, sacred vestments are employed and rituals are prescribed. Customs develop. Actions are highly formalized. Right ways and wrong ways are plainly marked out; illicit behaviors are distinguished from licit ones. Professional watchdogs supervise formal correctness. Moments of silence are observed. Concentration and intensity are indispensable. To attain them, drugs or special disciplines of spirit might be employed; ordinary humans, in the ordinary ups and downs of daily experience, cannot be expected to perform routinely at the highest levels of awareness.

Religions are built upon *ascesis*, a word that derives from the disciplines Greek athletes imposed upon themselves to give their wills and instincts command of their bodies; the word was borrowed by Christian monks and hermits. It signifies the development of character, through patterns of self-denial, repetition, and experiment. The type of character celebrated in the central rituals, more likely than not, reveals the unconscious needs of the civilization—extols the very qualities that more highly conscious formulations are likely to deny. Thus, the cults have a revelatory quality; they dramatize what otherwise goes unspoken.

Religions also channel the feeling most humans have of danger, contingency, and chance—in a word, Fate. Human plans involve ironies. Our choices are made with so little insight into their eventual effects that what we desire is often not the path to what we want. The decisions we make with little attention turn out to be major turning points. What we prepare for with exquisite detail never happens. Religions place us in the presence of powers greater than ourselves, and seek to reconcile us to them. The rituals of religion give these powers almost human shape, forms that give these powers visibility and tangible effect. Sports events in baseball, basketball, and football are structured so that "the breaks" may intervene and become central components in the action.

Religions make explicit the almost nameless dreads of daily human life: aging, dying, failure under pressure, cowardice, betrayal, guilt. Competitive sports embody these in every combat.

Religions, howsoever universal in imperative, do not treat rootedness, particularity, and local belonging as unworthy. On the contrary, they normally begin by blessing the local turf, the local tribe, and the local instinct of belonging—and use these as paradigms for the development of larger loyalties. "Charity begins at home." "Whoever says that he loves God, whom he does not see, but hates his neighbor, whom he does see, is a liar and the truth is not in him."

Religions consecrate certain days and hours. Sacred time is a block of time lifted out of everyday normal routines, a time that is different, in which different laws apply, a time within which one forgets ordinary time. Sacred time is intended to suggest an "eternal return," a fundamental repetition like the circulation of the human blood, or the eternal turning of the seasons, or the wheeling of the stars and planets in their cycles: the sense that things repeat themselves, over and over, and yet are always a little different. Sacred time is more like eternity than like history, more like cycles of recurrence than like progress, more like a celebration of repetition than like a celebration of novelty. Yet sacred time is full of exhilaration, excitement, and peace, as though it were more real and more joyous than the activities of everyday life—as though it were *really living* to be in sacred time (wrapped up in a close game during the last two minutes), and comparatively boring to suffer the daily jading of work, progress, history.

To have a religion, you need to have heroic forms to try to live up to: patterns of excellence so high that human beings live up to them only rarely, even when they strive to do so; and images of perfection so beautiful that, living up to them or seeing someone else live up to them, produced a kind of "*ah!*"

You need to have a pattern of symbols and myths that a person can grow old with, with a kind of resignation, wisdom, and illumination. Do what we will, the human body ages. Moves we once could make our minds will but our bodies cannot implement; disciplines we once endured with suppressed animal desire are no longer worth the effort; heroes that once seemed to us immortal now age, become enfeebled, die, just as we do. The "boys of summer" become the aging men of winter. A religion celebrates the passing of all things: youth, skills, grace, heroic deeds.

To have a religion, you need to have a way to exhilarate the human body, and desire, and will, and the sense of beauty, and a sense of oneness with the universe and other humans. You need chants and songs, the rhythm of bodies in unison, the indescribable feeling of many who together "will one thing" as if they were each members of a single body.

All these things you have in sports.

Sports are not Christianity, or Judaism, or Islam, or Buddhism, or any other of the world religions. Sports are not the civil religion of the United States of America, or Great Britain, or Germany, or the Union of Soviet Socialist Republics, or Ghana, or any other nation.

But sports are a form of religion. This aspect of sports has seldom been discussed. Consequently, we find it hard to express just what it is that gives sports their spirit and their power.

Athletes are not merely entertainers. Their role is far more powerful than that. People identify with them in a much more priestly way. Athletes exemplify something of deep meaning—frightening meaning, even. Once they become superstars, they do not quite belong to themselves. Great passions are invested in them. They are no longer treated as ordinary humans or even as mere celebrities. Their exploits and their failures have great power to exult—or to depress. When people talk about athletes' performances, it is almost as though they are talking about a secret part of themselves. As if the stars had some secret bonding, some Siamese intertwining with their own psyches.

. . .

The tales of *Gawain and the Green Knight*, the *Song of Roland*, the exploits of Ivanhoe—these are the ancient games in which human beings have for centuries found refreshment. The crowds who watched the jousts of old are still cheering, still quenching the dust in their throats with cold drinks between the acts, and still seeing enacted before their naked eyes myths of courage, brains, and skill.

We are so close to sports, so enmeshed in them, that we do not truly *see* them, we do not marvel. We overlook the wonder even the existence of sports should cause, let alone their persistence and their power. Long after the Democratic Party has passed into history, long after the United States has disappeared, human beings will still be making play fundamental to their lives.

Play is the most human activity. It is the first act of freedom. It is the first origin of law. (Watch even an infant at play, whose first act is marking out the limits, the rules, the roles: "This is the road. . . ." The first free act of the human is to assign limits within which freedom can be at play.) Play is not tied to necessity, except to the necessity of the human spirit to exercise its freedom, to enjoy something that is not practical, or productive, or required for gaining food or shelter. Play is human intelligence, and intuition, and love of challenge and contest and struggle; it is respect for limits and laws and rules, and high animal spirits, and a lust to develop the art of doing things perfectly. Play is what only humans truly develop. Humans could lives as animals (and often we do, governed by instinct), envying what seems to be the freedom of the wild, the soaring aloft of birds, the unfettered wanderings of jungle felines "born free." But animals are not free, not as humans are. Animals do not multiply cultures and languages, and forms of play, and organizational patterns. Animals play as they have for centuries, while humans ceaselessly invent, produce the multiple varieties of religion and play that establish on the soil of nature the realm of culture, the field of liberty. The religions we have, like the games we have, have issued forth from the historical response of humans to their own liberty.

In all these ways, religions and sports have much in common. Sports belong in the category of religion.

. . .

Sports constitute the primary lived world of the vast majority of Americans. The holy trinity—baseball, basketball, and football—together with tennis, bowling, skiing, golf, hiking, swimming, climbing (not to mention gambling, Monopoly, cards, and other forms of play), are not simply interludes but the basic substratum of our intellectual and emotional lives. Play provides the fundamental metaphors and the paradigmatic experiences for understanding the other elements of life. . . .

Television: The Shared Arena

Joshua Meyrowitz

Media has been an important theme within Canadian social theory, from Harold Innis' investigations of oral and literate cultures through Marshall McLuhan's probes into electronic media and society to the more recent postmodern investigations of Arthur Kroker and other contemporary communication theorists. In this article Meyrowitz explores what happens to culture and society when television, rather than physical sites such as town halls and even countries, becomes the arena for public life. While television seems to be all around us, its impacts have yet to be seen clearly. In what ways are we attached to, and detached from, each other by the shared experience of viewing television?

Meyrowitz discusses what television reveals about us even as we use it to identify with each other. How does it bind and separate us? How does its ubiquity in this era of electronic media undermine some of the central distinctions upon which our culture is based? Finally, how can Meyrowitz's research help us to understand and anticipate some of the consequences of widespread access to the Internet?

In 1950, only 9 percent of U.S. homes owned television sets. Little more than 25 years later, only 2 percent of households were without one. In a remarkably short time, television has taken a central place in our living rooms and in our cultural and political lives. On average, a U.S. household can now receive 30 channels; only 7 percent of homes receive six or fewer stations. Some 95 percent of homes own a colour TV, 63 percent own two or more sets, and 64 percent own a video cassette recorder.

Television is the most popular of the popular media. Indeed, if Nielsen research and other studies are correct, there are few things that Americans do more than they watch television. On average, each household has a TV on almost 50 hours a week. Forty percent of households eat dinner with the set on. Individually, Americans watch an average of 30 hours a week.

Source: Joshua Meyrowitz is Professor of Communication at the University of New Hampshire and the author of *No Sense of Place: The Impact of Electronic Media on Social Behaviour.* This article originally appeared in *The World & I* in July 1990.

We begin peering at TV through the bars of cribs and continue looking at it through the cataracts of old age.

Plato saw an important relationship between shared, simultaneous experience and a sense of social and political interconnectedness. Plato thought that his Republic should consist of no more than 5000 citizens because that was the maximum number of people who could fit in an arena and simultaneously hear the voice of one person. Television is now our largest shared arena. During the average minute of a typical evening, nearly a hundred million Americans are tuned in. While a book can usually win a place on the lists of the top 50 fiction or non-fiction bestsellers for the *year* with 115,000 hardcover sales, a prime-time network program with fewer than 15 million viewers for *each episode* is generally considered a failure.

Even the biggest bestsellers reach only a fraction of the audience that will watch a similar program on television. It took 40 years for *Gone with the Wind* to sell 21 million copies; 55 million people watched the first half of the movie on television in a single evening. The television mini-series "Roots" was watched, in part or whole, by approximately 130 million people in only eight days. Even with the help of the television promotion, fewer than 5 million copies of *Roots* sold in eight years.

The television arena, like a street corner or a marketplace, serves as an environment for us to monitor but not necessarily identify with. Reading a newspaper requires an investment of money and reading effort, and at least some minimal identification with its style and editorial policy. We have to reach out for it and embrace it—both literally and metaphorically. But with television, we simply sit back and let the images wash over us. While we usually select reading material that clearly reflects our own self-image, with TV we often feel we are passively observing what other people are like.

Most of us would feel uncomfortable stopping at a local store to pick up the current issue of a publication titled *Transvestite Times* or *Male Strippers' Review*, or a magazine on incest, child abuse, or adultery. But millions of viewers feel quite comfortable sharing their homes with transvestites, male strippers, and victims and perpetrators of incest, or almost anyone else who appears on "Donahue," "Oprah," or "Geraldo." Ironically, our personal dissociation with TV content allows for the most widespread sharing of similar experience in the history of civilization.

In the 1950s, many intellectuals were embarrassed to admit that they owned a television set, let alone that they spent any valuable time watching it. But the massive saturation of television into virtually every U.S. home now imbues the activity of watching television with multiple layers of social significance. One can watch popular programs not merely to see the program but to see what others are watching. To watch television may not be to stare into the eyes of America, but it is to look over its shoulder and see what Americans see. Watching television—with its often distorted versions of reality—does not allow us to keep our finger on the pulse of the nation so much as it allows us to keep our finger on the pulse the nation is keeping its finger on. With television, it somehow makes sense for a viewer to watch the tube avidly while exclaiming, "My God, I can't believe people watch this stuff!"

Even though many people watch it alone, television is capable of giving each isolated viewer a sense of connection with the outside world and with all the other people who are watching. During major television events—whether fictional or non-fictional—such as the final episode of "M*A*S*H" or the explosion of the *Challenger*, one is likely to find that more than one out of every two people one sees on the street the next day has had a similar experience the night before. Regardless of specific content, then, television often serves a social function similar to the weather: No one takes responsibility for it, often it is quite bad, but nearly everyone pays attention to it and sees it as a basis of common experience and conversational topics. Perhaps this is why even pay cable households spend more than half their viewing time watching "regular" network programming and why the most frequent use of VCRs is for time shifting of programs broadcast by network-affiliated stations.

For many people, someone or something that does not appear on television does not fully exist in the social sense. The Watergate scandals became "real" not when the *Washington Post* reported the stories but when network television news reported that the *Washington Post* reported the stories. Similarly, civil rights and anti-Vietnam War protests became social realities not when demonstrators took to the streets but when the protests were viewed on television. And although most of our early presidents were seen by only a few of the voters of their day, it is now impossible to imagine a serious candidate who would not visit us all on TV. And so it is that politicians, salespeople, protestors, and terrorists all design their messages in the hope of capturing the television eye.

TOO CLOSE TO THE SET

Despite its ubiquity, the impact of television is not yet seen very clearly. For one thing, most of us watch television too closely—not in the way that mothers warn their children about, but in the sense of evaluating television primarily on the basis of whether we like or don't like its programs. Even scholars tend to reduce the impact of television to its past and current programming and to the motives of the institutions that control it. The overwhelming majority of television research and criticism has focused on the nature of the programs, their imitative or persuasive power, their aesthetic value or bankruptcy, the range of meanings that viewers can draw from them, or their underlying economic and political purposes.

These are important but insufficient questions. The effects of a new communication technology cannot be understood fully by looking only at the medium's typical content and patterns of control. To see the limits of such an approach, we need only to consider what its use in the fifteenth and sixteenth centuries would have revealed about the impact of the printing press, then spreading through western culture. A content/institutionalized approach to printing probably would have led observers to conclude that books had two major effects: 1. the fostering of religion (most early books were religious in content) and 2. the strengthening of central religious and monarchal authorities (who controlled much of what was printed). The underlying, but ultimately more significant, long-term effects of the printing press—such as the growth of individual thinking and science and the spread of nationalism and constitutional systems—would remain invisible.

This is not to suggest that the short-term, surface effects are inconsequential. Just look at William Carter. He printed a pro-Catholic pamphlet in England in 1584 and was promptly hanged. Similarly, our current information environment is choked and narrowed by the way television is controlled.

The television business is not structured to deliver quality programming to viewers but rather to deliver viewers to advertisers. We are sold to advertisers in lots of a thousand. The real programming on television is the commercial. That is where the time, the money, and the competition are. That is where the most creative television "artists" (if we want to use that term) are working. The TV shows—whether news or entertainment or "infotainment"—are simply the bait.

We are misled when we are told that program ratings are part of an audience-centred, "democratic" process that allows us to "vote" for shows. In fact, we are usually offered choices among advertiser-friendly programs. This is why TV ratings systems rarely asked whether or why we like or dislike a show or what we would like to see instead. Most ratings simply measure how many of what type of people are there for the ads.

Even if networks can draw millions of viewers to a program, the last thing they want to do is put the audience in a mood that does not mix well with consumption of the advertised products. One of the most-watched programs in television history, "The Day After," for example, was an ad failure. After all, what companies would want their products associated with nuclear holocaust? In fact, as one vice-president of the network confided to me, the airing of "The Day After"—as bland a treatment of its subject as it was—almost led to a stockholder suit against ABC because the network could have made more money airing a rerun of a program such as "The Harlem Globetrotters Visit Gilligan's Island."

But to reduce television to a cultural nuisance or to a slickly disguised salesperson, as some analysts do, is to miss what is happening in our culture because of television and how it—not merely through its content but also as a certain form of shared experience—reshapes our attitudes and behaviours.

The effects of new media of communication arise not solely from their content but also from the new ways in which the medium packages and transmits information. Writing and print, for example, were able to foster the rise of individual thinking and science because they literally put information in the hands of individuals and because they allowed for the recording and wide-scale distribution of ideas that were too complicated to memorize (even by the people who came up with them). Even as William Carter swung from the gallows by regal decree, printing was quietly working against its apparent masters, ultimately secularizing the culture and encouraging the overthrow of monarchies. Similarly, the impact of television cannot be reduced to programs that come through the tube or to the institutions that control it. There are effects apart from, and even in opposition to, these forces.

The most significant long-term effects of television may also lie in its manner of packaging and transmitting information and in the ways that it undoes some of the systems of communication supported by print. Television has changed "who knows what about whom" and "who knows what compared to whom." As a result, it has changed the way we grow from childhood to adulthood, altered our sense of appropriate gender behaviour, shifted our perceptions of our political and other leaders, and affected our general sense of "them" and "us."

VIDEO NURSERY

As printing and literacy spread through western culture, literate adults discovered they could increasingly keep secrets from preliterate and semi-literate children. Adults used books to communicate among themselves without children overhearing. Clerics argued for the development of expurgated versions of the classics, and the notion of the innocence of childhood began to take hold, eventually spreading to the lowest classes with the growth of universal education.

Childhood was to be a time of innocence and isolation. Children were protected from the nasty realities of adult life. Unable to read, very young children had no access to the information available in books. Young children were presented with an idealized version of adult life. Children were slowly walked up the ladder of literacy with a new, somewhat less idealized view of adult life presented to them at each step of reading ability.

Television dilutes the innocence of childhood by undermining the system of information control that supported it. Television bypasses the year-by-year slices of knowledge given to children. It presents the same information to adults and to children of all ages. Children may not understand everything they see on television, but they are exposed to many aspects of adult life that their parents (and traditional children's books) would have once protected them from.

Parents often clamour for more and better children's television. But one could argue that there is no such thing as "children's television," at least not in the sense that there is children's literature. Children's literature is the only literature that children can read, and only children read it. In contrast, studies since the early days of television have found that children often prefer to watch programs aimed at adults. And adults often watch programs aimed at children—about a third of the audience for "Pee Wee Herman's Playhouse" is over eighteen.

At some point during the last decade, each of the following has been among the most popular programs in *all* age groups, including ages two to eleven: "Dallas," "The Muppets," "The Dukes of Hazzard," "Love Boat," "The A-Team," "Cheers," "Roseanne," and "The Golden Girls." Thus, children have been avid viewers of adult soap operas, and adults have found pleasure in a children's puppet show.

In both fictional and non-fictional programs, children learn that adults lie, cheat, drink too much, use drugs, and kill and maim each other. But perhaps the most dramatic revelation that television provides to young children is that parents struggle to control children.

Unlike books, television cannot be used easily by adults as a tool to discuss how to raise children. A parental advice book can be used by adults to communicate among themselves about what to tell and not to tell children. But the same conversation on television is usually overheard by thousands of children, who are thereby exposed to the very topics suggested for secrecy and to the "secret of secrecy" itself—the fact that adults are anxious about their parental roles and conspire to keep secrets from children.

Even seemingly innocent programs reveal significant secrets to children. When the first TV generation watched programs such as "Father Knows Best" and "Leave It to Beaver," for example, they learned that parents behaved one way in front of their children and another way when they were alone. In front of their children, the TV parents were calm, cool, and collected, but away from their kids, they were anxious and concerned about their parental behaviour. Because we often reduce the effects of TV to imitation, we forget that while the children *on* such programs were innocent and sheltered, the children *watching* the shows often saw how adults manipulated their behaviours to make it appear to their children that they knew best. This is a view that undermines traditional parental authority by making children less willing to take adult behaviour at face value. It is no wonder, perhaps, that the children who grew up watching "Father Knows Best" became concerned with the "credibility gap"; that is, the difference between what people proclaim publicly and what they say and feel privately.

Subsequent situation comedies, such as "One Day at a Time," shocked many viewers because the parents in the shows revealed their fears and anxieties about parenting in front of their children and because the child characters on the shows were no longer sheltered or innocent. But in terms of what *child* viewers learned about the concerns of parents, there was relatively little new information. The third phase of family shows, including "The Bill Cosby Show" and "Family Ties," offers a compromise between the two earlier family visions: The line between parents and children has been partly re-established, but the children are more sophisticated than early TV children and the parents are both less surefooted in front of their children and less conspiratorial away from them.

In a book culture, control over the flow of information is literally placed in parents' hands. Parents can easily give some books to children and withhold others. Parents can read one book while their children sit in the same room reading another. Television is not so co-operative. Parents often find it difficult to censor their children's viewing without censoring their own, and parents cannot always anticipate what will happen on TV the way they can flip through a book. A father may think he is giving his daughter a lesson in science as they watch the *Challenger* take off, only to discover that he has exposed her instead to adult hubris and tragedy.

Most television programs are accessible to children in a way that most book content is not. The visual/aural form of television allows children to experience many behaviours and events without the skill of decoding written sentences. And it is much simpler for children to wander off "Sesame Street" and slip beyond "Mr. Rogers' Neighborhood" into grown-up television than it is for children to buy or borrow books from a grown-up library or bookstore. Television takes our children across the globe before we as parents even give them permission to cross the street.

As children's innocence declines, children's literature and children's programming have changed as well. Some children's books now discuss sex and drugs and other once-taboo topics, and war and divorce have recently visited "Mr. Rogers' Neighborhood."

This does not mean that adults should abdicate their authority over children or even give up trying to control children's viewing of television. Adults are more experienced and more knowedgeable. But it does mean that the old support system for unquestioned adult authority has been undermined by television. In a television culture, children are more suspicious of adult authority, and many adults feel somewhat exposed, finding it more difficult to pretend to know everything in front of their children. The result is a partial blurring of traditional child and adult roles. Children seem older and more knowledgeable, and adults now reveal to their children the most childish sides of themselves, such as doubts, fears, and anxieties. Thus, we are seeing more adultlike children and more childlike adults, behaviour styles characteristic of preliterate societies.

GENDER BLENDER

Our society once tried to maintain a clear distinction between the male realm and the female realm. The Victorians spoke of the "two spheres": a public, male world of brutal competitions, rationality, and accomplishments; and a private, female world of home, intuition, and childrearing. Men were to suppress their emotions and women were to suppress their competitiveness. The ideal of separate spheres was quite strong in our society when television became the newest home appliance.

Yet even as television situation comedies and other programs featured very traditional gender roles in the two separate spheres, television, as a shared arena, was beginning to break down the distinction between the male and the female, between the public and private realms. Television close-ups reveal the personal side of public figures and events (we see tears well up in the eyes of the president; we hear male voices crack with emotion) just as most public events have become dramas that are played out in the privacy of our kitchens, living rooms, and bedrooms. Television has exposed even homebound women to most of the parts of the culture that were once considered exclusively male domains—sports, war, business, medicine, law, politics—just as it has made men more aware of the emotional dimensions and consequences of public actions.

When Betty Friedan wrote in *The Feminine Mystique* that women in 1960 felt a "schizophrenic split" between the frilly, carefree image of women in women's magazines and the important events occurring in "the world beyond the home," most of her examples of the latter were unwittingly drawn from the top television news stories of the year. By 1960, television was present in nearly 90 percent of U.S. homes. Similarly, other feminist writers have described changes in the 1960s by writing metaphorically of the "breaking of boundaries" (Gloria Steinem), "a sudden enlargement of our world" (Elizabeth Janeway), and of women having "seen beyond the bucolic peacefulness of the suburbs to the war zone at the perimeter" (Barbara Ehrenreich and Deirdre English). But these writers seem unaware of how closely their metaphors describe the literal experience of adding a television to a suburban household.

The fact that early TV programs generally portrayed active men and passive, obedient women had no more of an imitative effect on women viewers than the innocent child characters on "Father Knows Best" had on child viewers. Television, it is true, suggested to women how society thought they should behave, just as etiquette books had for centuries. But television did something else as well: It allowed women to observe and experience the larger world, including all-male interactions and behaviours. Indeed, there is nothing more frustrating than being exposed constantly to adventures, activities, and places that you are told are reserved for someone else. Television also demystified the male realm, making it and its inhabitant seem neither very special nor very intimidating. No wonder women have since demanded to integrate that realm.

Television's impact has been greatest on women because they have traditionally been more isolated. But men are affected as well, partly because women have demanded changes in their behaviour and partly because television emphasizes those traits traditionally ascribed to women: feelings, appearance, emotion. On television, "glorious victories" and "crushing defeats" are now conveyed through images of blood and limp bodies and the howls of survivors. Television has helped men to become more aware of their emotions and of the fact that emotions cannot be completely buried. Even at televised public hearings, it is hard to ignore the facial expressions, the yawns, the grimaces, the fatigue.

The way men react to public issues is also being subtly feminized. Men used to make fun of women for voting for candidates because of the candidates' appearance rather than their stands on the issues. But recent polls show that millions of men, as well as women, will now vote for a candidate they disagree with on the issues, if they "personally like" the candidate. About a third of Ronald Reagan's votes came from such supporters.

Television is one of the few public arenas in our culture where men routinely wear make-up and are judged as much on their personal appearance and "style" as on their "accomplishments." If it was once thought that women communicated and men accomplished, it is telling that our most successful recent president was dubbed the "Great Communicator" and was admired for his gentle voice and manner and his moist-eyed emotional appeals.

With television, boys and girls and men and women tend to share a great deal of similar information about themselves and the "other." Through TV close-ups, men and women see, in one month, many more members of the opposite sex at "intimate distance" than members of earlier generations saw in a lifetime. Further, unlike face-to-face interactions, in which the holding of a gaze may be construed as insulting or as an invitation to further intimacy, television allows one to stare and carefully examine the face, body, and movements of the other sex. Television fosters an easy and uninvolved intimacy.

Just as women have become more involved in the public realm, men are becoming more involved in the private realm, especially in the role of fathers. Traditional distinctions cannot be erased in a generation, of course. But dramatic changes have taken place in a remarkably short time. In 1950, only 12 percent of married women with children under six worked; by 1987, 57 percent did. Recent studies also show that men are now more likely to turn down overtime pay or travel and relocation offers in order to spend more time with their families.

In spite of its often sexist content, television, as an environment shared by both sexes, has made the membranes around the male and female realms more permeable. As a result, the nature of those two realms has been blurring. We are witnessing more career-oriented women and more family-oriented men; we are developing more work-oriented homes, and there is increasing pressure to make the public realm more family-oriented.

PRESIDENTIAL PIMPLES

Just as television tends to mute differences between people of different ages and sexes, so does it tend to mute differences between levels of social status. Although television is certainly an important weapon in the arsenal of leaders, if often functions as a double-edged sword. Unlike other media, television not only allows leaders to reach followers, it also allows followers to gain unprecedented access to the close-up appearance and gestures of leaders.

"Leadership" and "authority" are unlike mere power in that they depend on performance and appeal; one cannot lead or be looked up to if one's presence is unknown. Yet, paradoxically, authority is weakened by excess familiarity. Awe survives through "distant visibility" and "mystified presence." One of the peculiar ironies of our age is that most people who step forward into the television limelight and attempt to gain national visibility become too visible, too exposed, and are thereby demystified.

The speaker's platform once lifted politicians up and away from average citizens, both literally and symbolically. In newspaper quotes and reports, the politician—as flesh-and-bones person—was completely absent. And on radio, politicians were disembodied voices. But the television camera now lowers politicians to the level of the common citizen and brings them close for our inspection. In recent years, we have seen our presidents sweat, stammer, and stumble—all in living colour.

Presidential images were once much better protected. Before TV coverage of press conferences, newspapers were not even allowed to quote a president without his explicit permission. As late as the start of the Eisenhower administration, *The New York Times* and other publications had to paraphrase the president's answers to questions. In earlier administrations, journalists had to submit their questions in advance and were forbidden from mentioning which questions the president refused to answer. Presidential advisers frequently corrected presidents' answers during meetings with the press, and such assistance went unreported. In the face of a "crisis," our presidents once had many hours, sometimes even weeks or months, to consult with advisers and to formulate policy statements to be printed in newspapers. Now, standing before the nation, a president is expected to have all relevant information in his mind—without notes and without consultation with advisers. A president must often start a sentence before the end of the sentence is fully formed in his mind. Even a five-second pause for thought can seriously damage a leader's credibility. The apparent inarticulateness of all our recent presidents may be related more to the immediacy of television than to a decline in our leaders' mental abilities.

In language, the titles "president," "governor," and "senator" still call forth respect. But the close-up TV pictures of the persons filling those offices are rarely as impressive. We cannot help but notice the sweat on the brow, the nervous twitch, the bags under the eyes.

Television not only reduces our awe of politicians, it increases politicians' self-doubt and lowers self-esteem. A speaker's nervousness and mistakes usually are politely ignored by live audiences and therefore soon forgotten by the speaker as well. But with videotape, politicians have permanent records of themselves misspeaking or anxiously licking their lips. Television may be a prime cause of the complaints of indecisive lea-dership and hesitant "followership" that we have heard since the mid-1960s.

In the 1950s, many people were shocked that a genuine hero, Dwight Eisenhower, felt the need to hire a Hollywood actor to help with his television appearances. But now we are much more sophisticated—and more cynical. We know that one cannot simply *be* the president, but that one has to *perform* the role of "president." The new communication arena demands more control on the part of politicians, but it also makes the attempts at control more visible. Many citizens lived through twelve years of FDR's presidency without being aware that his legs were crippled and that he often needed help to stand. But we are now constantly exposed to the ways in which our presidents and presidential candidates attempt to manipulate their images to create certain impressions and effects.

The result is that we no longer experience political performances as naïve audiences. We have the perspective of stage hands who are aware of the constructed nature of the drama. Certainly, we prefer a good show to a bad show, but we are not fully taken in by the performances. Rather than being fooled, we are willingly entertained, charmed, courted, and seduced. Ironically, all the recent discussions of how effectively we are being manipulated may only point out how visible and exposed the machinations now are.

I am not suggesting that television has made us a fully informed and aware electorate. Indeed, relatively few Americans realize how selective an image of the world we receive through television news. When the same sort of occurrences take place in El Salvador and in Nicaragua, or in Poland and in Chile, they are often covered in completely different ways, often in keeping with pre-existing news narratives concerning each country. But regardless of the ways in which the content of television news is often moulded, television is having other effects due to its immediacy and visual nature.

Most of our information about other countries once came through the president and State Department, often after careful planning about how to present the information to the public. This allowed the government to appear to be in control of events and always to have a ready response. In many instances, we now experience events at the same moment as our leaders, sometimes before them. The dramatic images of the fall of the Berlin Wall and other changes in Eastern Europe were watched by the president, the secretary of state, and millions of other Americans at the same moment. The immediacy of television often makes leaders appear to be "standing on the sidelines" rather than taking charge or reacting quickly.

Television's accessible, visual nature also works to level authority. Average citizens gain the feeling that they can form their own impressions of Mikhail Gorbachev, Philippine "People Power," and other people and events without depending on official interpretations. Once formed, the mass perceptions constrain our leaders' presentation of events. Ronald Reagan found he needed to temper his talk of the "Evil Empire" as the public formed a positive perception of Gorbachev. And the televising of Filipinos facing down Marco's tanks made it difficult for Americans to accept our president's suggestion that the reported results of that country's election should stand because "there was cheating on both sides." President Reagan might have changed the rhetoric on these topics in any case, but the public's direct access to the television images made it appear that Reagan was following rather than leading the nation.

The speed of television affects authority in relation to domestic events as well. The videotape of the attempted assassination of Ronald Regan aired on television *before* a coded transmission about the event was received by Vice-President Bush aboard his airplane. Several years later, Reagan had no immediate reaction to the *Challenger* explosion, because millions of Americans saw the *Challenger* explode before he had a chance to watch it on videotape. In both cases, the gap between the experience of the event and a unified administration response made the administration appear temporarily impotent.

As our leaders have lost much control over the flow of information—both about themselves and political events—they have mostly given up trying to behave like the imperial leaders of the past. We now have politicians who strive to act more like the person next door, just as our real neighbours seem more worldly and demand to have a greater say in national and international affairs.

SHARED PROBLEMS

The recognition of television as a new shared arena solves a number of mysteries surrounding television viewing, including: why people complain so bitterly about TV content but continue to watch so much of it; why many Americans say they turn to television for "most" of their news even through the script for an average evening network news broadcast would fill only two columns of the front page of *The New York Times*; why people who purchase videotape machines often discover that they have little interest in creating "libraries" of their favourite television programs.

The shared nature of the television environment creates many new problems and concerns over media content. Content that would be appropriate and uncontroversial in books directed at select audiences often becomes the subject of criticism when presented on television. When television portrays the dominant and "normal" white, middle-class culture, minorities and subcultures protest their exclusion. Yet when television portrays minorities, many members of the majority begin to fear that their insular world is being "invaded." The nature of the portrayal of some groups becomes a catch-22. If homosexuals are portrayed in a negative and stereotypical manner, for example, gay rights groups protest. If homosexuals are portrayed as normal people who simply have a different sexual orientation, however, other viewers object to television "legitimizing" or "idealizing" homosexual life.

Similarly, television cannot exclusively present content deemed suitable only for young children because adult viewers demand more mature entertainment and news. Yet, when truly mature content is placed on television, many parents complain that the minds of child viewers are being defiled.

Without the segregation of audiences, a program designed for one purpose may have quite different effects. An informational program for parents on teenage suicide may not only help some parents prevent a death, it may also encourage a previously non-suicidal teenager to consider the option. Similarly, a program on how to outwit a burglar may make some home owners more sophisticated about protecting their home against professional criminals at the same time that it makes unsophisticated burglars more professional.

Even a choice between happy endings and realistic endings becomes controversial on television. When programs end happily, critics argue that serious issues are trivialized through 30- or 60-minute formulas for solving major problems. Yet when realistic endings are presented—a criminal escapes or good people suffer needlessly—critics attack television for not presenting young children with the ideals and values of our culture.

When looked at as a whole, then, it becomes clear that much of the controversy surrounding television programming is not rooted in television content per se but in the problems inherent in a system that communicates everything to all types of people at the same time.

As a shared environment, television tends to include some aspect of every facet of our culture. Fairy tales are followed by gritty portrayals of crime and corruption. Television preachers share the airwaves with female wrestlers. Poets and prostitutes appear on the same talk shows. Actors and journalists compete for Nielsen ratings. But there is little that is new about any of the information that is presented on television; what is new is that formerly segregated social arenas are blurred together. Information once shared only among people of a certain age, class, race, religion, sex, profession, or other subgroup of the culture has now been thrown into a shared, public forum—and few are wholly satisfied with the mishmash.

A substantial part of the social significance of television, therefore, may lie less in what is on television than in the very existence of television as a shared arena. Television provides the largest simultaneous perception of a message than humanity has ever experienced. Through television, Americans often gain a strange sort of communion with each other. In times of crisis—whether an assassination or a disaster—millions of Americans sit in the glow of their TV sets and watch the same material over and over again in an effort, perhaps, to find comfort, see meaning, and feel united with the other faceless viewers.

Even when video cassettes and other activities pull people away from broadcast and cable television, the shared arena is not destroyed. The knowledge of its existence functions in many ways like the knowledge of the "family home" where relatives can spontaneously gather at times of crisis or celebration. The shared arena does not have to be used every day to have a constant psychological presence.

MAJORITY CONSCIOUSNESS

The shared arena of television does not lead to instant physical integration or to social harmony. Indeed, the initial effect is increased social tension. Informational integration heightens the perception of physical, economic, and legal segregation. Television enhances our awareness of all the people we cannot be, the places we cannot go, the things we cannot possess. Through exposure to a wider world, many viewers gain a sense of being unfairly isolated in some pocket of it.

Shared experiences through television encourage members of formerly isolated and distinct groups to demand equal rights and treatment. Today's "minority consciousness," then, is something of a paradox. Many people take renewed pride in their special identity, yet the heightened consciousness develops from the ability to view one's group from the outside; that is, it is the result of no longer being fully in the group. The demand for full equality in roles and rights dramatizes the development of a mass "majority," a single large group whose members do not want to accept any arbitrarily imposed distinctions in roles and privileges. The diminutive connotation of the term *minority* does not seem to refer to the small number of people in the group, but rather to the limited degree of access the members feel they have to the larger society. The concept of minority as it is sometimes applied to women—the majority of the population—is meaningless in any other sense.

Ironically, many minority group members express their special desires to dissolve into the mainstream, to know what everyone else knows, to experience what everyone else experiences. When gays, blacks, Hispanics, women, the disabled, and others publicly protest for equal treatment under the law, they are not only saying: "I'm different and I'm proud of it," they are also saying, "I should be treated as if I'm the same as everyone else." As gay politician Harry Britt of San Francisco has said: "We want the same rights to happiness and success as the nongay." In this sense, many minorities proclaim their special identity in the hope of losing at least part of it.

Television makes it seem possible to have integration, but the social mechanisms are not always in place. The potential for gaining access to the male realm, for example, is much greater for some women than for others. The feminist movement has primarily advanced upper- and middle-class women—often through the hiring of lower-class women to clean house and mind children. For many segments of our society, television has raised expectations but provided few new opportunities.

The shared information environment fostered by television also does not lead to the identical behaviour or attitudes among all individuals. Far from it. What is increasingly shared is a similar set of options. The choice of dress, hairstyle, speech pattern, profession, and general style of life is no longer as strongly linked as it once was to traditionally defined groups.

Michel Foucault argued convincingly that the membranes around prisons, hospitals, military barracks, factories, and schools thickened over several hundred years leading up to the twentieth century. Foucault described how people were increasingly separated into distant spheres in order to homogenize them into groups with single identities ("students," "workers," "prisoners," "mentally ill," etc.). The individuals within these groups were, in a sense, interchangeable parts. And even the distinct identities of the groups were subsumed under the larger social system of internally consistent, linearly connected, and hierarchically arranged units. While Foucault, observed that modern society segregated people in their "special spheres" in order to homogenize individuals into components of a larger social machine, he did not observe the current, post-modern counter-process. As the membranes around spatially segregated institutions become more informationally permeable, through television and other electronic media, the current trend is toward integration of all groups into a relatively common experiential sphere—with a new recognition of the special needs and idiosyncrasies of individuals. Just as there is now greater sharing of behaviours among people of different ages, different sexes, and different levels of authority, there is also greater variation in the behaviours of people of the same age, same sex, and same level of authority.

A GLOBAL MATRIX

In many instances, the television arena is now international in scope. Over 400 million people in 73 countries watch the Academy Awards; "Live Aid" reached 1.5 billion people in 160 countries; Eastern European countries monitor western television; westerners watched Romanian television as capturing the TV station became the first goal of a revolution; the world watched as Chinese students in Tiananmen Square held English protest signs in front of western TV cameras. The shared arena of television is reinforced through worldwide phone systems, satellites, fax machines, and other electronic media.

But this larger sense of sharing with "everybody" is too wide and diffuse, too quickly changing, too insubstantial. Metaphors aside, it is not possible to experience the whole world as one's neighbourhood or village. Even discounting the numerous political, economic, and cultural barriers that remain, there is a limit to the number of people with whom one can feel truly connected. Electronic sharing leads to a broader, but also a shallower, sense of "us."

The effect of this is both unifying and fractionating. Members of the whole society (and world) are growing more alike, but members of particular families, neighbourhoods, and traditional groups are growing more diverse. On the macro level, the world is becoming more homogeneous, but on the micro level, individuals experience more choice, variety, and idiosyncrasy. We share more experiences with people who are thousands of miles away, even as there is a dilution of the commonality of experience with the people who are in our own houses and neighbourhoods. So the wider sense of connection fostered by electronic media is, ironically, accompanied by a greater retreat to the core of the isolated self. More than ever before, the post-modern era is one in which *everyone* else seems somewhat familiar—and somewhat strange.

As traditional boundaries blur—between regions, between nations, between east and west—there is a rise in factional and ethnic violence within areas that formerly seemed relatively homogeneous. Along with increased hope for world peace, the shared arena may stimulate new types of unrest. As the threat of world war recedes, we are faced with an increase in skirmishes, riots, and terrorism. Whether the era we are now entering will ultimately be viewed as a time of unprecedented unity or a period of unprecedented chaos remains to be seen.

Civic Participation in an Increasingly Complex and Demotivated World

J.S. Frideres

What are a citizen's civic responsibilities? How are citizens' political participation and community involvement essential to sustaining modern liberal democracies? Why are citizens opting out of traditional forms of civic engagement such as voting in elections, contributing to charitable causes, and volunteering? Are citizens less engaged, or are they simply using new means, such as the Internet, to create and sustain communities? Here Frideres addresses these issues in the context of how immigration, multiculturalism, and privatization are putting additional pressures on national unity.

  While these questions pertain to our society as a whole, Frideres is especially concerned with how they relate to youth, notably immigrant youth who, he argues, are disinclined to participate in the common good of building a nation. How do the main concerns of youth lead them to limit their political participation?

In the 19th century, Durkheim, Marx and Weber as well as others expressed concerns about the loss of solidarity, the weakening of Gemeinschaft and the increase in anomie. By the mid 20th century, this issue still remained a concern of social scientists. For example, Banfield (1958) analyzed the lack of social trust and civic engagement in a Sicilian village and coined the term "amoral familism" to designate the political culture of a community that was dysfunctional for the state. The evidence suggested the primacy of parochial over national considerations. As we move into the 21st century, that concern has been heightened and is being echoed by many social scientists writing about the erosion of community (Paxton, 1999; Misztal, 1996) and the loss of community identity and engagement (Dekker, 1999; Skocpol and Fiorina, 1999). Perspectives in addressing this issue range from the new communitarians (Etzioni, 2001) to more quantitative approaches that empirically assess citizen involvement (Norris, 1999; Mamakrishnan

Source: J. S. Frideres, "Civic Participation in an Increasingly Complex and Demotivated World." Reprinted with the permission of the author.

and Baldassare, 2004). Putnam's (2000, 2002) much applauded and criticized books address the issues of civic participation and the declining import of community and subsequent negative impact on nation building. He laments the decline of community and debates the connection between civic engagement, social trust, and social capital. His results show that there has been a dramatic disengagement from community life by the postwar generation of baby boomers and their offspring. His social view of the world suggests a lessening of the role of the state and the belief that today's society should foster the expression of individual interests while preserving the greater good of society. Put in operational terms, this means that the politics of the market and free competition among rational economic actors is the preferred method of developing a democratic society. It means that the state is no longer required to serve society's health and security needs, and thus individual for-profit and non-profit organizations are expected to take on an increasing portion of responsibility for social issues.[1]

Canadians may feel smug in this portrayal of American and other societies, fully aware that Canada is not like this. The question is are we not? As we view social participation in our society, e.g., voting, membership in voluntary associations, volunteerism, we find our assessment becomes less definitive. For example, we find that participation in Canadian federal elections was about 75% from World War II to the 1980's. However, the turnout has been steadily declining since that time and in the election of 2000, we find less than two thirds participated. What is even more dramatic is the decline among youth (18-24 year olds), who had a participation rate of 25% in 2000. We also find that people are less likely to vote at the provincial level (50% participation rate) and even less for municipal elections (30%). At each level, the participation rate for youth has become even lower than the overall rate. When we look at the civic and political engagement of immigrants and particularly immigrant youth, we find even lower rates of participation. Volunteerism has been decreasing every year for the past decade. The percentage of people who donate to charities is now down to less than one quarter of the population. And for those few that do give to charity, the amount is at an all time low of 0.68% of their taxable income. On the face of it, it looks like we exhibit all the classic attributes of a nation in trouble. So the question is, why are we finding that civic responsibility and/or political participation is decreasing at a rate that questions the sustainability of a viable democracy?

Ester and Vinken (2003) and others argue that the data is spurious and this line of reasoning is outmoded and simply wrong. They argue that traditional civil society theorists lack an awareness and understanding of contemporary forms of solidarity, connectedness and civic and political engagement. They argue that social scientists who deal with this issue simply miss novel forms of political participation and civic engagement.

They argue that today's citizen is just as involved in his/her community but through alternative forms of political action, civic engagement and social engagement. For example, they argue that the Internet is the new method of participating in civil society and engaging in political participation (Hill and Hughes, 1998). They argue through the use of the Internet, a "virtual community" is established that replaces the traditional notion of community as used in the social sciences. In summary, they argue that you cannot use indicators of civic involvement, community engagement or political action that were developed for past generations as good indicators for the current generation, as each new generation develops new strategies to become active citizens. The alternative argument is that the X generation is just as involved as the older generation but through different strategies. Hence, if you use the old paradigm, you will come to an erroneous conclusion.

This would seem to be a reasonable rival hypothesis to the proposition that we are at a crisis in developing our nation. So we should assess how viable this rival hypothesis is. Let's begin with reviewing the assumptions that this argument uses. First, the maintenance of the "good society" is based on the assumption that citizen participation is based on collective interests inspired by a sense of altruism and idealism (Dekker, 1995). Second, supporting this ethos is the family, the school, the neighbourhood, and religious and civic organizations, or as Tocqueville would call them, "the little schools of citizenship." Third, these schools of citizenship provide the essential psychological and symbolic functions in providing people with basic values and meaning structures. Fourth, if these institutions do not play a mediating (socializing/modeling) role between the individual and the society, the social fabric will be torn, resulting in the fragmentation of individual goals. As such the citizen becomes detached, non-affiliated and feels no personal loyalty to the wider community. In conclusion, as we assess the assumptions behind their argument, we find they agree with the basic traditional framework (Wolfe, 2000).[2]

So what is different? First of all, they argue that the traditional indicators of civic and political involvement are inappropriate. Unfortunately they do not provide alternative measures for traditional indicators such as voting, volunteering and civic participation. Second, they claim that the Internet is "independent" and devoid of control by national and transnational companies with an agenda. However, the data shows that the Internet and other alternative modes of communication are being taken over by transnational companies, which are re-establishing hegemonic control as they did with earlier modes of communication—newspapers, radio and television programs. Third, they argue that the use of the Internet gives the individual a sense of control, control over what they see, what they say and when they say it. In the end, however, Internet users' sense of control is specious, misplaced and continues to support the existing social structure.[3] In addition, users of the Internet are not the "free wheeling" and "exploring" individuals that are suggested by proponents of the rival hypothesis. In fact, researchers find that extensive use of the Internet is restrictive and limiting. As a result of seeing only part of the world, Internet users have a tendency to have a unique and unreal view of reality as the individual creates his/her fictive community. If you read only what you want to read, you develop a narrow and incomplete view of reality that does not allow you to evaluate all of the knowledge you need to make informed choices. Finally, the cost of the hardware and the transmission costs eliminate a large part of the population from being "users." Moreover, those who can afford the Internet tend to use it as a business or recreational tool to enhance their profit or leisure time and not as an "alternative" public participation tool. In a recent national Canadian study, it was found that use of the Internet was not age specific (Quinless, 2002) and once a person goes on-line, his/her frequency of Internet use is not determined by his/her age.[4]

In summary, contrary to the argument of Ester and Vinken and others, the users of the Internet are not contributing to enhanced civic responsibility or political participation. On the contrary, they are contributing to a furthering of fragmentation, degeneration of community, and further lack of civic and political participation.[5] Young people using the Internet have fewer analytical skills and are poor critical thinkers. In conclusion, we find that social cohesion, social connectedness, altruistic involvement in organizations that support the "common good" and moral principles are in steep decline in Canada. Young people are not being engaged in forms of civic engagement and political participation advancing the common good through the Internet. What they are advancing is their individual good as they see it through their narrowly focused perspectives.[6]

Having rejected the rival hypothesis, we now come back to our original question, "Why are Canadians becoming less involved in voting, less willing to become engaged at the political and civic level?"[7] Why are youth (and particularly immigrant youth) so cynical in participating in the "common good" of building a nation? First, they are cynical of a system that allows decision-makers to make and change rules when it is in their own best interest. The recent events of WorldCom, Enron and Group Action reveal that the elite are not held responsible for their actions and this is symptomatic of what Canadians are experiencing in their everyday lives. Second, young people are not being asked to become involved in decision making at a number of levels—community, regional, national, international (Nolin et al., 1997; Pammett and LeDuc, 2003). As a result, responsibility for both young and old Canadians is being rerouted to sub-communities, e.g., ethnic enclaves and even smaller units such as family. Third, Canadians have witnessed a number of situations in which a political party has won a majority of the votes but has actually ended up being the opposition. Moreover, the number of seats won is not always reflective of the percentage of votes and there are cases where a party has won 17 percent of the vote but has sent no one to the Legislative Assembly—all symptomatic of the democratic deficit.

Young Canadians look at the judicial system and find that older Canadians either dislike youth or love to incarcerate them. We have one of the highest rates of youth incarceration in the world. Police detain visible minority youth at a rate five to seven times higher than they do other Canadians. Incarceration rates for visible minorities are twice as high as they are for members of the community who are not visible minorities. These observations do not go unnoticed by young Canadians and they add to their lack of respect for the legal system and their unwillingness to participate in the larger community.

Finally, it should be noted that those in power positions today have developed ideologies and ethos that reflect the current bipolar value systems, e.g., good or bad, us versus them, traditional sex roles, adherence to rigid ingroup/outgroup boundaries, and generally have been intolerant of different ideologies and lifestyles. Moreover, we learn very quickly that when the rules of the system don't support the personal goals of the elite, there is little incentive for the elite to abide by them or sanctions against them breaking them (see C. Lasch, *The Revolt of the Elites and the Betrayal of Democracy*, 2003).

Governments (at all levels) now support a new world view (an ideological perspective) through the creation of a "fictional reality" that is devoid of any evidence. With the creation of a new world view by governments, decisions made by governments are made on the basis of whether or not they will support the world view or not. Policies and programs that do not "fit" the fictional reality are dismissed as irrelevant or inappropriate. This has become increasingly apparent since the 9/11 events and the imposition of various executive directives and laws. Being a visible minority today is a more critical burden than it was before 9/11. The new ethic of "user pay" and the primacy of the private sector, particularly in the areas of education and health, have convinced Canadians that their elected leaders are unaware of the issues facing them. The gap between the real world and the ideology espoused by government officials is becoming insurmountable.

Only since 1971, when the government implemented the Multi-culturalism Policy, has government shown a vision for the country and a goal that all Canadians could work toward achieving. However, since that time, governments have wandered aimlessly and without vision. Canadians have recognized this lack of vision and have retrenched to interact within their community (Blais et al., 2003).

Today a two-tiered level of action is taking place. On one level are those who are still convinced that civic responsibility and political engagement is meaningful. These are individuals who are the leaders in the various institutional spheres of society. These "leaders" are unable to relate or identify with the workers in this country, they are unaware of the implications of their present action on the future and they are concerned only with satisfying their own desires. Moreover, they are interested only in representing their interests (although when forced will argue their interests are also in the interest of the workers). At a second level are the workers who have no faith in government, are cynical about how it operates and have chosen to redirect their efforts on activities they can control, e.g., family, local community and ethnic enclave. Why case your vote if you know that it doesn't count? Why engage in civic responsibility if leaders have none? Until these individuals are convinced they can have a meaningful say in the political realm or in the civic realities, they will not be participating in volunteerism, voting and other traditional forms of political and civic engagement. Immigrants know that only four out of ten highly educated members of their community will find a job that uses their skills within one year. For visible minority persons, the poverty rate is nearly 40 percent (compared to the overall poverty rate in Canada of 21 percent)[8] (Jackson, 2001). In addition, those visible minorities who work earn about 79 cents for every dollar earned by a Canadian-born person who is not a member of a visible minority. Why would you cast your vote for a government that provides that kind of incentive? Why would you engage in community civic responsibility when Canadians with economic and political power don't feel it is their responsibility to ensure immigrants are integrated into Canadian society? The recent mining disaster in Westray (1992), where management was not found responsible for either the disaster or the deaths, is just one example of where corporate Canada has failed. When people do not feel valued, they are likely to withdraw, taking their talent, experience and expertise with them to apply it elsewhere (see Li, 2003, for a more elaborate discussion on this issue.)

Governments have traditionally focused on the provision of economic opportunity, without consideration of the specific needs of immigrants. Support for local ethnic communities to foster their development (Papillon, 2002) has not been pursued. Ensuring adequate representation of immigrant communities within community associations and other local advisory and political structures has been ignored. The provision of cultural and leisure opportunities for immigrant groups has not been addressed. The inescapable truth is that Canada, like other countries in the world, is gripped by the cancers of poverty and deprivation. It is infected by the malign forces of inequity and greed and contaminated with the conflict that rises from the mistrust that corporate and political leaders have created. It continues to be threatened by widespread and deepening political, social and economic disparities (Piper, 2004). In short, governments have not created guidelines for culturally sensitive planning and development policies and practices for immigrants and particularly for youth.

The continual gap between those who are governing and those who are governed threatens the sustainability of democracy and the framework of Canadian society. Yet those who are "leading" the country seem unaware or unwilling to acknowledge that such a gap has been created. And, even if they have recognized the gap, they are unwilling to introduce change. The elite today have more in common with other elite in Hong Kong, Berlin or London than with working Canadians. Their inability, or some might say unwillingness, to see this will continue to further decrease civic responsibility and political participation.

NOTES

1. At the same time, scholars such as Skocpol (1992, 2003) have argued that civil society in the United States "... has never flourished independent from active government and inclusive democratic politics" (p. 23). She demonstrates that voluntary civic associations have both pressured for the creation of public social programs, and worked with government to administer and expand social programs after they were established. However, all agree that the long-term sustainability of these organizations remains tenuous. The lack of stable funding and the trend toward "professionalization" leads to a concentration of power in "managerial administration" in these organizations.

2. The ideas of Ester and Vinken are suggestive and are most productive in the area related to how new forms of interaction (Internet) change the rules of engagement. The Internet does even the playing field and makes the traditional guardians of information and senders (government, religions, political parties) compete with others and their position in society is not taken as "given" to be true and the "last word." This loss of hegemonic control over information and the truth value by traditional socializing agents means that the new generation can re-establish what they think is right or wrong, who to listen to and who to ignore and who they choose to interact with; these are all important factors in the new realities of civic responsibilities and engagement.

3. Under current conditions, almost anyone can put their views on the Internet and no one knows the validity and/or reliability of these views.

4. Quinless (2002) showed that income, occupation and education are not positively linked to Internet usage. She found that males were much more likely to use the Internet than females.

5. While the use of the Internet could provide an opportunity for young people to become politically engaged and take on civic responsibility, there is little evidence they are doing so. It may well be true that the Internet could be used by the younger generation to gain and advance social capital, to build and take part in communities and to contribute to the common good, but it is not and there is no indication that it is going in this direction.

6. One area in which Canadians have used the Internet is in the area of developing identity. This is particularly true for the generation X immigrant group (those born

after 1965) (Wellman and Haythornthwaite, 2002). Evidence, sketchy as it is, suggests that a number of members of ethnic groups have utilized the Internet to obtain information about their heritage, establish and maintain contact with others from the "homeland" and develop structures to support their ethnic/national identity (Gray, 2001).

7. In a recent study, Orlando (2003) asked voters why they did not vote in the 2000 American presidential election. She found that 21% claimed they were "too busy," 15% claimed an illness or emergency interfered with their ability to vote, 12% were "not interested," ten percent were out of town. Seven percent had registration problems and four percent forgot. The remainder offered reasons such as "disliked all candidates," found voting inconvenient, had transportation problems, couldn't vote because of bad weather.

8. Unemployment rates for visible minority immigrants were significantly higher than the average in 1995 and the gaps were largest among the most highly educated groups. For those who arrived from 1986 to 1990, the rate was 35 percent, and it rose to 52 percent for immigrants who arrived from 1991 to 1996. For some visible minorities, e.g., black immigrants, the rate is more than 50 percent. Visible minority immigrants who were employed full-time earned $32,000 compared to the Canadian average of $38,000. More surprising is that one of every five visible minority immigrants with a university education were found in the bottom and poorest 20 percent of Canadians.

References

Banfield, E. 1958 *The Moral Basis of a Backward Society*, Chicago, The Free Press.

Blais, A., L. Massicotte, and A. Dohbrzynska 2003 Why is Turnout Higher in Some Countries than in Others? Ottawa, Elections Canada, www.elections.ca.

Dekker, P. 1995 "Participate in de civil society: versachtingen en empirishe bevindingen," in M. Wissenburg (ed), *Civil Politics and Civil Society*, pp. 121–150.

1999 *Vrijwilligerswerk vergeleken. Nederland in International en Historisch Perspectief. Civil Society en Vrijwilligerswerk III*, Den The Hague, Sociaal en Cultureel Planbureau.

Ester, P. and H. Vinken 2003 "Debating Civil Society," *International Sociology*, 18, 4: 659–680.

Etzioni, A. 2000 *The Monochrome Society*, Princeton, Princeton University Press.

Gray, C. 2001 *Cyborg Citizen: Politics in the Posthuman Age*, New York, Routledge.

Hill, A. and J. Hughes 1990 *Cyberpolitics: Citizen Activism in the Age of the Internet*, Lanham, MD, Rowman and Littlefield.

Jackson, A. 2002 "Poverty and Racism," www.ccsd.ca/perception/244/racism.htm.

Kunz, J., A. Milan and S. Schetagne 2000 *Unequal Access: A Canadian Profile of Racial Differences in Education, Employment, and Income*, Ottawa, Canadian Race Relations Foundation.

Li, P. 2003 Deconstructing Canada's Discourse of Immigrant Integration, Working Paper No. WP04-03, Edmonton, Prairie Centre of Excellence.

Mamakrishnan, S. and M. Baldassare 2004 *The Ties That Bind: Changing Demographic and Civic Engagement in California*, San Francisco, Public Policy Institute of California.

Misztal, B. 1996 *Trust in Modern Societies: The Search for the Bases for Social Order*, Cambridge, Polity Press.

Nolin, M., C. Chapman and K. Chandler 1997 *Adult Civic Involvement in the United States*, U.S. Department of Education, Office of Educational Research and Improvement, Washington DC.

Norriss, P. (ed) 1999 *Critical Citizens: Global Support for Democratic Government*, Oxford, Oxford University Press.

Orlando, A. 2004 "Why People Don't Vote," *The Change Agent*, Issue 18, p. 9.

Pammett, J. and L. DeDuc 2003 Explaining the Turnout Decline in Canadian Federal Elections: A New Survey of Non-voters, Ottawa, Elections Canada, www.elections.ca.

Papillon, M. 2001 *Immigration, Diversity and Social Inclusion in Canada's Cities*, Ottawa, Canadian Policy Research Networks Inc.

Paxton, P. 1999 "Is Social Capital Declining in the United States? A Multiple Indicator Assessment," *American Journal of Sociology*, 105: 88–127.

Piper, M. 2005 Learning in the 21st Century: Lessons from Lester Pearson, Ottawa, The Canadian Club, May 18.

Putnam, R. 2000 *Bowling Alone: The Collapse and Revival of American Community*, New York, Simon & Schuster.

——. 2002 *Democracies in Flux: The Evolution of Social Capital in Contemporary Society*, New York, Oxford University Press.

Quinless, J. 2006 Logging on to the Network: A Study of the Frequency of Home-Base Internet Use in Canada, Department of Sociology, University of Calgary.

Skocpol, T. 1992 *Protecting Soldiers and Mothers: The Political Origins of Social Policy in the United States*, Cambridge, Harvard University Press.

——. 2003 *Diminished Democracy: From Membership to Management in American Civic Life*, Oklahoma, University of Oklahoma Press.

Skocpol, T. and M. Fiorina (eds) 1999 *Civic Engagement in American Democracy*, Washington, DC, Brookings Institute Press.

Wellman, B. and C. Haythornthwaite 2007 *The Internet in Everyday Life*, Malden, MA, Oxford/Blackwell.

Globalisation

Anthony Giddens

Globalization has generally been topicalized by spokespersons within powerful political and business organizations, and depicted as a contemporary economic process affecting nations around the world. But is it all just fashionable or politically motivated talk? In this article Giddens challenges skeptics and radical advocates alike. He argues that the historical process of globalization is real and consequential for our lives, that it is a deeply political, technological, and cultural process, not simply economic, and that individually and collectively most of us feel powerless in the face of it.

But in what ways is it changing the very nature of our societies and our lives? What are the long-term consequences Giddens envisions from these changes? Finally, is it really possible to overcome the powerlessness most of us currently experience regarding it?

A friend of mine studies village life in central Africa. A few years ago, she paid her first visit to a remote area where she was to carry out her fieldwork. The day she arrived, she was invited to a local home for an evening's entertainment. She expected to find out about the traditional pastimes of this isolated community. Instead, the occasion turned out to be a viewing of *Basic Instinct* on video. The film at that point hadn't even reached the cinemas in London.

Such vignettes reveal something about our world. And what they reveal isn't trivial. It isn't just a matter of people adding modern paraphernalia—videos, television sets, personal computers and so forth—to their existing ways of life. We live in a world of transformations, affecting almost every aspect of what we do. For better or worse, we are being propelled into a global order that no one fully understands, but which is making its effects felt upon all of us.

Globalisation may not be a particularly attractive or elegant word. But absolutely no one who wants to understand our prospects at century's end can ignore it. I travel a lot to speak abroad. I haven't been to a single country recently where globalisation isn't being intensively discussed. In France, the word is *mondialisation*. In Spain and Latin America, it is *globalización*. The Germans says *Globalisierung.*

The global spread of the term is evidence of the very developments to which it refers. Every business guru talks about it. No political speech is complete without reference to it. Yet even in the late 1980s the term was hardly used, either in the academic literature or in everyday language. It has come from nowhere to be almost everywhere.

Given its sudden popularity, we shouldn't be surprised that the meaning of the notion isn't always clear, or that an intellectual reaction has set in against it. Globalisation has something to do with the thesis that we now all live in one world—but in what ways exactly, and is the idea really valid? Different thinkers have taken almost completely opposite views about globalisation in debates that have sprung up over the past few years. Some dispute the whole thing. I'll call them the sceptics.

According to the sceptics, all the talk about globalisation is only that—just talk. Whatever its benefits, its trials and tribulations, the global economy isn't especially different from that which existed at previous periods. The world carries on much the same as it has done for many years.

Most countries, the sceptics argue, gain only a small amount of their income from external trade. Moreover, a good deal of economic exchange is between regions, rather than being truly world-wide. The countries of the European Union, for example, mostly trade among themselves. The same is true of the other main trading blocs, such as those of Asia-Pacific or North America.

Others take a very different position. I'll label them the radicals. The radicals argue that not only is globalisation very real, but that its consequences can be felt everywhere. The global market-place, they say, is much more developed than even in the 1960s and 1970s and is indifferent to national borders. Nations have lost most of the sovereignty they once had, and politicians have lost most of their capability to influence events. It isn't surprising that no one respects political leaders any more, or has much interest in what they have to say. The era of the nation-state is over. Nations, as the Japanese business writer Kenichi Ohmae puts it, have become mere "fictions." Authors such as Ohmae see the economic difficulties of the 1998 Asian crisis as demonstrating the reality of globalisation, albeit seen from its disruptive side.

The sceptics tend to be on the political left, especially the old left. For if all of this is essentially a myth, governments can still control economic life and the welfare state remain intact. The notion of globalisation, according to the sceptics, is an ideology put about by free-marketeers who wish to dismantle welfare systems and cut back on state expenditures. What has happened is at most a reversion to how the world was a century ago. In the late nineteenth century there was already an open global economy, with a great deal of trade, including trade in currencies.

Well, who is right in this debate? I think it is the radicals. The level of world trade today is much higher than it ever was before, and involves a much wider range of goods and services. But the biggest difference is in the level of finance and capital flows. Geared as it is to electronic money—money that exists only as digits in computers—the current world economy has no parallels in earlier times.

In the new global electronic economy, fund managers, banks, corporations, as well as millions of individual investors, can transfer vast amounts of capital from one side of the world to another at the click of a mouse. As they do so, they can destabilise what might have seemed rock-solid economies—as happened in the events in Asia.

The volume of world financial transactions is usually measured in US dollars. A million dollars is a lot of money for most people. Measured as a stack of hundred-dollar notes, it would be eight inches high. A billion dollars—in other words, a thousand million—would stand higher than St Paul's Cathedral. A trillion dollars—a million million—would be over 120 miles high, 20 times higher than Mount Everest.

Yet far more than a trillion dollars is now turned over *each day* on global currency markets. This is a massive increase from only the late 1980s, let alone the more distant past. The value of whatever money we may have in our pockets, or our bank accounts, shifts from moment to moment according to fluctuations in such markets.

I would have no hesitation, therefore, in saying that globalisation, as we are experiencing it, is in many respects not only new, but also revolutionary. Yet I don't believe that either the sceptics or the radicals have properly understood either what it is or its implications for us. Both groups see the phenomenon almost solely in economic terms. This is a mistake. Globalisation is political, technological and cultural, as well as economic. It has been influenced above all by developments in systems of communication, dating back only to the late 1960s.

In the mid-nineteenth century, a Massachusetts portrait painter, Samuel Morse, transmitted the first message, "What hath God wrought?," by electric telegraph. In so doing, he initiated a new phase in world history. Never before could a message be sent without someone going somewhere to carry it. Yet the advent of satellite communications marks every bit as dramatic a break with the past. The first commercial satellite was launched only in 1969. Now there are more than 200 such satellites above the earth, each carrying a vast range of information. For the first time ever, instantaneous communication is possible from one side of the world to the other. Other types of electronic communication, more and more integrated with satellite transmission, have also accelerated over the past few years. No dedicated transatlantic or transpacific cables existed at all until the late 1950s. The first held fewer than 100 voice paths. Those of today carry more than a million.

On 1 February 1999, about 150 years after Morse invented his system of dots and dashes, Morse Code finally disappeared from the world stage. It was discontinued as a means of communication for the sea. In its place has come a system using satellite technology, whereby any ship in distress can be pinpointed immediately. Most countries prepared for the transition some while before. The French, for example, stopped using Morse Code in their local waters in 1997, signing off with a Gallic flourish: "Calling all. This is our last cry before our eternal silence."

Instantaneous electronic communication isn't just a way in which news or information is conveyed more quickly. Its existence alters the very texture of our lives, rich and poor alike. When the image of Nelson Mandela may be more familiar to us than the face of our next-door neighbour, something has changed in the nature of our everyday experience.

Nelson Mandela is a global celebrity, and celebrity itself is largely a product of new communications technology. The reach of media technologies is growing with each wave of innovation. It took 40 years for radio in the United States to gain an audience of 50 million. The same number was using personal computers only 15 years after the personal computer was introduced. It needed a mere 4 years, after it was made available, for 50 million Americans to be regularly using the Internet.

It is wrong to think of globalisation as just concerning the big systems, like the world financial order. Globalisation isn't only about what is "out there," remote and far away from the individual. It is an "in here" phenomenon too, influencing intimate and personal aspects of our lives. The debate about family values, for example, that is going on in many countries might seem far removed from globalising influences. It isn't. Traditional family systems are becoming transformed, or are under strain, in many parts of the world, particularly as women stake claim to greater equality. There has never before been a society, so far as we know from the historical record, in which women have been even approximately equal to men. This is a truly global revolution in everyday life, whose consequences are being felt around the world in spheres from work to politics.

Globalisation thus is a complex set of processes, not a single one. And these operate in a contradictory or oppositional fashion. Most people think of globalisation as simply "pulling away" power or influence from local communities and nations into the global arena. And indeed this is one of its consequences. Nations do lose some of the economic power they once had. Yet it also has an opposite effect. Globalisation not only pulls upwards, but also pushes downwards, creating new pressures for local autonomy. The American sociologist Daniel Bell describes this very well when he says that the nation becomes not only too small to solve the big problems, but also too large to solve the small ones.

Globalisation is the reason for the revival of local cultural identities in different parts of the world. If one asks, for example, why the Scots want more independence in the UK, or why there is a strong separatist movement in Quebec, the answer is not to be found only in their cultural history. Local nationalisms spring up as a response to globalising tendencies, as the hold of older nation-states weakens.

Globalisation also squeezes sideways. It creates new economic and cultural zones within and across nations. Examples are the Hong Kong region, northern Italy, and Silicon Valley in California. Or consider the Barcelona region. The area around Barcelona in northern Spain extends into France. Catalonia, where Barcelona is located, is closely integrated into the European Union. It is part of Spain, yet also looks outwards.

These changes are being propelled by a range of factors, some structural, others more specific and historical. Economic influences are certainly among the driving forces—especially the global financial system. Yet they aren't like forces of nature. They have been shaped by technology, and cultural diffusion, as well as by the decisions of governments to liberalise and deregulate their national economies.

The collapse of Soviet communism has added further weight to such developments, since no significant group of countries any longer stands outside. That collapse wasn't just something that just happened to occur. Globalisation explains both why and how Soviet communism met its end. The former Soviet Union and the East European countries were comparable to the West in terms of growth rates until somewhere around the early 1970s. After that point, they fell rapidly behind. Soviet communism, with its emphasis upon state-run enterprise and heavy industry, could not compete in the global electronic economy. The ideological and cultural control upon which communist political authority was based similarly could not survive in an era of global media.

The Soviet and the East European regimes were unable to prevent the reception of Western radio and television broadcasts. Television played a direct role in the 1989 revolutions, which have rightly been called the first "television revolutions." Street protests taking place in one country were watched by television audiences in others, large numbers of whom then took to the streets themselves.

Globalisation, of course, isn't developing in an even-handed way, and is by no means wholly benign in its consequences. To many living outside Europe and North America, it looks uncomfortably like Westernisation— or perhaps, Americanisation, since the US is now the sole superpower, with a dominant economic, cultural and military position in the global order. Many of the most visible cultural expressions of globalisation are American—Coca-Cola, McDonald's, CNN.

Most of the giant multinational companies are based in the US too. Those that aren't all come from the rich countries, not the poorer areas of the world. A pessimistic view of globalisation would consider it largely an affair of the industrial North, in which the developing societies of the South play little or no active part. It would see it as destroying local cultures, widening world inequalities and worsening the lot of the impoverished. Globalisation, some argue, creates a world of winners and losers, a few on the fast track to prosperity, the majority condemned to a life of misery and despair.

Indeed, the statistics are daunting. The share of the poorest fifth of the world's population in global income has dropped, from 2.3 per cent to 1.4 per cent between 1989 and 1998. The proportion taken by the richest fifth, on the other hand, has risen. In sub-Saharan Africa, 20 countries have lower incomes per head in real terms than they had in the late 1970s. In many less developed countries, safety and environmental regulations are low or virtually non-existent. Some transnational companies sell goods there that are controlled or banned in the industrial countries— poor-quality medical drugs, destructive pesticides or high tar and nicotine content cigarettes. Rather than a global village, one might say, this is more like global pillage.

Along with ecological risk, to which it is related, expanding inequality is the most serious problem facing world society. It will not do, however, merely to blame it on the wealthy. It is fundamental to my argument that globalisation today is only partly Westernisation. Of course the Western nations, and more generally the industrial countries, still have far more influence over world affairs than do the poorer states. But globalisation is becoming increasingly decentred—not under the control of any group of nations, and still less of the large corporations. Its effects are felt as much in Western countries as elsewhere.

This is true of the global financial system, and of changes affecting the nature of government itself. What one could call "reverse colonisation" is becoming more and more common. Reverse colonisation means that non-Western countries influence developments in the West. Examples abound—such as the latinising of Los Angeles, the emergence of a globally oriented high-tech sector in India, or the selling of Brazilian television programmes to Portugal.

Is globalisation a force promoting the general good? The question can't be answered in a simple way, given the complexity of the phenomenon. People who ask it, and who blame globalisation for deepening world inequalities, usually have in mind economic globalisation and, within that, free trade. Now, it is surely obvious that free trade is not an unalloyed benefit. This is especially so as concerns the less developed countries. Opening up a country, or regions within it, to free trade can undermine a local subsistence economy. An area that becomes dependent upon a few products sold on world markets is very vulnerable to shifts in prices as well as to technological change.

Trade always needs a framework of institutions, as do other forms of economic development. Markets cannot be created by purely economic means, and how far a given economy should be exposed to the world marketplace must depend upon a range of criteria. Yet to oppose economic globalisation, and to opt for economic protectionism, would be a misplaced tactic for rich and poor nations alike. Protectionism may be a necessary strategy at some times and in some countries. In my view, for example, Malaysia was correct to introduce controls in 1998, to stem the flood of capital from the country. But more permanent forms of protectionism will not help the development of the poor countries, and among the rich would lead to warring trade blocs.

The debates about globalisation I mentioned at the beginning have concentrated mainly upon its implications for the nation-state. Are nation-states, and hence national political leaders, still powerful, or are they becoming largely irrelevant to the forces shaping the world? Nation-states are indeed still powerful and political leaders have a large role to play in the world. Yet at the same time the nation-state is being reshaped before our eyes. National economic policy can't be as effective as it once was. More importantly, nations have to rethink their identities now the older forms of geopolitics are becoming obsolete. Although this is a contentious point, I would say that, following the dissolving of the Cold War, most nations no longer have enemies. Who are the enemies of Britain, or France, or Brazil? The war in Kosovo didn't pit nation against nation. It was a conflict between old-style territorial nationalism and a new, ethically driven interventionalism.

Nations today face risks and dangers rather than enemies, a massive shift in their very nature. It isn't only of the nation that such comments could be made. Everywhere we look, we see institutions that appear the same as they used to be from the outside, and carry the same names, but inside have become quite different. We continue to talk of the nation, the family, work, tradition, nature, as if they were all the same as in the past. They are not. The outer shell remains, but inside they have changed—and this is happening not only in the US, Britain, or France, but almost everywhere. They are what I call "shell institutions." They are institutions that have become inadequate to the tasks they are called upon to perform.

As the changes I have described in this chapter gather weight, they are creating something that has never existed before, a global cosmopolitan society. We are the first generation to live in this society, whose contours we can as yet only dimly see. It is shaking up our existing ways of life, no matter where we happen to be. This is not—at least at the moment—a global order driven by collective human will. Instead, it is emerging in an anarchic, haphazard, fashion, carried along by a mixture of influences.

It is not settled or secure, but fraught with anxieties, as well as scarred by deep divisions. Many of us feel in the grip of forces over which we have no power. Can we reimpose our will upon them? I believe we can. The powerlessness we experience is not a sign of personal failings, but reflects the incapacities of our institutions. We need to reconstruct those we have, or create new ones. For globalisation is not incidental to our lives today. It is a shift in our very life circumstances. It is the way we now live.

PART FIVE

Contested Terrains: Difference, Inequality, Control, and Change

Richard Edwards used the term *contested terrain* in his controversial study of the transformation of the workplace where, he argued, impersonal bureaucracies legitimate hierarchies and enhance the employer's control over workers.[1] The contested terrain, in Edwards' book, refers to the actual shop floor where workers' and owners' ongoing struggle for power results in conflict. We have extended his apt phrase to refer to the many other sites beyond the contemporary workplace which are characterized by change and dispute and which involve issues of difference, inequality, power, and control.

The prevailing sociological approach to conflict and change originates out of the work of Karl Marx and Max Weber and, more recently, feminism and post-structuralism. Under the label of conflict theory, these theorists formulate society as a dynamic order of interest groups competing for wealth, power, prestige, and other social and material rewards. These groups have differing amounts and types of power, unequal access to resources, and different ideas about how society should be organized. In this view, society is an ongoing negotiation between these interest groups, each of which is trying to improve its position, sometimes under the guise of concern for the common good.

Conflict theory challenges the customary view of conflicts as disruptions to the stability and continuity of social life by acknowledging that conflicts and changes are always both present and imminent. As Randall Collins points out, sometimes conflicts are overt. However, most of the time they persist as unequal social relationships and arrangements of domination and submission within stable institutions.[2]

In addition, many conflicts result from modern societies' commitments to change. Distinct from traditional societies—organized around, and committed to, the principles of stability and continuity—modern societies are organized around change and progress. Market economies, waves of technological invention, the rationalization of social life, and a multitude of scientific discoveries are all modern developments. Just as very few social arrangements exist because they actually are in everybody's interest (though most are justified in these terms), very few changes occur because they serve the common good or are equally beneficial. Sociology, notably conflict theory, reminds us that change is economically and politically driven and much change is managed in accordance with the interests of those groups who possess power and privilege.

Conflict is usually depicted as an undesirable interpersonal or political problem, and most responses to conflict, be they therapeutic or diplomatic, consist of strategies to reduce its effects or punish its advocates. However, sociologically, some kinds of conflict are inherent in the very makeup of human society and social life itself. Some forms are not only necessary, but are also valuable.

For example, conflict often arises when groups attempt to create and maintain social order. If submission to a ruling authority is one requirement for civil society, there must be, as well, those who have the power to enforce this authority. Whether this authority is granted or seized, there is a resulting imbalance of power. There are those who have the power to make and enforce the rules, and others who are subject to them.

In addition, sociologists have pointed out that society requires some degree of stability and cohesiveness, which emerges out of a generally shared agreement within a prevailing cultural system. However, in establishing and enforcing such a broad consensus, groups establish the conditions for intolerance towards difference and diversity. Meeting such societal needs has led to serious excesses, abuses, and inequities in modern societies such as ours which need to be remedied. But, if we understand obeying authority and some level of cohesiveness as essential for civil society, then the question becomes not "How can we get rid of the inequity?" but, "Are some forms of social inequality and control justifiable?" What kinds of social inequality can we collectively and willingly live with?

NOTES

1. Richard C. Edwards, *Contested Terrain: The Transformation of the Workplace in the 21st Century,* Basic Books, 1980.
2. Randall Collins, *Four Sociological Traditions,* Oxford University Press, 1994.

READING THIRTY

Woman's Place in Man's Life Cycle

Carol Gilligan

In this excerpt from her book *In a Different Voice*, Gilligan challenges major developmental psychologists with regard to their views on moral development and morality. Drawing on interviews with children and adults, Gilligan identifies differences in male and female worldviews and values, and in the kinds of relationships they develop to support these. A guiding question for the reader: What are the differences between the masculine and feminine systems of valuing that emerge from her research?

... At a time when efforts are being made to eradicate discrimination between the sexes in the search for social equality and justice, the differences between the sexes are being rediscovered in the social sciences. This discovery occurs when theories formerly considered to be sexually neutral in their scientific objectivity are found instead to reflect a consistent observational and evaluative bias. Then the presumed neutrality of science, like that of language itself, gives way to the recognition that the categories of knowledge are human constructions. The fascination with point of view that has informed the fiction of the twentieth century and the corresponding recognition of the relativity of judgment infuse our scientific understanding as well when we begin to notice how accustomed we have become to seeing life through men's eyes.

. . .

Psychological theorists have fallen ... into the same observational bias. Implicitly adopting the male life as the norm, they have tried to fashion women out of a masculine cloth. It all goes back, of course, to Adam and Eve—a story which shows, among other things, that if you make a woman out of a man, you are bound to get into trouble. In the life cycle, as in the Garden of Eden, the woman has been the deviant.

The penchant of developmental theorists to project a masculine image, and one that appears frightening to women, goes back at least to Freud (1905), who built his theory of psychosexual development around the experiences of the male child that culminate in the Oedipus complex. In the 1920s, Freud struggled to resolve the contradictions posed for his theory by the differences in female anatomy and the different configuration of the young girl's early family relationships. After trying to fit women into his masculine conception, seeing them as envying that which they missed, he came instead to acknowledge, in the strength and persistence of women's pre-Oedipal attachments to their mothers, a developmental difference. He considered this difference in women's development to be responsible for what he saw as women's developmental failure.

Having tied the formation of the superego or conscience to castration anxiety, Freud considered women to be deprived by nature of the impetus for a clear-cut Oedipal resolution. Consequently, women's superego—the heir to the Oedipus complex—was compromised: it was never "so inexorable, so impersonal, so independent of its emotional origins as we require it to be in men." From this observation of difference, that "for women the level of what is ethically normal is different from what it is in men," Freud concluded that women "show less sense of justice than men, that they are less ready to submit to the great exigencies of life, that they are more often influenced in their judgements by feelings of affection or hostility" (1925, pp. 257–258).

Thus a problem in theory became cast as a problem in women's development, and the problem in women's development was located in their experience of relationships. Nancy Chodorow (1974), attempting to account for "the reproduction within each generation of certain general and nearly universal differences that characterize masculine and feminine personality and roles," attributes these differences between the sexes not to anatomy but rather to "the fact that women, universally, are largely responsible for early child care." Because this early social environment differs for and is experienced differently by male and female children, basic sex differences recur in personality development. As a result, "in any given society, feminine personality comes to define itself in relation and connection to other people more than masculine personality does" (pp. 43–44).

In her analysis, Chodorow relies primarily on Robert Stoller's studies which indicate that gender identity, the unchanging core of personality formation, is "with rare exception firmly and irreversibly established for both sexes by the time a child is around three." Given that for both sexes the primary caretaker in the first three years of life is typically female, the interpersonal dynamics of gender identity formation are different for boys and girls. Female identity formation takes place in a context of ongoing relationship since "mothers tend to experience their daughters as more like, and continuous with, themselves." Correspondingly, girls, in identifying themselves as female, experience themselves as like their mothers, thus fusing the experience of attachment with the process of identity formation. In contrast, "mothers experience their sons as a male opposite," and boys, in defining themselves as masculine, separate their mothers from themselves, thus curtailing "their primary love and sense of empathetic tie." Consequently, male development entails a "more emphatic individuation and a more defensive firming of experienced ego boundaries." For boys, but not girls, "issues of differentiation have become intertwined with sexual issues" (1978, pp. 150, 166–167).

Writing against the masculine bias of psychoanalytic theory, Chodorow argues that the existence of sex differences in the early experiences of individuation and relationship "does not mean that women have 'weaker' ego boundaries than men or are more prone to psychosis." It means instead that "girls emerge from this period with a basis for 'empathy' built into their primary definition of self in a way that boys do not." Chodorow thus replaces Freud's negative and derivative description of female psychology with a positive and direct account of her own: "Girls emerge with a stronger basis for experiencing another's needs or feelings as one' own (or of thinking that one is so experiencing another's needs and feelings). Furthermore, girls do not define themselves in terms of the denial of preoedipal relational modes to the same extent as do boys. Therefore, regression to these modes tends not to feel as much a basic threat to their ego. From very early, then, because they are parented by a person of the same gender ... girls come to experience themselves as less differentiated than boys, as more continuous with and related to the external object-world, and as differently oriented to their inner object-world as well" (p. 167).

Consequently, relationships, and particularly issues of dependency, are experienced differently by women and men. For boys and men, separation and individuation are critically tied to gender identity since separation from the mother is essential for the development of masculinity. For girls and women, issues of femininity or feminine identity do not depend on the achievement of separation from the mother or on the progress of individuation. Since masculinity is defined through separation while femininity is defined through attachment, male gender identity is threatened by intimacy while female gender identity is threatened by separation. Thus males tend to have difficulty with relationships, while females tend to have problems with individuation. The quality of embeddedness in social interaction and personal relationships that characterizes women's lives in contrast to men's, however, becomes not only a descriptive difference but also a developmental liability when the milestones of childhood and adolescent development in the psychological literature are markers of increasing separation. Women's failure to separate then becomes by definition a failure to develop.

The sex difference in personality formation that Chodorow describes in early childhood appear during the middle childhood years in studies of children's games. Children's games are considered by George Herbert Mead (1934) and Jean Piaget (1932) as the crucible of social development during the school years. In games, children learn to take the role of the other and come to see themselves through another's eyes. In games, they learn respect for rules and come to understand the ways rules can be made and changed.

Janet Lever (1976), considering the peer group to be the agent of socialization during the elementary school years and play to be a major activity of socialization at that time, set out to discover whether there are sex differences in the games that children play. Studying 181 fifth-grade, white, middle-class children, ages ten and eleven, she observed the organization and structure of their playtime activities. She watched the children as they played at school during recess and in physical education class, and in addition kept diaries of their accounts as to how they spent their out-of-school time. From this study, Lever reports sex differences: boys play out of doors more often than girls do; boys play more often in large and age-heterogeneous groups; they play competitive games more often, and their games last longer than girls' games. The last is in some ways the most interesting finding. Boys' games appeared to last longer not only because they required a higher level of skill and were thus less likely to become boring, but also because, when disputes arose in the course of a game, boys were able to resolve the disputes more effectively than girls: "During the course of this study, boys were seen quarrelling all the time, but not once was a game terminated because of a quarrel and no game was interrupted for more than seven minutes. In the gravest debates, the final word was always, to 'repeat the play,' generally followed by a chorus of 'cheater's proof'" (p. 482). In fact, it seemed that the boys enjoyed the legal debates as much as they did the game itself, and even marginal players of lesser size or skill participated equally in these recurrent squabbles. In contrast, the eruption of disputes among girls tended to end the game.

Thus Lever extends and corroborates the observations of Piaget in his study of the rules of the game, where he finds boys becoming through childhood increasingly fascinated with the legal elaboration of rules and the development of fair procedures for adjudicating conflicts, a fascination that, he notes, does not hold for girls. Girls, Piaget observes, have a more "pragmatic" attitude toward rules, "regarding a rule as good as long as the game repaid it" (p. 83). Girls are more tolerant in their attitudes toward rules, more willing to make exceptions, and more easily reconciled to innovations. As a result, the legal sense, which Piaget considers essential to moral development, "is far less developed in little girls than in boys" (p. 77).

The bias that leads Piaget to equate male development with child development also colors Lever's work. The assumption that shapes her discussion of results is that the male model is the better one since it fits the requirements for modern corporate success. In contrast, the sensitivity and care for the feelings of others that girls develop through their play have little market value and can even impede professional success. Lever implies that, given the realities of adult life, if a girl does not want to be left dependent on men, she will have to learn to play like a boy.

To Piaget's argument that children learn the respect for rules necessary for moral development by playing rule-bound games, Lawrence Kohlberg (1969) adds that these lessons are most effectively learned through the opportunities for role-taking that arise in the course of resolving disputes. Consequently, the moral lessons inherent in girls' play appear to be fewer than in boys'. Traditional girls' games like jump rope and hopscotch are turn-taking games, where competition is indirect since one person's success does not necessarily signify another's failure. Consequently, disputes requiring adjudication are less likely to occur. In fact, most of the girls whom Lever interviewed claimed that when a quarrel broke out, they ended the game. Rather than elaborating a system of rules for resolving disputes, girls subordinated the continuation of the game to the continuation of relationships.

Lever concludes that from the games they play, boys learn both the independence and the organizational skills necessary for coordinating the activities of large and diverse groups of people. By participating in controlled and socially approved competitive situations, they learn to deal with competition in a relatively forthright manner—to play with their enemies and to compete with their friends—all in accordance with the rules of the game. In contrast, girls' play tends to occur in smaller, more intimate groups, often the best-friend dyad, and in private places. This play replicates the social pattern of primary human relationships in that its organization is more cooperative. Thus, it points less, in Mead's terms, toward learning to take the role of "the generalized other," less toward the abstraction of human relationships. But it fosters the development of the empathy and sensitivity necessary for taking the role of "the particular other" and points more toward knowing the other as different from the self.

The sex differences in personality formation in early childhood that Chodorow derives from her analysis of the mother-child relationship are thus extended by Lever's observations of sex differences in the play activities of middle childhood. Together these accounts suggest that boys and girls arrive at puberty with a different interpersonal orientation and a different range of social experiences. . . .

"It is obvious," Virginia Woolf says, "that the values of women differ very often from the values which have been made by the other sex" (1929, p. 76). Yet, she adds, "it is the masculine values that prevail." As a result, women come to question the normality of their feelings and to alter their judgments in deference to the opinion of others. In the nineteenth century novels written by women, Woolf sees at work "a mind which was slightly pulled from the straight and made to alter its clear vision in deference to external authority." The same deference to the values and opinions of others can be seen in the judgments of twentieth century women. The difficulty women experience in finding or speaking publicly in their own voices emerges repeatedly in the form of qualification and self-doubt, but also in intimations of a divided judgment, a public assessment and private assessment which are fundamentally at odds.

Yet the deference and confusion that Woolf criticizes in women derive from the values she sees as their strength. Women's deference is rooted not only in their social subordination but also in the substance of their moral concern. Sensitivity to the needs of others and the assumption of responsibility for taking care lead women to attend to voices other than their own and to include in their judgment other points of view. Women's moral weakness, manifest in an apparent diffusion and confusion of judgment, is thus inseparable from women's moral strength, an overriding concern with relationships and responsibilities. The reluctance to judge may itself be indicative of the care and concern for others that infuse the psychology of women's development and are responsible for what is generally seen as problematic in its nature.

Thus women not only define themselves in a context of human relationship but also judge themselves in terms of their ability to care. Women's place in man's life cycle has been that of nurturer, caretaker, and helpmate, the weaver of those networks of relationships on which she in turn relies. But while women have thus taken care of men, men have, in their theories of psychological development, as in their economic arrangements, tended to assume or devalue that care. When the focus on individuation and individual achievement extends into adulthood and maturity is equated with personal autonomy, concern with relationships appears as a weakness of women rather than as a human strength (Miller, 1976).

The discrepancy between womanhood and adulthood is nowhere more evident than in the studies on sex-role stereotypes reported by Broverman, Vogel, Broverman, Clarkson, and Rosenkrantz (1972). The repeated finding of these studies is that the qualities deemed necessary for adulthood—the capacity for autonomous thinking, clear decision-making, and responsible action—are those associated with masculinity and considered undesirable as attributes of the feminine self. The stereotypes suggest a splitting of love and work that relegates expressive capacities to women while placing instrumental abilities in the masculine domain. Yet looked at from a different perspective, these stereotypes reflect a conception of adulthood that is itself out of balance, favoring the separateness of the individual self over connection to others, and leaning more toward an autonomous life of work than toward the interdependence of love and care.

The discovery now being celebrated by men in mid-life of the importance of intimacy, relationships, and care is something that women have known from the beginning. However, because that knowledge in women has been considered "intuitive" or "instinctive," a function of anatomy coupled with destiny, psychologists have neglected to describe its development. In my research, I have found that women's moral development centers on the elaboration of that knowledge and thus delineates a critical line of psychological development in the lives of both of the sexes. The subject of moral development not only provides the final illustration of the reiterative pattern in the observation and assessment of sex differences in the literature on human development, but also indicates more particularly why the nature and significance of women's development has been for so long obscured and shrouded in mystery.

The criticism that Freud makes of women's sense of justice, seeing it as compromised in its refusal of blind impartiality, reappears not only in the work of Piaget but also in that of Kohlberg. While in Piaget's account (1932) of the moral judgment of the child, girls are an aside, a curiosity to whom he devotes four brief entries in an index that omits "boys" altogether because "the child" is assumed to be male, in the research from which Kohlberg derives his theory, females simply do not exist. Kohlberg's (1958, 1981) six stages that describe the development of moral judgment from childhood to adulthood are based empirically on a study of eighty-four boys whose development Kohlberg has followed for a period of over twenty years. Although Kohlberg claims universality for his stage sequence, those groups not included in his original sample rarely reach his higher stages (Edwards, 1975; Holstein, 1976; Simpson, 1974). Prominent among those who thus appear to be deficient in moral development when measured by Kohlberg's scale are women, whose judgments seem to exemplify the third stage of his six-stage sequence. At this stage morality is conceived in interpersonal terms and goodness is equated with helping and pleasing others. This conception of goodness is considered by Kohlberg and Kramer (1969) to be functional in the lives of mature women insofar as their lives take place in the home. Kohlberg and Kramer imply that only if women enter the traditional arena of male activity will they recognize the inadequacy of this moral perspective and progress like men toward higher stages where relationships are subordinated to rules (stage four) and rules to universal principles of justice (stages five and six).

Yet herein lies a paradox, for the very traits that traditionally have defined the "goodness" of women, their care for and sensitivity to the needs of others, are those that mark them as deficient in moral development. In this version of moral development, however, the conception of maturity is derived from the study of men's lives and reflects the importance of individuation in their development. Piaget (1970), challenging the common impression that a developmental theory is built like a pyramid from its base in infancy, points out that a conception of development instead hangs from its vertex of maturity, the point toward which progress is traced. Thus, a change in the definition of maturity does not simply alter the description of the highest stage but recasts the understanding of development, changing the entire account.

When one begins with the study of women and derives developmental constructs from their lives, the outline of a moral conception different from that described by Freud, Piaget, or Kohlberg begins to emerge and informs a different description of development. In this conception, the moral problem arises from conflicting responsibilities rather than from competing rights and requires for its resolution a mode of thinking that is contextual and narrative rather than formal and abstract. This conception of morality as concerned with the activity of care centers moral development around the understanding of responsibility and relationships, just as the conception of morality as fairness ties moral development to the understanding of rights and rules.

This different construction of the moral problem by women may be seen as the critical reason for their failure to develop within the constraints of Kohlberg's system. Regarding all constructions of responsibility as evidence of a conventional moral understanding, Kohlberg defines the highest stages of moral development as deriving from a reflective understanding of human rights. That the morality of rights differs from the morality of responsibility in its emphasis on separation rather than connection, in its consideration of the individual rather than the relationship as primary, is illustrated by two responses to interview questions about the nature of morality. The first comes from a twenty-five-year-old man, one of the participants in Kohlberg's study:

> [*What does the word morality mean to you?*] Nobody in the world knows the answer. I think it is recognizing the right of the individual, the rights of other individuals, not interfering with those rights. Act as fairly as you would have them treat you. I think it is basically to preserve the human being's right to existence. I think that is the most important. Secondly, the human being's right to do as he pleases, again without interfering with somebody else's rights.
>
> [*How have your views on morality changed since the last interview?*] I think I am more aware of an individual's rights now.

I used to be looking at it strictly from my point of view, just for me. Now I think I am more aware of what the individual has a right to.

Kohlberg (1973) cites this man's response as illustrative of the principled conception of human rights that exemplifies his fifth and sixth stages. Commenting on the response, Kohlberg says: "Moving to a perspective outside of that of his society, he identifies morality with justice (fairness, rights, the Golden Rule), with recognition of the rights of others as these are defined naturally or intrinsically. The human's being right to do as he pleases without interfering with somebody else's rights is a formula defining rights prior to social legislation" (pp. 29–30).

The second response comes from a woman who participated in the rights and responsibilities study. She also was twenty-five and, at the time, a third-year law student:

[*Is there really some correct solution to moral problems, or is everybody's opinion equally right?*] No, I don't think everybody's opinion is equally right. I think that in some situations there may be opinions that are equally valid, and one could conscientiously adopt one of several courses of action. But there are other situations in which I think there are right and wrong answers, that sort of inhere in the nature of existence, of all individuals here who need to live with each other to live. We need to depend on each other, and hopefully it is not only a physical need but a need of fulfillment in ourselves, that a person's life is enriched by cooperating with other people and striving to live in harmony with everybody else, and to that end, there are right and wrong, there are things which promote that end and that move away from it, and in that way it is possible to choose in certain cases among different courses of action that obviously promote or harm that goal.

[*Is there a time in the past when you would have thought about these things differently?*] Oh, yeah, I think that I went through a time when I thought that things were pretty relative, that I can't tell you what to do and you can't tell me what to do, because you've got your conscience and I've got mine.

[*When was that?*] When I was in high school. I guess that it just sort of dawned on me that my own ideas changed, and because my own judgment changed, I felt I couldn't judge another person's judgment. But now I think even when it is only the person himself who is going to be affected, I say it is wrong to the extent it doesn't cohere with what I know about human nature and what I know about you, and just from what I think is true about the operation of the universe, I could say I think you are making a mistake.

[*What led you to change, do you think?*] Just seeing more of life, just recognizing that there are an awful lot of things that are common among people. There are certain things that you come to learn promote a better life and better relationships and more personal fulfillment than other things that in general tend to do the opposite, and the things that promote these things, you would call morally right.

This response also represents a personal reconstruction of morality following a period of questioning and doubt, but the reconstruction of moral understanding is based not on the primacy and universality of individual rights, but rather on what she describes as a "very strong sense of being responsible to the world." Within this construction, the moral dilemma changes from how to exercise one's rights without interfering with the rights of others to how "to lead a moral life which includes obligations to myself and my family and people in general." The problem then becomes one of limiting responsibilities without abandoning moral concern. When asked to describe herself, this woman says that she values "having other people that I am tied to, and also having people that I am responsible to. I have a very strong sense of being responsible to the world, that I can't just live for my enjoyment, but just the fact of being in the world gives me an obligation to do what I can to make the world a better place to live in, no matter how small a scale that may be on." Thus while Kohlberg's subject worries about people interfering with each other's rights, this woman worries about "the possibility of omission, of your not helping others when you could help them."

The issue that this woman raises is addressed by Jane Loevinger's fifth "autonomous" stage of ego development, where autonomy, placed in a context of relationships, is defined as modulating an excessive sense of responsibility through the recognition that other people have responsibility for their own destiny. The autonomous stage in Loevinger's account (1970) witnesses a relinquishing of moral dichotomies and their replacement with "a feeling for the complexity and multifaceted character of real people and real situations" (p. 6). Whereas the rights conception of morality that informs Kohlberg's principled level (stages five and six) is geared to arriving at an objectively fair or just resolution to moral dilemmas upon which all rational persons could agree, the responsibility conception focuses instead on the limitations of any particular resolution and describes the conflicts that remain.

Thus it becomes clear why a morality of rights and non-interference may appear frightening to women in its potential justification of indifference and unconcern. At the same time, it becomes clear why, from a male perspective, a morality of responsibility appears inconclusive and diffuse, given its insistent contextual relativism. Women's moral judgments thus elucidate the pattern observed in the description of the developmental differences between the sexes, but they also provide an alternative conception of maturity by which these differences can be assessed and their implications traced. The psychology of women that has consistently been described as distinctive in its greater orientation toward relationships and interdependence implies a more contextual mode of judgment and a different moral understanding. Given the differences in women's conceptions of self and morality, women bring to the life cycle a different point of view and order human experiences in terms of different priorities.

. . .

References

Broverman, I., Vogel, S., Broverman, D., Clarkson, F., and Rosenkrantz, P. "Sex-role Stereotypes: A Current Appraisal." *Journal of Social Issues* 28 (1972): 59–78.

Chodorow, Nancy. "Family Structure and Feminine Personality." In M. Z. Rosaldo and L. Lamphere, eds., *Woman, Culture and Society.* Stanford: Stanford University Press, 1974.

Chodorow, Nancy. *The Reproduction of Mothering.* Berkeley: University of California Press, 1978.

Edwards, Carolyn P. "Society Complexity and Moral Development: A Kenyan Study." *Ethos* 3 (1975): 505–527.

Freud, Sigmund. *The Standard Edition of the Complete Psychological Works of Sigmund Freud,* trans. and ed. James Strachey. London: The Hogarth Press, 1961.

Freud, Sigmund. *Three Essays on the Theory of Sexuality* (1905). Vol. VII.

Freud, Sigmund. "Some Psychical Consequences of the Anatomical Distinction Between the Sexes" (1925). Vol. XIX.

Holstein, Constance. "Development of Moral Judgment: A Longitudinal Study of Males and Females." *Child Development* 47 (1976): 51–61.

Kohlberg, Lawrence. "The Development of Modes of Thinking and Choices in Years 10 to 16." Ph.D. Diss., University of Chicago, 1958.

Kohlberg, Lawrence. "Stage and Sequence: The Cognitive-Development Approach to Socialization." In D. A. Goslin, ed., *Handbook of Socialization Theory and Research.* Chicago: Rand McNally, 1969.

Kohlberg, Lawrence. "Continuities and Discontinuities in Childhood and Adult Moral Development Revisited." In *Collected Papers on Moral Development and Moral Education.* Moral Education Research Foundation, Harvard University, 1973.

Kohlberg, Lawrence. *The Philosophy of Moral Development.* San Francisco: Harper and Row, 1981.

Kohlberg, L., and Kramer, R. "Continuities and Discontinuities in Child and Adult Moral Development." *Human Development* 12 (1969): 93–120.

Lever, Janet. "Sex Differences in the Games Children Play." *Social Problems* 23 (1976): 478–487.

Loevinger, Jane, and Wessler, Ruth. *Measuring Ego Development.* San Francisco: Jossey-Bass, 1970.

Mead, George Herbert. *Mind, Self, and Society.* Chicago: University of Chicago Press, 1934.

Miller, Jean Baker. *Toward a New Psychology of Women.* Boston: Beacon Press, 1976.

Piaget, Jean. *The Moral Judgment of the Child* (1932). New York: The Free Press, 1965.

Piaget, Jean. *Structuralism.* New York: Basic Books, 1970.

Simpson, Elizabeth L. "Moral Development Research: A Case Study of Scientific Cultural Bias." *Human Development* 17 (1974): 81–106.

Stoller, Robert J. "A Contribution to the Study of Gender Identity." *International Journal of Psycho-analysis* 45 (1964): 220–226.

Woolf, Virginia. *A Room of One's Own*. New York: Harcourt, Brace and World, 1929.

The World of the Professional Stripper

Chris Bruckert

"Things are not as straightforward as they seem." With this, Bruckert begins her study of the world of strip clubs. Her approach draws on Marxism, feminism, and symbolic interactionism. While both acknowledging and resisting the moral assumptions and everyday perceptions with which we usually interpret the setting and what occurs within it, the author asks us to attend to the perspective of those working within the industry. As such, the setting is less about entertainment and sex than about how stripping is a job like any other.

At the same time, the stripper is in a disreputable occupation and engaged in labour of which the public generally disapproves. How do these workers handle the stereotypical assumptions, disapproval, stigma, and moral censure associated with their jobs? If a dancer's livelihood depends on "recreating social relations" how does she go about doing this? Finally, in what ways is stripping "women's work"?

If you wandered into one of the over 200 strip clubs in Ontario, you might notice the dim lighting, the pool tables and video games, the continually running pornographic movies, and the smell of stale beer. You might notice that this is clearly a "male space" that is, somewhat ironically, defined by the presence of (some) women. Women in scant attire "hanging out," women sitting and listening with apparently rapt attention to men, women at some phase of undress dancing on stage, women in champagne rooms[1] dancing for, or talking with, (clothed) men who are sitting only inches away from their naked bodies. At first glance the scene appears so imbued with the markers of gendered oppression, objectification, and exploitation that analysis is hardly necessary. Nonetheless, things are not as straightforward as they seem. From the perspective of the women "deep" in conversation or

Source: Chris Bruckert, "The World of the Professional Stripper," adapted from *Taking It Off; Putting It On: Women in the Strip Trade*, Women's Press, 2002 [as printed in Merle Jacobs (editor), *Is Anyone Listening? Women, Work and Society*, Toronto: Women's Press: 2002: 155–172.*] Reprinted by permission of Canadian Scholars' Press Inc. and/or Women's Press.

dancing on the stage, strip clubs are not about entertainment, or immorality, or sex. They are about work.

In this [article] we explore the work of strippers through the lens of feminist labour theory. Using an approach informed by Marxism, symbolic interactionism, and feminism allows us to shift between analytic levels and consider the intersection and tension between market economy, social and gender relations, regulatory frameworks, dominant discourses, labour processes, and work site practices. When we step outside of morally loaded assumptions and attend to the understanding of industry workers, it is quickly apparent that strippers' work is both similar to and markedly different from other working-class women's labour.[2]

FROM ENTERTAINMENT TO SERVICE

The trajectory of labour of Ontario's strippers over the last three decades speaks to the unique position of strip clubs as both commercial enterprises embedded in the market economy and, at the same time, the product and focus of dynamic social processes including moral and legal regulation. It also illustrates how broader labour market trends and economic shifts not only position clubs to exploit strippers but also condition the nature of that exploitation. In the mid and late 1970s strippers were entertainers who, in exchange for wages,[3] performed five sets of four songs (three fast, one slow floor show) during their six-hour shift. During the 1980s and into the 1990s Canada experienced periods of recession, a general stagnation of fiscal growth, and high rates of unemployment (Phillips, 1997:64). During this period of economic restructuring, manufacturing jobs were displaced, the service sector expanded exponentially, and women's labour market position was destabilized (Luxton and Corman, 2001). For working-class women labour market reorganization resulted in a move into labour-intensive consumer service sector employment characterized by low pay, low capital-labour ratio, limited job security, poor working conditions, and non-standard labour arrangements such as part-time, casual, and seasonal work. In principle protected through labour legislation, in practice marginal, non-unionized workers in this sector have limited recourse to legal protection and are susceptible to a range of exploitive practices (Duffy and Pupo, 1997). In addition, the vanishing social safety net compounded the vulnerability, economic need, and domestic responsibilities of this social strata. As a result, workers were not only, by default, increasingly employed in the service sector but situated to embrace work in the growing non-standard labour market, including casual and flexible self-account work, as an income-generating strategy that allowed them to fulfill their many social and personal obligations.

It was in this context of economic decline and dwindling options for working-class women that the new industry innovation of table dancing[4] was introduced in the early 1980s and used to justify cutting dancers' pay to $30 or $40 a day. At the same time shifts were increased from six to eight hours, and bar fees[5] were implemented. By the early 1990s as the economy continued to spiral downward threatening even the "bad jobs" in the service sector, clubs went from exploiting workers to the full appropriation of their labour. Many dancers found their pay eliminated, as they were offered the option of working for "tips" or not at all. In short, in Ontario between 1980 and 2000 stripping was "deprofessionalized,"[6] dancers were redefined as service providers, wages were reduced, and the labour requirements were substantially increased.

Today, while some dancers continue to work "on-schedule" earning between $35 and $45 for an eight-hour shift, most work as "freelancers"[7] receiving no financial compensation from the club. Under either arrangement dancers are expected to pay the established bar fee of between $10 and $20, follow house rules, remain in the bar for a predetermined period of time—hanging out and "looking like a hooker" (Debbie)[8]—and perform between one and five three-song "sets" on stage. Similar to other subcontracting relations (i.e., electricians) exotic dancers are responsible for furnishing tools, in this case music, costumes, and transportation. In exchange for labour, fees, and compliance with the expectations of the club, the bar provides the labour site—the physical space (bar, chairs, champagne rooms) and other coordinated and necessary labour by disk jockeys, bartenders, servers, and doormen. This setting is, of course, crucial. Without it, a dancer cannot solicit the private dances that constitute her income.[9]

In spite of receiving no, or minimal, pay the workers' labour and general deportment remains under the control of management, who establish the house rules governing attire and behaviour, expectations of stage shows, interactions with customers, and services offered in the champagne rooms. Compliance is realized through economic sanctions in the form of fines and by the club's power to deny access to customers. A dancer who is defined as troublesome, who complains "too much," who doesn't follow the house rules, or who leaves with a customer may be suspended or barred permanently. "Troublesome" dancers also risk being blacklisted. This can have dire consequences, since the marked dancer will be unable to pursue her trade anywhere in the city. Put this way, today strippers are in a contradictory space—on the one hand they are managed like employees and subject to disciplinary regimes if they fail to comply, while on the other hand they are denied the pay and protection generally associated with employment.

The exploitative nature of managerial attempts to extract maximum labour power notwithstanding, there have also been positive implications for workers in the shift from entertainment to service. With deprofessionalization and lower labour costs, a new industry standard of continuous stages and lots of "girls" emerged. These changes in turn meant new employment opportunities and an opening up of the labour market as the demand increased. They also conditioned the relationship between management and dancers in new ways. Their limited commitment to a particular labour site affords individual dancers greater levels of autonomy and allows them to determine, within particular confines, where, when, and how much they work. Since the club no longer pays workers but exchanges fees and labour site access for free labour, the ability of management to control labour has been somewhat eroded. This is exacerbated by the managerial need for a stable work force and their subsequent hesitancy to alienate the dancers on whom they rely. This is particularly true for women with considerable organizational assets (i.e., a "sexy" appearance, a client list).

Moreover, the new organization not only conditions labour relations but also interacts with class to shape the labour site itself. In the past, the nature of the entertainment-based industry compelled dancers to work full-time and travel "the circuit." These conditions effectively excluded many women workers who embraced other "respectable" social roles: children, partners, school commitments, other jobs. Today, dancers can opt to work full-time, part-time, or occasionally; and either never, or only periodically, go "on the road" in response to particular financial difficulties. In real terms this, coupled with the impoverishment of women workers in Canada generally, opened up the industry to reputable working-class women and women from middle-class families whose eroding economic position (coupled with ideological changes regarding the meaning of nudity) has rendered morally suspect labour increasingly tenable. Tina, a sole-support mother, started working as a stripper when after years of steady employment she found herself:

> On welfare for seven months. And it was hard and ... I saw those, ah, those ads [in the newspaper]. And one day I decided to, to go, to try it y'know. But it was scary. I was 29 years old and I didn't know what was going on there.

Like Tina, these new workers need to overcome their own stereotypical assumptions about strip clubs; however, those who effectively deconstruct the dominant discourses sometimes remain in the occupation for considerable periods of time. In turn as these new workers bring to the labour site their own class culture and investment in respectability, these values have become embedded in the industry structure itself. Today, the markers of rough working-class culture—practices (partying, drugs), appearance (cut-off jean shorts, tattoos), values (being "solid"), and language (talking tough)—are either absent from strip clubs or are limited to one token "rough bar."[10]

In 1973, amendments to the provincial Liquor Control Act expanded the definition of "theatre" (Ontario, 1973) and made it possible in Ontario to combine alcohol and nudity in a legal commercial endeavour. Since then, the trajectory of strip clubs in the province reveals how the complex interplay between market economy and labour structure shapes the labour process in marginal spheres at the same time as the labour process shapes the class origin of the available employee base. Workers are then positioned in a contradictory class location: they are both independent entrepreneurs who manage their own business—thus, are the bourgeois— and employees who sell (or in the case exchange) their labour power and who rely on, and must comply with the expectations of, an individual capitalist—in this case, the proletariat.

THE JOB

When we shift our focus and apply the feminist labour lens to the question of labour practices another set of questions emerge. What does the work entail? What skills and competencies are workers expected to bring to the labour site? What strategies do dancers employ to negotiate the occupational hazards? What are the particular challenges of the job? In addition, by retaining the focus on class, we are also positioned to ask: How does strippers' labour compare to that of working-class women more generally?

The Stage

. . .

A woman working in a strip club as a stripper first and foremost has to *act* like a *stripper*; whether she is on the stage or not, she is always *performing*. This involves both the ceremony common to visible employees of "playing [her] condition to realize it" (Goffman, 1959:76) and the fact that the dancer is allowed some creativity, although, like actors generally, she is required to assume a role that is not her own, nor of her making (Henry and Sims, 1970). To entertain, she has to "do a stage." This public erotic labour involves the ability to perform for, but also interact with, the audience, whose very presence legitimates the work.[11] In addition, a stripper's act requires a degree of comfort with nudity, a willingness to expose herself physically, and a self-assured and confident presentation-of-self. Many strippers develop a strong stage presence and are often competent dancers, proficient not only in the standard stripper "moves" but able to incorporate, and execute (in very high heels) their own eclectic mix of ballet, jazz, acrobatics, aerobics, and posing. On stage a dancer must continue to smile or at least assume the appropriate sexually vacant expression—"I think about doing laundry or watch the TV" (Debbie)—in the face of apathy and, sometimes, taunts.

These kinds of verbal comments touch not only on her performance but, in light of the gendered appearance imperative, on her value as a woman. In short, she needs to develop the capacity to distance herself from the negative evaluation of the audience.

Although obscured by the performance component and nudity, stage shows are physically demanding labour. And, like so much physical labour, it can be dangerous.[12] In addition to the risks inherent to dancing in stiletto heels, there is the threat of infectious disease. While many dancers take protective measures,[13] the dressing rooms, washrooms, stage, and pole are, at least in some clubs, not particularly well-maintained. The work is also exhausting and technically difficult: "Pole work is a lot of hanging upside down, it's a lot of balance, muscle technique. It's hard to look sexy when you're upside down and all the blood's rushing to your head!" (Diane). Put another way, the "moves" can only be erotic if they appear effortless and natural, a feat that necessitates practice, skill, and considerable muscle development.[14] In short, constructing sexuality is not natural or easy but hard *work*; however, the more effective the illusion, the more sexual the portrayal, the more the *work* is invisible.

The question becomes: How is erotic labour understood and negotiated by participants? Perhaps the most telling finding was how few comments were made by interviewees about sexuality. It appeared to be largely incidental. While dominating public consciousness, nudity, sexual presentations, and interactions are normalized within the cultural environment of the strip club, so that the erotic nature of the labour is essentially a non-issue for participants....

Perhaps more important still are the meanings scripted onto the labour. A dancer engages with the indicators of sexuality, and these links to the erotic appear to define her job as a *stripper*. However, this explicitly erotic labour operates at the level of the visible body. It is not about sex but nudity and the visual presentations of the erotic: "You manipulate your body in a certain way and you throw a sexual aspect to it" (Debbie). Put another way, dancers engage in surface acting where "the body not the soul is the main tool of the trade. The actor's body evokes passion in the *audience*, but the actor is only *acting* if he has the feeling" [emphasis in original] (Hochschild, 1983:37). The eroticized setting, available props, and their own expectations may ensure that the audience defines the entertainers as sexual, but the experience of workers is markedly different.

. . .

The Floor

As previously noted, today's dancers must continually negotiate two discrete, and sometimes conflicting, jobs during their work day. The quasi-contractual obligation is to perform strip-tease shows and "hanging out"—tasks for which she receives not a paycheck but attains access to customers. As self-account service workers, all or most of the worker's income is directly paid by customers in fee-for-service arrangements. In order to "make her money" dancers must first solicit and sell their private dances by convincing "a guy that he really wants a dance" (Debbie). Here labour practices are constrained not only by house rules but also by individual inclination. Some dancers flatly refuse to approach customers: "Some girls go around and ask, 'hi baby how you doing' and start shaking their things in front of him. No! I don't like that at all. I just wait for them. If they want me bad enough, they'll come and get me, they'll signal me or tell the waitress" (Rachel). Others "work the floor"—socializing and engaging promising looking customers in conversation. The most aggressive hustlers greet all customers. At a minimum they "give them the eye, just like you would in a bar" (Debbie).

Having "sold" her service, the dancer accompanies the patron to the champagne room where she seeks to maximize her income by employing a variety of special skills. "Once they come and get me, they're screwed. They're stuck with me, and I'm gonna keep them and siphon out every last dime I can get" (Rachel). While this may entail dirty dancing, more frequently dancers employ "straight" strategies to maximize income:

> I don't stop [dancing] until they tell me to stop and then I tell them how much. I don't do one dance and then sit ... I used to do that, one dance and that's it. Then you don't get another dance. So I just keep dancing. (Sally)

In the champagne room a dancer needs to encourage the customer, retain his attention and good will, and yet remain firmly in control of the situation. The challenges have increased with the media and public discourses throughout the lap-dancing debates. Apparently, customers frequently equate surface presentations of sexuality with actual sexuality, so that dancers are wrongly presumed to be, if not prostitutes, then highly promiscuous. Today "99% by the customers, oh yes, 'You must have a price.' ... the way society is, they're allowed to expect it" (Marie). This means that an individual dancer is required to cope with customers' anticipation of sexual fulfillment while she labours in an environment where she is presumed to be, but cannot be, sexually available.

Not surprisingly given the physical space and discursive parameters, making money also renders dancers vulnerable to physical or sexual aggression. As a result, they must remain vigilantly attentive to clues that identify potentially dangerous patrons (body language, conversation, approach, intoxication). Dancers also routinely rely on each other for protection—"In the champagne room we're all watching each other's back" (Debbie)—and most of the more experienced dancers have perfected strategies that maximize their control of the interaction. . . .

Emotional Labour

While erotic labour, either on the public stage or in the relatively private champagne rooms, appears to define strippers' labour, in practice strippers are increasingly required not only to engage in the surface acting essential for the selling and providing of private entertainment services but also to provide an interpersonal social service that necessitates a unique set of skills and strategies. Here Arlie Hochschild's (1983) concept of emotional labour has resonance.[15] Many customers are only marginally interested in nude entertainment whether it is on the public stage or in a private champagne room. Instead these men come to strip clubs because they "want someone to talk to" (Rachel) and will "spend a couple of hundred bucks and they sit there and talk to a girl that's nice to them and makes them feel good for a few hours" (Diane).

For the dancer this parody of social relations necessitates "playing a game. . . . It depends on the guy, the drippier you are, the more money you'll make. The more you laugh at his jokes, the more money you'll make" (Sally). In essence the dancer presents a cynical performance (Goffman, 1959:18), instrumentally and consciously playing to the expectations of an audience of one.

. . .

Essentially, a dancer's livelihood depends on her ability to recreate social relations and "treat them like they're people. You don't just treat them like they're a ten dollar bill" (Rachel). Interactions are routinized charades where dancers create the illusion of a novel interaction with a "special" person. In short, strippers' daily labour involves not only continual performance—playing the role of a stripper—but also adopting other personas, in effect playing a number of roles, within a particular spectrum of possibilities, consecutively and sometimes concurrently:

> I used to give every guy a different age depending on what they wanted. I also gave different stories, but that's complicated to keep track of. Sometimes I acted really young and walk[ed] around the club in a skirt being cutesie. You don't even have to look that young, just act young. It's really weird. Different guys want different things. (Sarah)

Like other direct service workers, a stripper has to be able to manage her emotions and anger in the face of ignorant and trying customers. However, there is something more—she participates in a financial interaction that masquerades as a social relationship with its sense of reciprocity: "I should probably have my PhD in psychology by now for all the problems I've listened to and all the advice I've given" (Rachel). Social relationships are normally defined by mutual concern. In the strip club, however, the appearance of concern becomes a commodity that is purchased: "I feel guilty when they tell me things. Because personally I don't give a shit. But I have to pretend I do" (Jamie). Notably, unlike the professionals to whom Rachel compares herself, a dancer has neither the language nor the professional training on which to rely to guide them through the interaction; instead, she has to improvise as she continually reinvents herself and adapts her performance.

Although talking to customers appears to be a rather innocuous activity, many dancers express exasperation: "You have to go sit down with the guy and blah blah blah blah blah blah blah blah. I hate that" (Tina). In fact, the most distress was voiced by research participants about this activity. On reflection, this is not surprising. As capitalism expands and the service industry swells to include the supply of emotional and interpersonal services (for men) in a commercial imitation of authentic social relationships,[16] the boundaries are being blurred and the product is not only the service but the server herself. For strip industry workers this means they are alienated not just from their bodies—through their physical capacity to work or their labour power—nor from their surface sexual self-presentation in a way that was normal in burlesque theatres. They are alienated from something more—their social selves:

> Temporarily you're someone you're not, just for this guy, just so you can get his money. If he wants to believe something then you just play right along with it. "Ya I'm from wherever" and make yourself up to be something you're not. (Sally)

The result is a disassociated sense of self, so that "I pretend I'm somebody else and I get all glamorous and I go into work. I'm a completely different person in the club, a completely different person" (Debbie). Workers are very explicit about the need to distance and separate their different selves: "I have a very distinct difference between my job and my life, and I find if I mix the two of them that I can't keep it straight" (Ann). This assumption of separate identity is in part facilitated by the use of stage names so that "on stage I'm Kim so that's not me either" (Alex). It would appear that, as new areas of social and interpersonal life are transformed into services to be bought, the alienation inherent to the labour process in modern capitalist societies is also extended into a new arena.

STRIPPING IS WOMEN'S WORK

To summarize the discussion so far, when we abandon morally loaded assumptions, explore labour structure and practices, and "normalize" the labour of strippers by making links to the "reputable" work of working-class women, similarities start to emerge. Today strippers are contractual own-account workers who experience the same sorts of issues confronting other working-class women in Canada, including a non-supportive work environment, exploitation and oppression by owners and managers, non-standard labour arrangements, lack of security, and minimal protection by the state. As part of the burgeoning consumer service sector they, like many other direct service workers, do a job that requires erotic and emotional labour. The job itself is physically exhausting, emotionally challenging, and definitely stressful.[17] Success is contingent on the development of complex skills and competencies including performance, construction of sexuality, sales, and finely tuned interpersonal skills. The very existence of these skills belies the customary focus on deviance rather than work process in much of the literature. Of course that these skills are largely dismissed, or rendered invisible, is not unique to the strip trade but characterizes many working-class women's jobs (Gaskell, 1986). It does, however, affirm once again the relative and subjective nature of what is defined as skills.

It is women's work in another way as well. Traditionally women were expected to provide men with nurturance, care, and support. Dancers provide this service for men who "want someone to listen to their problems" (Sarah) on the market on a fee-for-service basis. Suspending momentarily what it says about the state of alienation in advanced capitalist society that men are prepared to pay $10 for every four minutes[18] they spend in the company of a woman ($85 per hour if they take the flat rate[19]), we can appreciate that this is fully consistent with the move of capital into the types of services traditionally performed in the home. In the context of intimate relations this empathetic support is not experienced as particularly challenging; within the labour market it proves to be difficult, emotionally taxing labour that requires both surface and deep acting and the implementation of complex skills. Like so much of the labour women do, it is obscured, even to participants, by the context in which it occurs and the taken-for-granted nature of the competencies. That is to say not only is the labour structured so that work is interspersed with social interaction but emotions do not "fit" into the language of work, so that while the strippers are fully aware that "it's hard on your head after a while" (Diane), they are, nonetheless, sometimes not fully cognizant of this as *labour* activity.

BUT NOT A JOB LIKE ANY OTHER!

Recognizing this work as labour, we must exercise caution. While we can legitimately make links to more reputable labour sites for almost every aspect of the dancer's work, few jobs require this combination of skills and necessitate that the worker operate in such a complex and emotionally taxing labour environment. In this last section we attend to specificity and consider stripping as a *marginal* labour activity and reflect on the implications for workers.

First, we need to consider that stripping is a stigmatized labour location.[20] While participation in the paid labour force is a taken-for-granted imperative for most Canadians, the nature of an individual's work is something they are presumed to choose. "Choosing" a labour market location that is on the margins of legality, morality, or propriety can have profound implications, as the stigma of labour location is transformed into a stigma of the worker (Polsky, 1969)....

These workers must contend with moral righteousness and stereotypical assumptions in interpersonal relations and in a range of social and economic areas from housing—"some places don't rent to strippers" (Diane)—to finance—"it's hard to get credit in a bank" (Marie)—that are generally assumed to operate outside of moral consideration. Put another way, *working* as a stripper becomes *being* a stripper, an identity marker with very real implications in the lives of women in the industry and that shapes the worker's experience of the wider world.[21] While most dancers deconstruct the discourses and challenge the assumptions that underlie the stigma (prostitution, drug abuse, immorality) and effectively manage their personal and social identities, they must, nonetheless, continually engage in social and personal exchanges where their labour location is understood to be definitive.

The implications of participating in "disreputable" labour extend beyond questions of identity and social interaction. Dancers must also negotiate a web of state regulatory practices unknown to employees in more "reputation" occupations. Throughout the 1980s and into the early 1990s, in response to claims made by community groups that linked strip clubs to increased crime and vice, municipalities throughout Ontario began to regulate the industry through severe zoning restrictions, banning clubs from residential areas, restricting the clubs to commercial zones, and stipulating no strip-club parameters around churches and schools. They also introduced licencing that required the newly designated "exotic entertainment parlour attendants" to purchase annual licences under threat of fines and even imprisonment.[22] The nature of the licencing is revealing and speaks to the moral subtext of these strategies....

In principle, these controls are intended to regulate the industry in the interests of broader society; in practice, they not only stigmatize and marginalize workers but also further restrict the employment options of women workers: some clubs are "zoned" out of existence, while live entertainment ceases to be economically viable for smaller clubs in light of the hefty annual fees.

CONCLUSION

If I have done my job well, it should be clear to the reader that the work women perform in strip clubs, is *hard* work. In order to be able to practice her trade, a dancer has to appear periodically on stage, dance, and remove her clothes for a roomful (or worse, *not* a roomful) of men "for free." In order to "make her money," she has to present herself as an attractive "sexy" woman, sell her service to an individual patron, and retain his attention by engaging in erotic and/or emotional labour while carefully maintaining physical and psychological boundaries. In the champagne room, her naked body may well be inches from her client, but she is continually being monitored by the manager, the doorman, other dancers, and the police. Like a rape victim, if she is inappropriately touched, she is held responsible and sanctioned. All the while she has to cope with the particular stress of working in a leisure site as well as deal with the chaotic environment and interpersonal conflicts that abound. When she leaves the labour site, she continues to engage with the stigmatized nature of her occupation, managing her social and personal identity as well as coping with the stereotypical assumptions of her friends, intimate partners, and the state agencies with whom she interacts. In other words, while we can legitimately make links to more reputable labour sites for almost every aspect of the dancer's work, there are few jobs that require this combination of skills and necessitate that the worker operate in such a complex and emotionally taxing labour environment. Furthermore, the implication of stigma means that the labour has far-reaching costs in the worker's personal life.

At the same time, the implications of having a "job like no other" are not all bad. Unlike most workers who provide traditional women's work on the open market, a stripper is well compensated for her labour. Furthermore, not only does the job offer her a flexibility and autonomy seldom available to working-class women, it allows her to develop competencies that are useful outside the labour site—assertiveness, boundary maintenance, and interpersonal skills. In addition, although her work may leave her frustrated and angry, it also affords her a broader vision, enhanced self-esteem, good body image, comfort with her sexuality, and confidence—all worthwhile attributes and ones that many women continue to struggle to realize.

NOTES

1. These cubicles, measuring perhaps three feet by five feet each, are equipped with two (most often vinyl) chairs facing each other, an ashtray and a ledge to hold drinks. While the cubicles are usually hidden from the general view of the club, they are open to be monitored by anyone passing down the aisle between them.

2. This chapter is based on data gathered during a year of participant observation in a southern Ontario strip club, fifteen in-depth, semi-structured interviews with

women working as strippers, and a series of interviews with other industry employees including managers, doormen, bartenders, waitresses, and disk jockeys. For a more detailed description of the methodology or for a further development of the arguments see Bruckert (2002).

3. Wages ranged from $275 to $600 a week in the late 1970s.

4. Table dances are a one-song strip show performed at the patron's table. Today, in spite of the advertised availability of $5 table dances, these are rare. Most dancers simply refuse to remove their clothes in the middle of the bar. At any rate, most patrons are easily persuaded to enjoy the privacy afforded by the champagne rooms, where for $10, the stripper either dances on a stool or sits in close proximity to the customer and moves—a dance in name only.

5. Dancers are required to pay bar fees or "DJ fees" of between $10 and $20 per shift. In practice this means that the dancer must "pay to work there" (Jamie). Depending on the club these fees compensate the disk jockey and sometimes the bartender, who also receive no pay in the traditional sense.

6. This redefining of labour as semi-skilled is consistent with the trend towards deskilling that Braverman (1974) identified as characteristic of twentieth-century capitalism. That deskilling is ideologically and economically useful (for capitalists) is revealed when we realize that throughout the 1980s and 1990s, at the same time as skills were being denied, employers in mainstream sectors of the labour market were establishing inordinate educational requirements (Rinehart, 1996:78). It would appear that labour-dependent personal service industries capitalize on existing age, gender, and racial stratifications by hiring marginal workers and then justify their low wages through reference to their marginal status (Reiter, 1991:148).

7. DERA (Dancer's Equal Rights Association of Ottawa) estimates that one in four Ottawa dancers are "on-schedule" (DERA, 2001).

8. On-schedule dancers are booked for eight-hour shifts, while freelancers must remain for a minimum period—usually four or five hours—established by the bar.

9. To perceive these arrangements as anomalous risks reaffirming marginality by locating it outside of established labour practices. In fact, in the way it is organized, stripping is comparable with the non-stigmatized service occupation of realty. Like strippers, real estate agents are in such a paradoxical relation to their "employers" that the term is hardly appropriate. Realtors are actively recruited by brokers; they are hired, and they can be fired. But since they receive no direct financial remuneration for their labour from their employer, the relationship is nuanced. In exchange for legal protection and access to the necessary legitimizing context (including the use of the name, licence, and insurance), means of production (phone services, office space, and technical support), the realtor commits her/himself to a particular brokerage firm (including providing "free" labour staffing the office).

10. There was a particular irony here. While the dominant discourse increasingly defines stripping as immoral, the clubs and workers are becoming progressively more committed to respectability: "they think of it as a business now, y'know, the newer generation; it's more like a business instead of just the stereotyped thing that people used to do. The girls are keeping their money. A lot less drugs" (Rachel). Furthermore, young women from the rough working-class, who wear the markers with pride, are being marginalized within the industry. It is precisely these women whose employment options are restricted and who are the most exploited population of workers.

11. Her agency is noteworthy. Far from being solely an object of the male gaze, it is the dancer who establishes the interaction with the audience and determines the pace, actions, and movement of the show. The audience's reading of her sexualized form does not erase her authorship. We see this clearly when a dancer enacts a fine parody as she plays with her own and her audience's sexuality, although she is usually quite careful, given the economic-power dynamic, not to let the audience in on the joke.

12. While not all working-class jobs are manual, physically challenging jobs are overwhelmingly working class. Consequently, the labour sold frequently has a socially unacknowledged (though recognized by the wage-labourers themselves) youth imperative and uncompensated costs in terms of health and well-being (Dunk, 1991; Houtman and Kompier, 1995:221).

13. These include bringing their own towels to sit on and sometimes their own cleaning materials.

14. In addition, creating an erotic persona necessitates countless hours of labour in appearance, clothes, make-up, and sometimes tanning salons or plastic surgeries.

15. Hochschild (1983) argues that rather than simply selling her mental and physical labour, the modern service workers must now engage in emotional labour. This requires the worker, in exchange for a wage, to "induce or suppress feeling in order to sustain the outward countenance that produces the proper state of mind in others" (1983:7) and engage in "deep acting" by re-creating personal experiences in a commercial setting. Such a worker must manage her feelings not just for private social relations (which we all do), but as a commodity to benefit the corporation that pays her wage. The process, which requires her to transform her smile into a *sincere* smile, cannot avoid creating a sense of alienation from feelings (Hochschild, 1983:21).

16. It is possible that capitalism is responding to the market and exploiting men's insecurity in the changing gender relations that characterize the latter half of the twentieth century. With the erosion of male power that "is based on the compliance of women and the economic and emotional services which women provided" (Giddens, 1992:132), men struggle with the new expectations and their own need for intimacy (Giddens, 1992:180).

17. For dancers "role overload," identified as a key contributor to workplace stress (Levi et al., 1986:55) is normal. Dancers have to constantly negotiate two separate, sometimes conflicting, jobs during their work day. The quasi-contractual obligation of the stripper is to perform strip-tease shows and "hang out"– "looking like a hooker" (Debbie)–tasks for which she receives not a paycheck but the opportunity to "make her money"; that is, to take the chance to utilize the profitable skills of soliciting and playing the game. Her job not only requires her to fulfill a number of roles at the same time but also to continually manage the emotional and sexual demands of patrons. She must try to maximize her income while simultaneously engaging in boundary maintenance to protect her emotional and physical space. In addition dancers are subject to the stress shared by other labourers engaged in emotional work (Adelmann, 1995:372) as well as the particular stressors shared by entertainers–performance anxiety and a fear of even minor physical injury that can effectively curtail their career (Sternbach, 1996): "I can't work with black eyes, I can't work with big scars across my face" (Jessie).

18. These prices were in effect in 1999.

19. These prices were in effect in 1999.

20. Of course other occupations are also stigmatized—morticians, custodians, and used car salespeople, to name a few.

21. It is also a "sticky" stigma infecting those around the dancer as well (Goffman, 1963:30), so that her family may be, or may perceive themselves to be, stigmatized. Certainly, those who share her labour site are. It is also sticky in the sense of enduring even after participation in the industry has ceased. The almost inevitable linguistic designation of *ex*-strippers in the media speaks to an understanding that participation in the trades legitimates continued assumptions of immorality.

22. In 2001 there is considerable provincial disparity. Some municipalities, such as London and Kitchener, require clubs, but not attendants, to purchase licences. In municipalities that continue to licence dancers, costs can be quite high. In Windsor, dancers must pay $225 plus administration and photo fees annually.

References

Adelmann, P. (1995). Emotional labour as a potential source of job stress. In S. Sauter & L. Murphy (Eds.), *Organizational risk factors for job stress*. Washington: American Psychological Association.

Braverman, H. (1974). *Labour and monopoly capital: The degradation of work in the twentieth century*. New York: Monthly Review Press.

Bruckert, C. (2002). *Taking it off, putting it on: Women in the strip trade*. Toronto: Women's Press.

DERA. (2001). Mission statement. Ottawa: Dancers Equal Rights Association of Ottawa Carleton. (n.p.)

Duffy, A. & N. Pupo. (1997). *Part-time paradox*. Toronto: McClelland & Stewart.

Dunk, T. (1991). *It's a working man's town: Male working class culture in northwestern Ontario*. Montreal: McGill-Queen's University Press.

Gaskell, J. (1986). Conceptions of skill and work of women: Some historical and political issues. In R. Hamilton & M. Barrett (Eds.), *The politics of diversity: Feminism, Marxism and nationalism*. London: Verso.

Giddens, A. (1992). *The transformation of intimacy*. Palo Alto, CA: Stanford University Press.

Goffman, E. (1959). *The presentation of self in everyday life*. New York: Doubleday.

_____. (1963). *Stigma*. Upper Saddle River, NJ: Prentice Hall.

Henry, W. & J. Sims. (1970). Actors' search for a self. *Trans-Action* 7–11.

Hochschild, A. (1983). *The managed heart: Commercialization of human feeling*. Berkeley, CA: University of California Press.

Houtman, I. & M. Kompier. (1995). Risk factors and occupational risk groups for work stress in the Netherlands. In S. Sauter & L. Murphy (Eds.), *Organizational risk factors for job stress*. Washington: American Psychological Association.

Levi, L., M. Frankenhauser, & B. Gardell. (1986). The characteristics of the workplace and the nature of its social demands. In S. Wolf & A. Finestone (Eds.), *Occupational stress: Health and performance at work*. Littleton, MA: PSG Publishing.

Luxton, M. & Corman, J. (2001). *Getting by in hard times: Gendered labour at home and on the job*. Toronto: University of Toronto Press.

Ontario. (1973). An act to amend the Liquor Licence Act. (Chapter 68, 69). *Statutes of the Province of Ontario*. Toronto: Thatcher.

Phillips, P. (1997). Labour in the new Canadian political economy. In W. Clement (Ed.), *Understanding Canada: Building the new Canadian political economy.* Montreal: McGill-Queen's University Press.

Polsky, N. (1969). *Hustlers, beats and others.* Garden City, NJ: Anchor Press.

Reiter, E. (1991). *Making fast food.* Montreal: McGill-Queen's University Press.

Rinehart, J. (1996) *The tyranny of work: Alienation and the labour process.* 3rd ed. Toronto: Harcourt Brace.

Sternbach, D. (1995). Musicians: A neglected working population in crisis. In S. Sauter & L. Murphy (Eds.), *Organizational risk factors for job stress.* Washington: American Psychological Association.

Ordering Choice: Women, Disability and Medical Discourse

Tanya Titchkosky

Titchkosky's paper presents a careful textual analysis of an ordinary newspaper article about a pregnant woman and an abnormal fetus. The article opens by calling attention to a problem. But what kind of problem? Something's wrong. But what's wrong? Titchkovsky draws on interpretive, feminist, and disability studies theory to make explicit the interpretive structures and decisions through which women and bodily anomalies are made meaningful in the newspaper article.

Titchkosky concludes her paper by pointing out that her analysis both assumes and demonstrates there are alternative ways of knowing and relating to bodies, mothers, and anomalies than the prevailing medical one presented in the ordinary newspaper article. What exactly is the alternative implied in Titchkosky's analysis? What does she mean by the "language game" of medical science and technology? How is the reader encouraged to think differently about disability?

The story begins by telling of a woman who knowingly remains pregnant with a disabled fetus.

> "The first clue appeared on the ultrasound, a hint in the microscopic clenched fists' awkward inward curl and in the umbilical cord's missing artery that something was amiss.
>
> The doctor called. It could be Down syndrome ... Only amniocentesis—examining a sample of fluid from the womb—could solve the mystery."

Sources: Tanya Titchkosky, "Ordering Choice: Women, Disability and Medical Discourse." Reprinted with permission of the author.

Margaret Philp, "Of Human Bondage: How the System Martyrs Parents of Disabled Kids." *The Globe and Mail*, February 16, 2002: F4–5. Reprinted with permission from *The Globe and Mail*.

This depiction of something amiss serves as the opening lines of a newspaper article titled "Of Human Bondage: How the System Martyrs Parents of Disabled Kids," written by Margaret Philp (*Globe and Mail*, February 16, 2002, pp. F4-5). That something is amiss is delivered to pregnant women and readers alike in one of the most familiar and ordinary of ways; the doctor will call, and we are already expecting this. This paper[1] examines the extraordinary social consequences of this ordinary orientation to bodily anomaly.

Situated at the intersection of disability studies and feminist theory, this paper conducts an analysis of a way that women and disability are made meaningful in everyday life. The paper analyzes a seemingly ordinary newspaper story of disability in order to show the ways that the text organizes the reader's consciousness so as to reproduce women as the embodiment of uncontrollable subjectivity and reproduce disability as nothing other than devalued luck.

Of Human Bondage: How the System Martyrs Parents of Disabled Kids

The first clue appeared on the ultrasound, a hint in the microscopic clenched fists' awkward inward curl and in the umbilical cord's missing artery that something was amiss.

The doctor called. It could be Down syndrome or some other genetic fluke, the simple biological mistake of a spare chromosome that can render a child mentally or physically shortchanged. Only amniocentesis—examining a sample of fluid from the womb—could solve the mystery.

Adela Crossley searched her soul. Her first child, Jason, was a strapping boy who seldom fell ill. If her second child turned out to be cursed with a debilitating genetic defect, would she opt for an abortion?

No, she decided, "Francine was conceived in love." Adela, 34, now says, "and we loved her with all our hearts even before she was born. The fact she was born disabled, didn't make her any less of a child." She refused the amnio, leaving the doctors to shake their heads as successive ultrasounds revealed ever more starkly the subtle deformities of a fetus with a severe chromosomal disorder.

Francine Crossley was born with an enormous bump on the side of her head and a flopping rag doll of a body weighing little more than four pounds. In seconds, she was in the intensive-care ward, where three days later she was diagnosed with Trisomy-18, which medical texts describe as "incompatible with life." Most babies die before their first birthdays, the victims of heart failure or an infection their faulty immune systems are helpless to fight.

"The pediatrician told us she was going to be extremely disabled mentally and physically to the point that she would not know us," Adela recalls. "Her quality of life would be impaired significantly and most likely, she would never walk or talk."

> The decision by so many women to delay having children and the onward march of medical science has resulted in a dramatic growth in the number of children born with severe disabilities.
>
> Statistics Canada found in 1999 that 3.1 million people—12.5 percent of the population over the age of 12—suffered from a "long-term disability or handicap," compared with 2.85 million or 11.6 per cent just two years earlier. In 1996, the National Longitudinal Survey of Children and Youth found that 436,000 youngsters—9.3 per cent of the population—were regarded as having special needs, from a learning disability to something as severe as a debilitating degenerative disease.
>
> These are just estimates. Even the expert data collectors at Statscan [Statistics Canada] won't have a solid figure until they see the results from a major survey on the subject conduct as part of last year's national census.
>
> But it's no secret that delivery room heroics now save thousands of disabled babies who once died shortly after birth. They also are creating a population of shell-shocked parents stunned to discover that, when it comes to rearing such children, they are largely on their own.
>
> *An ordinary newspaper article on women and disability (Margaret Philp, Globe and Mail, February 16, 2002, pp. F4-5).*

METHODOLOGICAL BACKGROUND: INTERPRETING INTERPRETATION

I regard the newspaper article I analyze here as exemplary: it is very ordinary and sensible, it is typical and seems reasonable, and thus this text provides the possibility of analyzing it for, in Judith Butler's (1997, p. 13) words, the "conditions of its emergence." The text has come to make an appearance in the midst of a variety of conditions, such as the normative order of literate culture, comodification of knowledge, as these intersect with a popular press industry that includes trained and accomplished authors such as Philp.

Still, there is more. There is the matter of the text itself as a context for inquiry. What conditions the meaning of this newspaper story of martyred parents and disabled children? How does a story about martyrdom take as its starting point the technological discovery of bodily anomaly presumably located within a woman's body?

I address the conditions of the text's emergence through an analysis that asks how the text is empowered by, and simultaneously empowers, the genre, or language game, of medicine.[2] In the context of the text resides the ongoing activity of not only making up the meaning of people and issues, but also constructing relations between these people and those issues.[3] This analysis is grounded in sociologically-oriented feminist and disability theorizing that flows from the linguistic turn influenced by the traditions of phenomenological and hermeneutic inquiry. These traditions dedicate themselves to an analysis of the order of interpretation asking how best to address that which appears as a reality built up through interpretation. The unique approach offered here requires a constant search for the making of meaning, which as Stiker (1999, p. 1) suggests aims to "... enlarge the understanding we already have." Thus, the experience of reading this writing can be consulted for the flashes of meaning that establish interpretive relations between women and disability.

VICTIMS OF LOVE: THE CONSTITUTION OF MEDICAL ISSUES

Establishing the Problem

The introductory tag lines, "Of Human Bondage: How the System Martyrs Parents of Disabled Kids," "Victims of Love" and "Care and Woes," are followed by a short block of text that introduces not only disability but also the author: "... Margaret Philp finds that many Canadian families are on the verge of despair ..." (Philp, 2002, p. F4). Her exposé of this despair begins with a depiction of an ultrasound test. Again:

> The first clue appeared on the ultrasound, a hint in the microscopic clenched fists' awkward inward curl and in the umbilical cord's missing artery that something was amiss.
> The doctor called.

Philp defers to the authoritative image of the ultrasound.[4] This image "voices over" Philp's own voice and displaces her as author. Still, the beginning of the text seems almost personal in its poetic description of an abnormal happening. There is a writer of course. But in the face of the startling clues of something amiss, it seems almost natural that the "really wrong" will appear as if it is outside of human acts of noticing and interpreting. The clue to something really wrong is made to appear *as if* it is not produced by human interpretation, but is instead produced by the stark and undeniable technological *authority* which has the ability to *see* inside the body—the ultrasound image. Displacing the author's voice offers the reader an up-close look at something going wrong, while simultaneously making it appear as if this desire to see and to know exists in the starkness of the *wrong* and not in the relation between writer and reader.

The problem is not that of a system that martyrs parents of disabled children, at least not so far. Nothing is attended to outside of the fetus and the technology geared for reading it for "microscopic clues" of abnormality. These clues are depicted as belonging solely to the domain of the ultrasound which is a common, yet political, diagnostic imaging procedure (Cook, 1996, p. 76; Hartouni, 1998, p. 208; Stabile, 1998, pp. 187-188). Imaging hints of difference represent the birth of disability as abnormality conceived of as a medical problem.

Medicine is laboring to separate women from their bodies. There is no mother mentioned in Philp's depiction of the ultrasound image and it is difficult to remember that "the division between woman and fetus is historically unprecedented" (Stabile, 1998, p. 172). In the face of the authoritative technoscience image of something wrong, it is relatively easy to forget that the image maker, like the embryo, and like its story, has a history (Sawicki, 1999, pp. 191-193). Moreover, medicine must further enact a separation between bodies and any sense of a wider social context. As Irving Zola (1977, p. 62) reminds us, constituting a medical trouble requires that technology be used to "... locate the source of trouble as well as the place of treatment primarily in individuals and makes the etiology of the trouble asocial and impersonal."

Readers are provided with a text of some "thing" objectively wrong and are not provided with images that reflect culturally specific ways to imagine wrongness. Without signs of medical decision-making and interpretation, it becomes difficult to remember that imaging something is a way of forming the decision to look. As Rod Michalko (1998, p. 40) says, any look is "... essentially a social act. Looking and noticing is located within a social web of interests, purposes, hopes, fears, anxieties, and so on." The ordinary phraseology for noticing something wrong—"The first clue appeared on the ultrasound"—accomplishes the extraordinary work of displacing not only the author but the entire sociality that grounds contemporary ways of looking at bodies. A coherent problem is imaged and exposed; the means of doing so are not.

Something Amiss

The term "ultrasound" is used in such a way as to establish the power of the god trick (Haraway, 1991, p. 189) and through this trick the reader comes to an unquestioned sense of something amiss. Technology appears to produce its results without the need for the reader to imagine even a technician operating the body-imaging-machine, let alone to imagine the presence of an "actual" woman's body (Smith, 1990, p. 5). Then, the doctor calls.

It does not matter who this doctor is, nor does it matter where the doctor is, nor does it matter when the doctor is calling. The doctor's name and gender are not given, nor is the reader given any sense of the doctor's training, speciality, or social location. An anonymous agent (Sawicki, 1999, p. 194), the "doctor" who "calls," is the only necessary identity marker. The only thing the reader needs to remember is that when medicine calls, there is a good (medical) reason for it to do so, and this reason has been provided for by Philp's account of the ultrasound image with its authoritative discovery of abnormality. No question about it, something is amiss, some-thing is wrong.

The socially abstract and disembodied doctor normalizes the notion that some human problems are beyond a need to attend to their social organization and production, thus reproducing the belief that some technologies and practices of medicine are not grounded in decisive and, thereby, oriented human action. This sensibility allows the reader to experience "The doctor called" as obvious while not experiencing the doctor's particularity as missing. This is not a strange way to speak of doctors. It makes sense. While bodily uncertainty or pain may make this call most welcome, the call nonetheless is a masterful organizer of uncertainty.

Seeing Abnormality as Devalued Form of Life

The doctor called. It could be Down syndrome or some other genetic fluke, the simple biological mistake of a spare chromosome that can render a child mentally or physically shortchanged. Only amniocentesis—examining a sample of fluid from the womb—could solve the mystery.

This represents the spirit of what the doctor said when the doctor called. The medical call of abnormality transforms the ultrasound "image" into the doctor's "actually seeing" of abnormality. "Clearly, technoscience knowledge confers power upon those who control it ..." (Goggin and Newell, 2003, p. 11). The doctor's call mediates what she or he has seen. Still, what is wrong is a mysterious problem to be unraveled by an amniocentesis test.[5]

Addressing this wrong entails two parts. First, medicine wants to solve the mystery as to what type of disability it is dealing with, for example, "Down syndrome or some other genetic fluke."[6] In the words of Nikolas Rose (1994, p. 50), "Our modern medical experience is, first, constituted in certain *dividing practices*. ..." The amniocentesis test will allow for medicine to divide one type of problem from another but, for now, the diagnosis of a *type* of disability remains a *mystery* and medicine is depicted as guessing at what "it could be. ..." Nonetheless, the consequences of disability are not mysterious; someone (and not medicine) will be, as Philp puts it, "shortchanged." Medicine is depicted as already successfully actualizing a dividing practice—the contours of the shortchanged life are seen as separate and distinct from the contours of normal fetal development, and the woman is separated out so radically that she is yet to make an appearance. It is now the self-proclaimed task of medicine to "... locate subjects in different relations to the decisions and actions made about their problems ..." (Rose, 1994, p. 52).

The second part of dealing with the problem involves articulating these consequences to whomever the doctor calls. Ultrasound in hand, the doctor calls the possessor of the womb that is carrying a fetus that displays an apparent abnormality. The doctor gives a partial diagnosis by regarding the abnormality as *caused*—Genes? Biological fluke? A caused abnormality, even at the microscopic level, grounds the doctor's ability to also provide a prognosis—this fetus represents a shortchanged life. The fetus is interpretively transformed into a child, but also a clue of abnormality is transformed into a living-lack. This prognosis is related to medicine's "social vocation" to determine where embodied differences fit and where they do not (Rose, 1994, p. 52).

Reading this, it is easy to not be startled by the text's radical movement from a description of the technological traces of a microscopic abnormality to the depiction of a child's form of life: a disabled one, a shortchanged one, a devalued one. Moving from an image of abnormal body parts to the imagining of an entire shortchanged form of life and its life-course is presented within Philp's text as a normal-fact and not as a decision based upon an interpretive relation to disability. The reader is given the not-so-mysterious articulation of disability: troublesome-lack. Nonetheless, the medical mystery continues to reside in the yet undetermined prognosis—just how shortchanged is this fetus? According to the text, this mystery can be solved by an amniocentesis test.

That a medical test can be used to foretell life-chances, or forms of life, makes sense only insofar as both the reader and writer conform to a set of taken-for-granted beliefs. Both reader and writer must accept the efficacy of such tests; both must believe in and expect a straightforward equation between type of disability and type of life; both must subscribe to the belief that disability is a fluke condition of a nameable type within which there are few variations. The primary shared belief required of readers and one actualized by the text is that a named problem condition is not merely a prediction of a form of life, but is also a reasonable depiction of a life conditioned by such problems.[7] All this meaning is being made under the technologically enhanced illusion that disability is unrelated to human interpretation. The clarification of these meanings becomes part of the doctor's call.

The text serves as a way to govern embodiment (Titchkosky, 2003b, 2002; Pratt and Valverde, 2002; Foucault, 1988, p. 16ff). Writer and reader can govern their relation to disability by the understanding that disability is an objectively given lack that is assembled *as if* it is completely unrelated to the symbolic realm of social meaning—unrelated to systems that offer poor supports, unrelated to medical techniques and knowledge, and unrelated to alternative interpretations that might be held by pregnant women or parents.

Governing Objectivity and Subjectivity

The yet un-named but objectified problem reported by Philp, unencumbered by any mention of a woman's body, or an actual doctor, is made to exist on its own, and made to appear completely within the purview of medical technology, practices, and beliefs. But when the doctor calls, she or he calls a woman to whom this problem apparently belongs, and the meaning of disability expands well beyond the issues of diagnosis and prognosis.

> Adela Crossley searched her soul. Her first child, Jason, was a strapping boy who seldom fell ill. If her second child turned out to be cursed with a debilitating genetic defect, would she opt for an abortion?
>
> No, she decided, "Francine was conceived in love." Adela, 34, now says, "and we loved her with all our hearts even before she was born. The fact she was born disabled, didn't make her any less of a child." She refused the amnio, leaving the doctors to shake their heads as successive ultrasounds revealed every more starkly the subtle deformities of a fetus with a severe chromosomal disorder.

Crossley must be the bodily container (Purdy, 1990) upon which the ultrasound was performed and who will be delivered the diagnosis, prognosis, and treatment issues that accompany a medical noticing of abnormal fetal development—she will be delivered the issues of deformity, disorder, defect, disability. Adela Crossley is whom the doctor, of unknown gender, name, experience, or place, has called. The disembodied doctor (objectivity) calls the embodied and socially located woman (subjectivity) who possesses the womb that carries a problem in order to recommend further examinations.

Through Crossley, illness has become a failing and disability a curse that is located in a context described only as a mother's love. As much as a cliché as this sounds, Crossley becomes the occasion to articulate complex and messy relations between disability (all that is bad and not strappingly normal) and women as the subjects who bear this objective wrong. "The fact she was born disabled, didn't make her any less of a child." Child and disability are separated and somehow this separation, like the medical separation between fetus and woman, makes sense. Crossley is depicted as splitting disability and personhood and, on the basis of this split, all sorts of other soul searching considerations enter into the meaning of disability.[8] The neater technologically enhanced medical distinction between normal and abnormal *conditions*, resulting in functional or shortchanged lives, becomes messy and unclear upon the advent of an imagined personhood whose meaning inevitably exceeds the sensibility offered by medical discourse.

Orienting to the medical meaning of disability as if it is both "natural" and "objective" conceals medicine's active interpretive relation to disability and makes Crossley appear all the more troublesome, indeed mistaken, in her thoroughly subjectivized stance. Crossley's reasoning seems attached only to a surfeit of subjectivity—"love." She refuses to regard the ultrasound's sound version of the fetus as something that can be totally contained by a medical conception of disability as organic lack of function and abnormality. In the process of intersecting with extra-medical institutions and contexts (such as motherhood), diagnosis, prognosis, and treatment of disability become more complex.

But, like the doctors, readers are positioned so as to only shake their heads at Crossley's overly subjectivized stance. The juxtaposition of the singular patient, saturated with her own subjectivity, and the many doctors, abstracted from any form of subjectivity beyond their function, provides the reader with a sense of the woman's decision as strange—Crossley will knowingly birth a disabled fetus. Moreover, she is humanizing that which medicine insists is best regarded as a form of life that departs from humanity.[9]

Victimage

The tacit meaning of "Victims of Love" now begins to crystallize. Some fetuses are cursed by a faulty genetic make-up. Technology is heralded as discovering this curse. Undoing this curse is difficult because the problem is in fetuses at the microscopic level manifesting in clenched fists and in hints of deformities and deviations. Currently, elimination of such fetuses is the only sure medical treatment. But, there are times when the curse is not undone. Because of some women's failure to be mastered by the call of medicine they are victimized by their own (subjective) orientation toward pregnancy and thus the healthy management of an orderly and normal life is put at risk. A second level of victimization then emerges as the fetus is forced to live with its curse because of the mother's decisions. Under the medicalization of life, mistaken or risky bodies are made manifest as lives victimized by an impaired sensibility, a sensibility unable to fully subject itself to the ruling authority of the day. A third level of victimization lies in this: the birth of such a being is the birth of a shortchanged life in regards to which both parent and medicine will have to offer compensatory measures. If we do not heed the call of medicine and willingly submit to it as the best way to order our lives, we all pay for it in the end. The price we pay in not accepting our subjectitude serves to remind all of us that a renewed acceptance of the total medicalization of our lives is rational, necessary, normal, and good.

CONCLUSION

The story of this victimage continues and Crossley ends up functioning in the text as merely one example of the many women who have "delayed" having children. However, due to space constraints, I end here by collecting how medicine interpretively orders women and disability. The dramatic medicalized depiction of Crossley's decision to give birth to a particular child does the work of preparing the reader to accept the truth-claim that the birth of disability *in general* is just as dramatically disturbing as are the dramatically increasing numbers of disabled youngsters that the article goes on to (mistakenly) statistically document. Disability transmogrified as matter gone wrong helps to produce and is produced by women regarded as those who makes things matter wrongly. In the words of Merleau-Ponty (1964, p. 98), "Every science secretes an ontology; every ontology anticipates a body of knowledge." Our ways of knowing, including medical technology, organize ways of being, and being a woman seems particularly subject to the necessity of knowing how we are going to manage Her excessiveness.

The newspaper story's way of knowing "secretes" disability as a being whose characteristic feature is devalued difference–abnormalcy. This way of constructing disability anticipates the knowledge that women who willingly give birth to disability are not only derelict in their duty, but monstrously mistaken in their choices (Shildrick, 2002; Braidott, 1997). This way of knowing is so certain in its understanding of disability that it can only shake its head, flabbergasted, at those who dare exercise women's right to choose in the face of disability.

Disability as the devalued under-life of all valued aspects of life comes with a further detrimental consequence. Disability, like all other forms of human life, is born of woman. Woman is made responsible, yet again. Today, disability is born of woman either because of her uncontrolled relation to the fetus or because of her "natural" ignorance. Had she known; considered the risk; had a few more tests ... had we been better able to contain her, disability would not be born. "But it is not simply that the feminine is presented only as a lack ... it is also the site of an unruly excess that must be repressed" (Shildrick, 2002, p. 105). Woman imagined as able to respond to (e.g., love) this devalued under-life puts into question woman's own tie to a reasonable version of humanity. Woman's lack is nonetheless clenched by her own surfeit of subjectivity and thus technoscience always needs to recover its hold on her.

My current analysis of the newspaper text both presupposes and demonstrates that there are alternative ways of coming to know embodiment. An alternative way of knowing (steeped in hermeneutics) begins from the assumption that there are, in fact, a plurality of ways of developing relations with bodies, mothers, and anomalies and that all ways, even medicine's, involves interpretive decisions and ontological commitments. A hermeneutic way of knowing, or science of understanding our interpretations of interpretations, requires the inquirer to begin from the assumption that we are all enmeshed in multiple and conflicting webs of interpretation. Texts call out for us to be certain kinds of readers, they act upon our consciousness in the act of reading itself, organizing and governing what we may and may not come to know.

With this alternative way of knowing and its focus on the activity of interpretation, disability, as the despised under-figure of a "natural and normal" body can become the discursive space where we confront culture as it makes our embodied possibilities out of the limits that it has constructed disability and women to be.

NOTES

1. This paper is derived in part from a larger paper, "Clenched Subjectivity: Women, Disability and Medicine" (forthcoming, *DSQ*: Special issue on Disability and Technology, Goggin and Newell (Eds.)), which is part of a larger unpublished book-length manuscript, *Reading Disability Differently*, that I am currently completing.

2. On genre see Bakhtin (1986) and Smith's development in "Telling the Truth after Postmodernism" (1999: 96–130). On language game, see Wittgenstein (1976). For a self-reflective discussion of language *as* the problematic of our relation to both agency and our being an effect of subjection, see Butler (1997: 1–30) and Bonner (2001).

3. I am speaking here of the issue of "governmentality," or as Foucault puts it, the conduct of conduct which governs our governing of our selves (Martin et al., 1988: 16ff). On governmentality see also Pratt and Valverde (2002), Rose (1999), Ruhl (2002). For an analysis of some of the ways that embodiment is governed as this is reflected in disability discourse, see Corker (1998), Corker and Shakespeare (2002), Davis (2002), Stiker (1997), Titchkosky (2003b).

4. Ultrasound is today regarded as the authoritative text/image of many different aspects of pregnancy. See, for example, Carol Stabile and Valerie Hartouni's articles in *The Visible Woman: Imaging Technologies, Gender, and Science* (1998). Ultrasound is a technology that uses pulsating sound waves to produce an image of tissue. In relation to pregnant woman who have access to such health care procedures, ultrasound imaging has become routine. A "... scanner moves over the abdomen of the woman, sound waves penetrate the uterus; these waves bounce back to a monitor that produces the image, which can be captured immediately by the Polaroid camera attached to the machine" (Cook, 1996: 74).

5. Amniocentesis testing is a procedure that involves collecting a sample of amniotic fluid by inserting a needle through the pregnant woman's abdomen. Medicine conducts amniocentesis tests only in the light of a medical version of risk, such as the age of the woman, history of disability, illness, and, of course, an abnormal ultrasound image. While medicine conducts this test in the face of such risk, the test itself has its own risks, such as spontaneous abortion (Berebue, 1996: 40-94).

6. That the problem of disability is located at the level of the gene is an important issue that is just beginning to receive theoretic attention. See Wilson (2002).

7. A belief in the efficacy of medical tests on women's bodies and fetuses to reveal not only problems but forms of life is, as Ruth Hubbard (1997:197ff) argues, empirically questionable and even fallacious. Moreover, such tests are completely unable to foretell what *others will make of embodied human differences*. See Rapp (2002:129ff) for how disability imagery is actually lived in women's lives in a variety of ways.

8. See Corker and Shakespeare (2002), Corker (2001, 1998), Michalko (2002), Overboe (1999), or Titchkosky (2001) on theorizing the meaning of the split between personhood and disability, and between impairment and disability.
9. That this appears unruly and even monstrous is steeped in an understanding of woman as dangerous; woman blurs and confounds the clear-cut categories of self and other especially in matters of maternity. Margrit Shildrick (2002: 30ff) explicates the mother/other/monster cultural conceptions and shows how mothers "as a highly discursive category, have often represented both the best hopes and the worst fears of societies faced with an intuitive sense of their own instability vulnerabilities."

References

Bakhtin, Mikhail Mikhailovich. (1986). *Speech Genres and Other Late Essays.* Trans. Vern W. McGee. Austin: University of Texas Press.

Balsamo, Anne. (1999). "Forms of Technological Embodiment: Reading the Body in Contemporary Culture" in Margrit Shildrick and Janet Price (Eds.), *Feminist Theory and the Body: A Reader.* New York: Routledge.

Berube, Michael. (2000). "Biotech Before Birth: Review of Testing Women, Testing the Fetus: The Social Impact of Amniocentesis in America, by Rayna Rapp (Routledge 1999)" in *Tikkun.* Vol. 15 (3): 73–75.

Bonner, Kieran M. (2001). "Reflexivity and Interpretive Sociology: The Case of Analysis and the Problem of Nihilism" in *Human Studies* Vol. 24: 267–292.

Braidott, Rosi. (1997). "Mothers, Monsters, and Machines" in Katie Conboy, Nadia Medina, and Sara Stanbury (Eds.). *Writing on the Body: Female Embodiment and Feminist Theory.* New York: Columbia University Press. 59–79.

Butler, Judith. (1993). *Bodies That Matter.* New York: Routledge.

Butler, Judith. (1997). *The Psychic Life of Power.* Stanford: Stanford University Press.

Cook, Kay. (1996). "Medical Identity: My DNA/Myself" in Sidonie Smith and Julia Watson (Eds.), *Getting a Life: Everyday Uses of Autobiography.* Minneapolis: University of Minnesota Press. 63–88.

Corker, Mairian. (1998). "Disability Discourse in a Postmodern World" in Tom Shakespeare (Ed.), *The Disability Reader.* London: Cassel. 221–233.

Corker, Mairian and Tom Shakespeare (Eds.), (2002). "Mapping the Terrain" in *Disability/Postmodernity: Embodying Disability Theory.* London: Continuum. 1–17.

Davis, Lennard J. (2002). *Bending Over Backwards: Disability, Dismodernism & Other Difficult Positions.* New York: New York University Press.

Foucault, Michel. (1988). "Technologies of the Self" in Luther Martin, Huck Gutman, and Patrick H. Hutton (Eds.), *Technologies of the Self: A Seminar with Michel Foucault.* Amherst: The University of Massachusetts Press. 16–49.

Goggin, Gerard, and Christopher Newell. (2003). *Digital Disability: The Social Construction of Disability in the New Media.* Oxford: Rowman and Littlefield Publishers, Inc.

Haraway, Donna. (1990). "A Manifesto for Cyborgs: Science, Technology, and Socialist Feminism in the 1980's" in Linda J. Nicholson (Ed.), *Feminism/Postmodernism.* New York: Routledge.

Hartouni, Valerie. (1998). "Fetal Exposures: Abortion Politics and the Optics of Allusion" in Paula A. Treichler, Lisa Cartwright, and Constance Penley (Eds.), *The Visible Woman: Imaging Technologies, Gender, and Science.* New York: New York University Press. 198–216.

Hubbard, Ruth. (1997). "Abortion and Disability" in Lennard Davis (Ed.), *The Disability Studies Reader.* New York: Routledge. 187–200.

Martin, Luther, Huck Gutman, and Patrick H. Hutton (Eds.). (1988). *Technologies of the Self: A Seminar with Michel Foucault.* Amherst: The University of Massachusetts Press.

Merleau-Ponty, Maurice. (1964). *The Primacy Of Perception and Other Essays on Phenomenological Psychology, the Philosophy of Art, History and Politics.* Evanston: Northwestern University Press.

Michalko, Rod. (2002). *The Difference that Disability Makes.* Philadelphia: Temple University Press.

Michalko, Rod. (1998). *The Mystery of the Eye and the Shadow of Blindness.* Toronto: University of Toronto Press.

Overboe, James. (1999). "'Difference in Itself': Validating Disabled People's Lived Experience." *Body and Society.* Vol. 5 (4): 17–29.

Philp, Margaret. (2002). "Of Human Bondage: How the System Martyrs Parents of Disabled Kids." *Globe and Mail.* (February 16, 2002): F4–5.

Pratt, Anna, and Mariana Valverde. (2002). "From Deserving Victims to 'Masters of Confusion': Redefining Refugees in the 1990s" in *The Canadian Journal of Sociology.* Vol. 27 (2): 135–161.

Purdy, L.M. (1990). "Are Pregnant Woman Fetal Containers?" *Bioethics* 4 (4): 273–291.

Rapp, Rayna. (1993). "Accounting for Amniocentesis" in Shirley Lindenbaum and Margaret Lock (Eds.), *Knowledge and Practice: The Anthropology of Medicine and Everyday Life.* Berkeley: University of California Press.

Rapp, Rayna. (2000). *Testing Women, Testing the Fetus: The Social Impact of Amniocentesis in America.* New York: Routledge.

Rose, Nikolas. 1994. "Medicine, History and the Present" in Colin Jones and Roy Porter (Eds.), *Reassessing Foucault: Power, Medicine and the Body (Studies in the Social History of Medicine).* London: Routledge. 48–82.

Ruhl, Lealle. (2002). "Dilemmas of the Will: Uncertainty, Reproduction, and the Rhetoric of Control." *Signs: Journal of Women in Culture and Society.* Vol. 27 (3): 641–663.

Sawicki, Jana. (1999). "Disciplining Mothers: Feminism and the New Reproductive Technologies" in Margrit Shildrick and Janet Price (Eds.), *Feminist Theory and the Body: A Reader.* New York: Routledge.

Shildrick, Margrit. (2002). *Embodying the Monster: Encounters with the Vulnerable Self.* London: Sage Publications.

Shildrick, Margrit, and Janet Price. (1996). "Breaking the Boundaries of the Broken Body" in *Body and Society.* Vol. 2 (4): 93–13.

Smith, Dorothy E. (1999). *Writing the Social: Critique, Theory, and Investigations.* Toronto: University of Toronto Press.

Smith, Dorothy E. (1990). *The Conceptual Practices of Power: A Feminist Sociology of Knowledge.* Toronto: University of Toronto Press.

Stabile, Carol. (1998). "Shooting the Mother: Fetal Photography and the Politics of Disappearance" in Paula A. Treichler, Lisa Cartwright, and Constance Penley (Eds.), *The Visible Woman: Imaging Technologies, Gender, and Science.* New York: New York University Press. 171–197.

Stiker, Henri-Jacques. (1997). *A History of Disability.* Foreword by David T. Mitchell. Ann Arbor: University of Michigan Press.

Titchkosky, Tanya. (2002). "Cultural Maps: Which Way to Disability?" in Mairian Corker and Tom Shakespeare (Eds.), *Disability/Postmodernity: Embodying Disability Theory.* London: Continuum. 145–160.

Titchkosky, Tanya. (2001). "Disability—A Rose By Any Other Name? People First Language in Canadian Society." *Canadian Review of Sociology and Anthropology.* Vol. 38 (2): 125–140.

Titchkosky, Tanya. (2003a). *Disability, Self and Society*. Toronto: University of Toronto Press.

Titchkosky, Tanya. (2003b). "Governing Embodiment: Technologies of Constituting Citizens with Disabilities." *Canadian Journal of Sociology*. Vol. 28 (4): 517–542.

Weir, Lorna. (1998). "Pregnancy Ultrasound in Maternal Discourse" in Margrit Shildrick and Janet Price (Eds.), *Vital Signs: Feminist Reconfigurations of the Bio/logical Body*. Edinburgh: Edinburgh University Press. 78–101.

Wilson, James C. (2002). "(Re)Writing the Genetic Body-Text: Disability, Textuality, and the Human Genome Project." *Cultural Critique* Vol. 50: 23–39.

Wittgenstein, Ludwig. (1976). *Philosophical Investigations*. Trans. G. E. M. Anscombe. Oxford: Basil Blackwell and Mott, Ltd.

Zola, Irving Kenneth. (1977). "Healthism and Disabling Medicalization" in Ivan Illich et al. (Eds.), *Disabling Professions*. London: Marion Boyars Publishers Ltd.

Media Magic: Making Class Invisible

Gregory Mantsios

Most of us are sure of the existence of social inequality and social class in our society. Yet most of us are usually unsure about how income and wealth are distributed or where we are situated within the class system. Of what social class do you consider yourself a member? Upon what do you base this characterization? What ideas, attitudes, perceptions, or lifestyle choices do you share with other members of your social class?

In this article, Mantsios documents different ways social class becomes either confounded or rendered invisible within the media, and thus effectively removed from public view and debate. How does the media influence our perceptions of the wealthy, the poor, and ourselves? How are class differences depicted on television or in the movies? What function is served by making social class "invisible," particularly in the American mass media, which Mantsios is investigating?

Of the various social and cultural forces in our society, the mass media is arguably the most influential in molding public consciousness. Americans spend an average twenty-eight hours per week watching television. They also spend an undetermined number of hours reading periodicals, listening to the radio, and going to the movies. Unlike other cultural and socializing institutions, ownership and control of the mass media is highly concentrated. Twenty-three corporations own more than one-half of all the daily newspapers, magazines, movie studies, and radio and television outlets in the United States.[1] The number of media companies is shrinking and their control of the industry is expanding. And a relatively small number of media outlets is producing and packaging the majority of news and entertainment programs. For the most part, our media is national in nature and single-minded (profit-oriented) in purpose. This media plays a key role in defining our cultural tastes, helping us locate ourselves in history, establishing our national identity, and ascertaining the range of national and social possibilities. In this essay, we will examine the way the mass media shapes how people think about each other and about the nature of our society.

Source: Gregory Mantsios, "Media Magic: Making Class Invisible," *Race, Class, and Gender in the United States: An Integrated Study*, 4th Ed. St. Martin's Press, 1998. Reprinted with permission from the author.

The United States is the most highly stratified society in the industrialized world. Class distinctions operate in virtually every aspect of our lives, determining the nature of our work, the quality of our schooling, and the health and safety of our loved ones. Yet remarkably, we, as a nation, retain illusions about living in an egalitarian society. We maintain these illusions, in large part, because the media hides gross inequities from public view. In those instances when inequities are revealed, we are provided with messages that obscure the nature of class realities and blame the victims of class-dominated society for their own plight. Let's briefly examine what the news media, in particular, tells us about class.

ABOUT THE POOR

The news media provides meager coverage of poor people and poverty. The coverage it does provide is often distorted and misleading.

The Poor Do Not Exist

For the most part, the news media ignores the poor. Unnoticed are forty million poor people in the nation—a number that equals the entire population of Maine, Vermont, New Hampshire, Connecticut, Rhode Island, New Jersey, and New York combined. Perhaps even more alarming is that the rate of poverty is increasing twice as fast as the population growth in the United States. Ordinarily, even a calamity of much smaller proportion (e.g., flooding in the Midwest) would garner a great deal of coverage and hype from a media usually eager to declare a crisis, yet less than one in five hundred articles in the *New York Times* and one in one thousand articles listed in the *Reader's Guide to Periodic Literature* are on poverty. With remarkably little attention to them, the poor and their problems are hidden from most Americans.

When the media does turn its attention to the poor, it offers a series of contradictory messages and portrayals.

The Poor Are Faceless

Each year the Census Bureau releases a new report on poverty in our society and its results are duly reported in the media. At best, however, this coverage emphasizes annual fluctuations (showing how the numbers differ from previous years) and ongoing debates over the validity of the numbers (some argue the number should be lower, most that the number should be higher). Coverage like this desensitizes us to the poor by reducing poverty to a number. It ignores the human tragedy of poverty—the suffering, indignities, and misery endured by millions of children and adults. Instead, the poor become statistics rather than people.

The Poor Are Undeserving

When the media does put a face on the poor, it is not likely to be a pretty one. The media will provide us with sensational stories about welfare cheats, drug addicts, and greedy panhandlers (almost always urban and Black). Compare these images and the emotions evoked by them with the media's treatment of middle-class (usually white) "tax evaders," celebrities who have a "chemical dependency," or wealthy businesspeople who use unscrupulous means to "make a profit." While the behavior of the more affluent offenders is considered an "impropriety" and a deviation from the norm, the behavior of the poor is considered repugnant, indicative of the poor in general, and worthy of our indignation and resentment.

The Poor Are an Eyesore

When the media does cover the poor, they are often pictured through the eyes of the middle class. For example, sometimes the media includes a story about community resistance to a homeless shelter or storekeeper annoyance with panhandlers. Rather than focusing on the plight of the poor, these stories are about middle-class opposition to the poor. Such stories tell us that the poor are an inconvenience and irritation.

The Poor Have Only Themselves to Blame

In another example of media coverage, we are told that the poor live in a personal and cultural cycle of poverty that hopelessly imprisons them. They routinely center on the Black urban population and focus on perceived personality or cultural traits that doom the poor. While the women in these stories typically exhibit an "attitude" that leads to trouble or a promiscuity that leads to single motherhood, the men possess a need for immediate gratification that leads to drug abuse or an unquenchable greed that leads to the pursuit of fast money. The images that are seared into our mind are sexist, racist, and classist. Census figures reveal that most of the poor are white not Black or hispanic, that they live in rural or suburban areas not urban centers, and hold jobs at least part of the year.[2] Yet, in a fashion that is often framed in an understanding and sympathetic tone, we are told that the poor have inflicted poverty on themselves.

The Poor Are Down on Their Luck

During the Christmas season, the news media sometimes provides us with accounts of poor individuals or families (usually white) who are down on their luck. These stories are often linked to stories about soup kitchens or other charitable activities and sometimes call for charitable contributions. These "Yule time" stories are as much about the affluent as they are about the poor: they tell us that the affluent in our society are a kind, understanding, giving people—which we are not.* The series of unfortunate circumstances that have led to impoverishment are presumed to be a temporary condition that will improve with time and a change in luck.

Despite appearances, the messages provided by the media are not entirely disparate. With each variation, the media informs us what poverty is not (i.e., systemic and indicative of American society) by informing us what it is. The media tells us that poverty is either an aberration of the American way of life (it doesn't exist, it's just another number, it's unfortunate but temporary) or an end product of the poor themselves (they are a nuisance, do not deserve better, and have brought their predicament upon themselves).

By suggesting that the poor have brought poverty upon themselves, the media is engaging in what William Ryan has called "blaming the victim." The media identifies in what ways the poor are different as a consequence of deprivation, then defines those differences as the cause of poverty itself. Whether blatantly hostile or cloaked in sympathy, the message is that there is something fundamentally wrong with the victims—their hormones, psychological makeup, family environment, community, race, or some combination of these—that accounts for their plight and their failure to lift themselves out of poverty.

But poverty in the United States is systemic. It is a direct result of economic and political policies that deprive people of jobs, adequate wages, or legitimate support. It is neither natural nor inevitable: there is enough wealth in our nation to eliminate poverty if we chose to redistribute existing wealth or income. The plight of the poor is reason enough to make the elimination of poverty the nation's first priority. But poverty also impacts dramatically on the nonpoor. It has a dampening effect on wages in general (by maintaining a reserve army of unemployed and underemployed anxious for any job at any wage) and breeds crime and violence (by maintaining conditions that invite private gain by illegal means and rebellion-like behavior, not entirely unlike the urban riots of the 1960s). Given the extent of poverty in the nation and the impact it has on us all, the media must spin considerable magic to keep the poor and the issue of poverty and its root causes out of the public consciousness.

ABOUT EVERYONE ELSE

Both the broadcast and the print news media strive to develop a strong sense of "we-ness" in their audience. They seek to speak to and for an audience that is both affluent and like-minded. The media's solidarity with affluence, that is, with the middle and upper class, varies little from one medium to another. Benjamin DeMott points out, for example, that the *New York Times* understands affluence to be intelligence, taste, public spirit, responsibility, and a readiness to rule and "conceives itself as spokesperson for a readership awash in these qualities."[3] Of course, the flip side to creating a sense of "we," or "us," is establishing a perception of the "other." The other relates back to the faceless, amoral, undeserving, and inferior "underclass." Thus, the world according to the news media is divided between the "underclass" and everyone else. Again the messages are often contradictory.

The Wealthy Are Us

Much of the information provided to us by the news media focuses attention on the concerns of a very wealthy and privileged class of people. Although the concerns of a small fraction of the populace, they are presented as though they were the concerns of everyone. For example, while relatively few people actually own stock, the news media devotes an inordinate amount of broadcast time and print space to business news and stock market quotations. Not only do business reports cater to a particular narrow clientele, so do the fashion pages (with $2,000 dresses), wedding announcements, and the obituaries. Even weather and sports news often have a class bias. An all news radio station in New York City, for example, provides regular national ski reports. International news, trade agreements, and domestic policies issues are also reported in terms of their impact on business climate and the business community. Besides being of practical value to the wealthy, such coverage has considerable ideological value. Its message: the concerns of the wealthy are the concerns of us all.

The Wealthy (as a Class) Do Not Exist

While preoccupied with the concerns of the wealthy, the media fails to notice the way in which the rich as a class of people create and shape domestic and foreign policy. Presented as an aggregate of individuals, the wealthy appear without special interests, interconnections, or unity in purpose. Out of public view are the class interests of the wealthy, the interlocking business links, the concerted actions to preserve their class privileges and business interests (by running for public office, supporting political candidates, lobbying, etc.). Corporate lobbying is ignored, taken for granted, or assumed to be in the public interest. (Compare this with the media's portrayal of the "strong arm of labor" in attempting to defeat trade legislation that is harmful to the interests of working people.) It is estimated that two-thirds of the U.S. Senate is composed of millionaires.[4] Having such a preponderance of millionaires in the Senate, however, is perceived to be neither unusual nor antidemocratic; these millionaire senators are assumed to be serving "our" collective interests in governing.

. . .

The Middle Class Is Us

By ignoring the poor and blurring the lines between the working people and the upper class, the news media creates a universal middle class. From this perspective, the size of one's income becomes largely irrelevant: what matters is that most of "us" share all intellectual and moral superiority over the disadvantaged. As *Time* magazine once concluded, "Middle America is a state of mind."[5] "We are all middle class," we are told, "and we all share the same concerns": job security, inflation, tax burdens, world peace, the cost of food and housing, health care, clean air and water, and the safety of our streets. While the concerns of the wealthy are quite distinct from those of the middle class (e.g., the wealthy worry about investments, not jobs), the media convinces us that "we [the affluent] are all in this together."

The Middle Class Is a Victim

For the media, "we" the affluent not only stand apart from the "other"— the poor, the working class, the minorities, and their problems—"we" are also victimized by the poor (who drive up the costs of maintaining the welfare roles), minorities (who commit crimes against us), and by workers (who are greedy and drive companies out and prices up). Ignored are the subsidies to the rich, the crimes of corporate America, and the policies that wreak havoc on the economic well-being of middle America. Media magic convinces us to fear, more than anything else, being victimized by those less affluent than ourselves.

The Middle Class Is Not a Working Class

The news media clearly distinguishes the middle class (employees) from the working class (i.e., blue collar workers) who are portrayed, at best, as irrelevant, outmoded, and a dying breed. Furthermore, the media will tell us that the hardships faced by blue collar workers are inevitable (due to progress), a result of bad luck (chance circumstances in a particular industry), or a product of their own doing (they priced themselves out of a job). Given the media's presentation of reality, it is hard to believe that manual, supervised, unskilled, and semiskilled workers actually represent more than 50 percent of the adult working, population.[6] The working class, instead, is relegated by the media to "the other."

In short, the news media either lionizes the wealthy or treats their interests and those of the middle class as one and the same. But the upper class and the middle class do not share the same interests or worries. Members of the upper class worry about stock dividends (not employment), they profit from inflation and global militarism, their children attend exclusive private schools, they eat and live in a royal fashion, they call on (or are called upon by) personal physicians, they have few consumer problems, they can escape whenever they want from environmental pollution, and they live on streets and travel to other areas under the protection of private police forces.[7]

The wealthy are not only a class with distinct life-styles and interests, they are a ruling class. They receive a disproportionate share of the country's yearly income, own a disproportionate amount of the country's wealth, and contribute a disproportionate number of their members to governmental bodies and decision-making groups—all traits that William Domhoff, in his classic work *Who Rules America* , defined as character-istic of a governing class.[8]

This governing class maintains and manages our political and economic structures in such a way that these structures continue to yield an amazing proportion of our wealth to a minuscule upper class. While the media is not above referring to ruling classes in other countries (we hear, for example, references to Japan's ruling elite),[9] its treatment of the news proceeds as though there were no such ruling class in the United States.

Furthermore, the news media inverts reality so that those who are working class and middle class learn to fear, resent, and blame those below, rather than those above them in the class structure. We learn to resent welfare, which accounts for only two cents out of every dollar in the federal budget (approximately $10 billion) and provides financial relief for the needy), but learn little about the $11 billion the federal government spends on individuals with incomes in excess of $100,000 (not needy), or the $17 billion in farm subsidies, or the $214 billion (twenty times the cost of welfare) in interest payments to financial institutions.

Middle-class whites learn to fear African Americans and Latinos, but most violent crime occurs within poor and minority communities and is neither interracial nor interclass. As horrid as such crime is, it should not mask the destruction and violence perpetrated by corporate America. In spite of the fact that 14,000 innocent people are killed on the job each year, 100,000 die prematurely, 400,000 become seriously ill, and 6 million are injured from work-related accidents and diseases, most Americans fear government regulation more than they do unsafe working conditions.

Through the media, middle-class—and even working-class—Americans learn to blame blue collar workers and their unions for declining purchasing power and economic security. But while workers who managed to keep their jobs and their unions struggled to keep up with inflation, the top 1 percent of American families saw their average incomes soar 80 percent in the last decade.[10] Much of the wealth at the top was accumulated as stockholders and corporate executives moved their companies abroad to employ cheaper labor (56 cents per hour in El Salvador) and avoid paying taxes in the United States. Corporate America is a world made up of ruthless bosses, massive layoffs, favoritism and nepotism, health and safety violations, pension plan losses, union busting, tax evasions, unfair competition, and price gouging, as well as fast buck deals, financial speculation, and corporate wheeling and dealing that serve the interests of the corporate elite, but are generally wasteful and destructive to workers and the economy in general.

It is no wonder Americans cannot think straight about class. The mass media is neither objective, balanced, independent, nor neutral. Those who own and direct the mass media are themselves part of the upper class, and neither they nor the ruling class in general have to conspire to manipulate public opinion. Their interest is in preserving the status quo, and their view of society s fair and equitable comes naturally to them. But their ideology dominates our society and justifies what is in reality a perverse social order—one that perpetuates unprecedented elite privilege and power on the one hand and widespread deprivation on the other. A mass media that did not have its own class interests in preserving the status quo would acknowledge that inordinate wealth and power undermines democracy and that a "free market" economy can ravage a people and their communities.

NOTES

1. Lee, M., & Solomon, N. (1990). *Unreliable sources*. New York: Lyle Stewart, p. 71. See also Bagdikian, B. (1990). *The media monopoly*. Boston: Beacon Press.
2. Department of Commerce, Bureau of the Census. Poverty in the United States: 1992. *Current population reports, consumer income*. Series P60–185, pp. xi, xv, 1.
3. Demott, B. (1990). *The imperial middle*. New York: William Morrow, p. 123.
4. Barnes, F. (1990, January 5). The zillionaires club, *The New Republic*, p. 24.
5. *Time*. (1979, January 5), p. 10.
6. Navarro, V. (1992, March 23). The middle class—A useful myth, *The Nation*, p. 1.
7. Anderson, C. (1974). *The political economy of social class*. Englewood Cliffs, NJ: Prentice-Hall, p. 137.
8. Domhoff, W. (1967). *Who rules America*. Englewood Cliffs, NJ: Prentice-Hall, p. 5.
9. Lee & Solomon, *Unreliable sources*, p. 179.
10. *Business Week*. (1992, June 8), p. 86.

   * Eds. Note-American households with incomes of less than $10,000 give an average of 5.5 percent of their earnings to charity or to a religious organization, while those making more than $100,000 a year give only 2.9 percent. After changes in the 1986 tax code reduced the benefits of charitable giving, taxpayers earning $500,000 or more slashed their average donation by nearly one-third. Furthermore, many of these acts of benevolence do not help the needy. Rather than provide funding to social service agencies that aid the poor, the voluntary contributions of the wealthy go to places and institutions that entertain, inspire, cure, or educate wealthy Americans—art museums, opera houses, hotels, orchestras, ballet companies, private hospitals, and elite universities. (Robert Reich, "Secession of the Successful," *New York Times Magazine*, February 17, 1991, 43.)

   * Eds. Note-In 92 percent of the murders nationwide, the assailant and the victim are of the same race (46 percent are white/white, 46 percent are black/black), 5.6 percent are black on white, and 2.4 percent are white on black. FBI and Bureau of Justice Statistics, 1985-1986 quoted in Raymond S. Franklin, *Shadows of Race and Class*, University of Minnesota Press, Minneapolis, 1991, p. 108.)

The Super-Rich

Stephen Haseler

General statistics on income and wealth distribution provide a broad measure of the levels of inequality which exist in a society overall. But they also conceal the existence of a despairing underclass as well as an elite group of "super-rich" individuals and families who own and control vast amounts of wealth and property. In this article Haseler provides an overview of who the super-rich are and the resources they control. He argues that the current generation of super-rich is decidedly different from earlier generations of the wealthy in that they "owe no loyalty to community or nation." What are some of the consequences of increasing globalization, higher profits, and more millionaires and billionaires? What does Haseler mean when he implies that there ought to be a "change in the political climate of the Western world"? On a different note, can one ever be too rich?

THE SUPER-RICH

The end of the forty years of Cold War was more than the political triumph of the West over the Soviet Union. It was also more than the victory of freedom and pluralism over command communism. When the Berlin Wall cracked open and the iron curtain fell a new form of capitalism came into its own—global capitalism—and with it new global elite, a new class.

This new class already commands wealth beyond the imagination of ordinary working citizens. It is potentially wealthier than any super-rich class in history (including the robber barons, those 'malefactors of great wealth' criticised by Teddy Roosevelt, and the nineteenth-century capitalists who inspired the opposition of a century of Marxists). The new class of super-rich are also assuming the proportions of overlordship, of an overclass—as powerful, majestic and antidemocratic as the awesome, uncompromising imperial governing classes at the height of the European empires.

Source: Stephen Haseler, *The Super-Rich: The Unjust New World of Global Capitalism*, 2000, Palgrave Macmillan, NY. Reproduced with permission of Palgrave Macmillan.

The awesome new dimension of today's super-rich—one which separates them sharply from earlier super rich—is that they owe no loyalty to community or nation. The wealthy used to be bounded within their nations and societies—a constraint that kept aggregations of wealth within reason and the rich socially responsible. Now, though, the rich are free: free to move their money around the world. In the new global economy super-rich wealth (capital) can now move their capital to the most productive (or high profit, low cost) haven, and with the end of the Cold War—and the entry into global economy of China, Russia, Eastern Europe and India—these opportunities have multiplied. The super-rich are also free to move themselves. Although still less mobile than their money, they too are becoming less rooted, moving easily between many different locations.

MILLIONAIRES

Mobility is made possible by the lack of a need to work—a 'lifestyle' normally fixed in one nation or location for many years at a time. It is this escape from the world of work which effectively defines the super-rich. The lowest-ranking dollar *millionaire* household can, depending upon the interest and inflation rate, secure an *unearned* annual income of, say, $60 000 per year, which is almost double that of the median annual income of American families and four times that of the median income of British households.[1]

These millionaires are by no means lavishly well-off, particularly if they are in three- or four-people families or households. However they are financially independent—as one commentary put it, then can 'maintain their lifestyle for years and years without earning even one month's pay.'[2] It has been estimated that in 1996 there were as many as six million dollar millionaires in the world, up from two million at the end of the Cold War.[3] Over half of those—estimates claim about 3.5—are to be found in the United States.[4]

MULTIMILLIONAIRES

However these dollar millionaires find themselves at the *very* lower reaches of the world of the super-rich. Their homes and pensions are included in the calculations that make them millionaires, they often work—if not for a living, then for extras—and their lifestyles are often not particularly extravagant or sumptuous. They are, in fact, poor cousins in comparison with the more seriously rich families and individuals who are now emerging in the global economy. Official US statistics report that around a million US households—the top 1 per cent of total US households—possess a *minimum* net worth of over $2.4 million each and an average of $7 million each. In Britain the top 1 per cent have an average of around $1.4 million each.

The top half a per cent of US households, about half a million people, are staggeringly rich. This group has a *minimum* net worth of $4.7 million and an *average* of over $10 million each, which could produce an unearned annual income of over $600 000. In Britain the top half a per cent, around 48 000 households, have on average something like $2 million each—a fortune that can produce, again depending on interest and inflation rates, an unearned annual income of around $120 000 before tax and without working.[5]

These households are the truly super-rich, whose net worth, much of it inherited, is the source of considerable economic power and produces an income (mainly unlinked to work) that allows, even by affluent Western standards, extraordinarily sumptuous lifestyles. Estimates vary about the world-wide number of such super-rich families and individuals, but over two million in the plus $2.5 million category and over one million in the over $4.7 million (average $10 million) category would seem reasonable.[6]

Although huge amounts of the money of these multimillionaires are held outside the United States, in Europe, Asia and Latin America, this tells us nothing about the nationality of the holders.[7] In a sense these super-rich multimillionaires are the world's true global citizens—owing loyalty to themselves, their families and their money, rather than to communities and territorial boundaries—but reasonable estimates suggest that over half of them are American, and that most of the rest are European, with—certainly until the 1998 crash in Asia—a growing contingent from Asia.[8]

Their money is highly mobile, and so are they themselves, moving between their various homes around the world—in London, Paris and New York; large houses in the Hamptons in the United States, in the English and French countryside, and in gated communities in sun-belt America, particularly Florida, southern California and Arizona, and for the global super-rich the literal mobility of yachts in tropical paradises not scarred by local poverty.

MEGA-RICH AND BILLIONAIRES

Amongst multimillionaires there is a sharpish distinction to be made between those at the lower end—say the $20 million net worth households—and those at the higher end—say the $500 million plus households. The distinction is one of power, not lifestyle. From most perspectives the income from $20 million (say $1 million)—about 70 000 US households in 1994—can, at least on the face of it, produce the same kind of lifestyle as income from the net worth of the more serious multimillionaires (there is arguably a limit to the number of homes, yachts and cars that can be enjoyed and consumed in a lifetime).[9] $50 million in net worth, however, simply does not command as much economic power—over employment, over small businesses—than do the resources of the big time multimillionaires, much of whose money is tied up in big transnational corporations.

At the very top of this mega-rich world are the dollar billionaires, those who command over $1000 million in net worth, a fortune that can secure an unearned annual income, depending on inflation and interest rates, of $50 million a year before tax—staggeringly well over 1000 times more than the average US income. In 1997 estimates of the number of these ultra-super-rich individuals varied from 358 to 447 world-wide, and the number is growing fast, virtually doubling during the few years of the post Cold War era.[10]

WHO ARE THE BILLIONAIRES?

The 400 or so billionaires in the world are a varied lot. In one sense they are like the rest of us (and like those who will read this book). They are overwhelmingly Western, primarily American or European, and male, but they represent no single ethnic group, no single social background, and certainly possess no single business or financial secret for acquiring these awesome fortunes.

Many of the billionaires, though, would not be in the mega-rich category without an inheritance—which remains the most well-trodden route to great multimillion dollar wealth. Of the top 400 wealthiest people in the United States, 39 made the list through inheritance alone and many of the others had some inheritance to help get them started.[11] The British queen, Elizabeth Windsor, is perhaps the most famous example of such massive unearned wealth. In 1997 Phillip Beresford (*The Sunday Times'* 'Rich List', (*Sunday Times* 6 April 1997) put her net worth at a staggering $10.4 thousand million in 1992 (double the 1997 figure for top-listed Joseph Lewis). However, after she took a rival 'rich list' to the Press Complaints Commission over its valuation of her assets, *The Sunday Times'* Wealth Register excluded from its calculations the royal art collection, which, had it been included, would have given her a $16 billion figure, making her the world's wealthiest woman and the second wealthiest person in the world, with half the net worth of the Sultan of Brunei but more than the Walton family.[12]

In contrast to the inheritors, there are some 400 'self-made' mega-rich men (there are no women). Yet even these men of merit have not necessarily made their inordinate fortunes through extraordinary amounts of work and talent—certainly not its continuous application. Many of the self-made mega-rich are certainly talented and creative (and often ruthless), but many of them have become mega-rich through one-off bursts of insight or risk or luck.

William (Bill) Gates is seen as 'self-made', very much the American entrepreneurial hero. His vast resources—*Newsweek* calls him 'the Croesus of our age'—have been built upon the meritorious image of having run a successful company which provides a real service, a real addition to human understanding and communication. His huge net worth—he was listed in 1997 by *Forbes* magazine as the richest American at $36.4 billion—is based upon the value of his shares in his company Microsoft. It was Gates' original burst of imagination that created his fortune—the initial stock offering in 1986 of 100 Microsoft shares cost $2100 but by the first trading day in August 1997 this had risen to 3600 shares at $138.50 each. Gates' personal share of the company rose from $234 million to $37.8 billion in the same period.[13] Certainly Gates has managed the company and taken many crucial decisions. Yet as Microsoft grew he needed the more 'routine' skills of thousands of major company directors—such as managerial aptitude and the ability to stave off competition. As with all established businesses, less and less risk and less and less creativity was needed (and a junior hospital doctor probably put in more hours).

Paul Raymond is a different type of self-made billionaire. Described by academic John Hills as Britain's richest man—in 1995 he placed him ahead of Joseph Lewis—Raymond's fortune is thought to be well over £1.65 billion. Having founded Raymond's revue bar in the Soho district of London, with topless dancers, he made his money by investing in soft pornography and property.[14] Like Gates he had the talent to spot a coming market—albeit one that was less elevating and educational. And also like Gates, and the other mega-rich, once the original burst of inventiveness (perhaps amounting only to the one great insight) was over the rest of his working life has consisted of simply managing his empire and watching his money grow. . . .

COMPARISONS

This group of late twentieth-century billionaires not only dwarf their 'ordinary' super-rich contemporaries but also the earlier race of mega-rich 'robber barons' who were so identified with the burgeoning capitalism of the early twentieth-century. In terms of resources at their personal command, in 1997 William Gates, was three times richer than John D. Rockefeller (Standard Oil) was in 1918, Warren Buffet was over ten times richer than Andrew Carnegie (Steel) was in 1918, and it was estimated that in 1992 the British queen was ten times richer than Henry Ford (automobiles) was in 1918, although some of these early-twentieth-century super-rich probably commanded a greater percentage of their nations' resources.[15]

The resources at the disposal of these super-rich families— a huge pool of the globe's wealth—are truly astounding, beyond the wildest imaginings of most of the affluent Western middle classes. These high net worth individuals (HNWI's, as they are depicted in the financial services sector that serves them) accounted for almost $17 trillion in assets in 1996.

The power—that is, command over resources—of the world's super-rich is normally expressed in raw monetary figures, but the sheer, egregious extent of these private accumulations of wealth can also be given some meaning by making comparisons.... Eighty four of the world's richest people have a combined worth greater than that of China.[16] So the wealth of just one of these super-rich individuals is equal to that of about 12.5 million of his fellow humans.

Just as awe-inspiring is the fact that the total wealth of the world's few hundred billionaires equals the combined income of 45 per cent of the planet's population.[17] It is also somewhat sobering to realise that the *individual* wealth of the world's billionaires can exceed the gross national product of whole nations.[18] The world's ten richest billionaires all individually possess more in wealth than the GNP of many nation-states. The world's richest individual, the Sultan of Brunei, weighing in at over $45 billion, commands more resources than the combined GNP of 40 nation-states. To give his wealth some form of reality, it is larger than the GNP of the Czech republic (population 10.3 million); while William Gates commands more resources than the GNP of Africa's oil-rich giant, Nigeria (with a population of 111.3 million); the Walton family commands over $27.6 billion, more than the GNP of Vietnam (peopled by 73.5 million); Paul Sacher and the Hoffman family command over $13 billion, more than the GNP of Bulgaria (population 8.4 million); Karl and Theo Albrecht command over $8 billion, more than the GNP of Panama (with its 2.6 million inhabitants); Joseph Lewis, the highest ranking mega-rich British citizen, commands just under $5 billion, which gives him more control over resources than his country of residence, the Bahamas.[19]

Another way of grasping the huge personal agglomerations of wealth in the modern global economy is to compare income levels. On 1997 interest-rate figures, and assuming that all assets are not income producing, the Sultan of Brunei could easily receive from his assets something in the region of $3 billion a year as income—compared with an average of $430 per person in the 49 lowest-income nation-states, $2030 per person in the 40 middle-income nation-states, $4260 in the 16 upper-middle income states and $24 930 in the 25 highest income economies....

Get the world's top three mega-rich (dollar billionaire) people into one room and you would have assembled command over more resources than the GNP of Israel; the top four and you would tie with Poland, the top ten and you would beat Norway and South Africa. Europe's richest

20 families command around $113 billion, a little more than the whole Polish economy; America's richest 10 ($158 billion) and Britain's richest 1000 families ($156 billion) together command more resources than the GNP of the entire Russian Federation.[20]

If the top 200 or so billionaires could ever be assembled together then the command over assets, in that one room, would outrank the GNP of each of Australia, the Netherlands, Belgium, possibly even Brazil; and with 400 or so billionaires the one gathering would outrank Britain and almost overtake France!

It is these kinds of statistic that bring into sharp focus the economic power limitations of elected presidents and prime ministers (and other public sector officials)—who also have to share their economic power with cabinets and parliaments—compared with the economic power of the unelected mega-rich, whose only accountability is to the market. Such economic power was on display when the American media billionaire Ted Turner decided to donate $1 billion to the United Nations and 'to put on notice ... every rich person in the world ... that they're going to be hearing from me about giving money'.[21] For a Western politician to move a billion dollars in the direction of the UN would have involved months and months of negotiating and a bruising campaign.

All of our four categories of the world's super-rich (the 'ordinary' millionaires with up to $2.5 million, those with $2.5-5 million, those with $5-1000 million, and the billionaires with over $1000 million) have a combined net worth of $17 trillion, more than double the GNP of the United States.[22]

Just as awe-inspiring is the proportion of national wealth of the Western nations held by their own passport-holding super-rich.[23] In 1995 in the US the amount of wealth (total net worth) held by 90 per cent of American households—everyone under the top 10 per cent—came to only 31.5 per cent, whereas the top 10 per cent of American households own 69.5 per cent of the US. More striking still, the top 1 per cent of Americans hold 35.1 per cent of US wealth, and the top half a per cent of households (500 000 households), those with a minimum net worth of $4.7 million, own 27.5 per cent of the US.

In Britain too the super-rich also own a huge proportion of the net worth of their country.[24] In 1992 the top 10 per cent of Britons owned half of the country's marketable wealth (for the top US 10 per cent the 1995 figure was a whopping 69.5 per cent). The wealthiest 5 per cent of Britons owned around 37 per cent of Britain's marketable wealth. The top 1000 super-rich families in Britain own about $160 billion worth of wealth, about the same average (0.16 billion each) as the top half a per cent in the US; Britain's top 100 command $89 billion, its top 50 own $69 billion and the top 20 own $42 billion.[25]

Among the 1997 British 'top twenty' Joseph Lewis (finance) was estimated to have a net worth of $4.8 thousand million; Hans Rausing (food packing) came just behind with $4.72 thousand million; David Sainsbury (retailing) and Garfield Weston and family (food production) third with $4 thousand million each; Richard Branson (airline, retailing and entertainment), Sir Adrian and John Swire (shipping and aviation) and the Duke of Westminster (landownership) all joint fifth with $2.72 thousand million each; Lakshimi and Usha Mittal (steel) eighth with $2.4 thousand million; and Joe and Sir Anthony Bamford (construction equipment) and Viscount Rothermere (newspapers) joint ninth with $1.92 thousand million.

A particular feature of the British super-rich scene is the concentration in very few hands of land ownership. Britain—or rather the land area known as the United Kingdom—is, quite literally, owned by a very small caste; as is the capital city, London. It remains a poignant commentary on wealth concentration that large tracts of London are owned by just a few individuals. The Duke of Westminster, through the Grosvenor Estate, owns around 200 acres of Belgravia and 100 acres of Mayfair—a dynastic inheritance created by the seventeen[th]-century marriage of Cheshire baronet Thomas Grosvenor to Mary Davies, the '12 year old heiress to a London manor that at the time included 200 acres of Pimlico'. Viscount Portman owns 110 acres north of Oxford Street. Lord Howard de Walden's four daughters, through a holding company, own 90 acres of Marylebone. Elizabeth Windsor, the queen, remains the 'official' owner of 150 acres of 'crown estates' in central London, as the eight crown estates in central London, as the eight crown estates commissioners address their annual report to her. Andrew Lycett has argued that although 'millions of pounds are exchanged every week in leasehold property deals ... London still has no sizable new landowners' with the exception of the Sultan of Brunei and Paul Raymond.[26]

RICHER STILL, YET RICHER

And the super-rich are getting richer. The former vice chairman of the US Federal Reserve Board said in 1997 that 'I think when historians look back at the last quarter of the twentieth century the shift from labour to capital, the almost unprecedented shift of money and power up the income pyramid, is going to be their number one focus.'[27] The figures are indeed dramatic. In the US the top half a per cent rose from 23 per cent to 27.5 per cent between 1989 and 1995. The next half a per cent rose from 7.3 per cent to 7.6 per cent in same period. However the next 9 per cent fell from 37.1 per cent [to] 33.2 per cent, while the lowest 90 per cent fell from 32.5 per cent to 31.5 per cent. As the most reliable and scholarly analysis put it, the evidence shows 'a statistically significant increase in the share of household net worth held by the wealthiest half a per cent of [US] households from 1992 to 1995'.[28]

There are no figures available for the British top half a per cent, but tax authority figures—which do not include the considerable amounts of offshore money held by the British-passport-holding rich—suggest that whereas the top 1 per cent of the population were losing ground between 1950 and 1980, during the Thatcherite, globalising 1980s and 1990s their share of the wealth of Britain stabilised.[29]

And the assets held by the world-wide super-rich (the HNWIs) are expected to continue to grow. One assessment portrays them as more than doubling (from $7.2 trillion in 1986 to $15.1 trillion in 1995), and they are projected to grow from the 1996 level of $17 trillion to $25 trillion (up by more than 50 per cent) by the new millennium. . . .

AN OVERCLASS?

If the new global super-rich do not amount to an old-style ruling class, they are certainly becoming an overclass: the mirror image of the more publicised urban underclass—separated from the rest of us, with distinct interests that differ from those of the mass of the peoples of Western societies.

In a very real sense the new super-rich are becoming removed from their societies. This is happening physically. The higher levels of the super-rich have always lived apart: within their walled estates or in wealthy ghettos in the centre of Manhattan, London and other cities. They have always owned possessions that have singled them out. Today, of course, mere diamonds, helicopters and expensive cars no longer signify the apex of great wealth. Now it is the luxury yacht (normally personally designed by John Banneman), the personal aeroplane—the Sultan of Brunei has a Boeing 747—(normally supplied by Grumanns), and one of two of the highest valued paintings that signify someone has reached the top. . . .

Of course one test of loyalty to a society is a willingness to pay its taxes, particularly if they are not onerous. Yet increasingly the super-rich are dodging the taxes of their countries of origin. In 1997 the *New York Times* reported that

> nearly 2,400 of the Americans with the highest incomes paid no federal taxes in 1993, up from just 85 individuals and couples in 1977. While the number of Americans who make $200,000 or more grew more than 15 fold from 1977 to 1993, the number of people in that category who paid no income taxes grew 28 fold or nearly twice as fast, according to a quarterly statistical bulletin issued by the IRS.[30]

So difficult was it for the US authorities to collect taxes from the super-rich that Congress introduced a new tax altogether—the Alternative Minimum Tax—to catch them.[31] With the American 'middle classes'—the middle income groups—paying a larger percentage of their earnings in taxes (including sales taxes, property taxes and social security payroll taxes), tax evasion and avoidance is becoming a growing cause of economic inequality and social fracture. . . .

'THE WORLD IS IN THE HANDS OF THESE GUYS'

The emergence of this global overclass not only raises the question of equality—or inequality—but also of power. Supporters of this new 'free market' global capitalism tend to celebrate it as a force for pluralism and freedom; yet so far these egregious aggregations of assets and money have placed in very few hands enormous power and influence over the lives of others. Through this accumulation of assets and money the super-rich control or heavily influence companies and their economic policies, consumer fashions, media mores, political parties and candidates, culture and art.

What is more the resources at the disposal of many of these super-rich individuals and families represent power over resources unattained by even the most influential of the big time state politicians and officials—'the panjamdrums of the corporate state' who populated the earlier, more social democratic era, and who became targets of the new capitalist right's criticism of the abuse of political power.

In the new capitalist dispensation it is the global super-rich who are 'lords of humankind', or 'lords of the Earth' like Sherman McCoy in Tom Wolfe's all too apt social satire on Wall Street, *Bonfire of the Vanities*, wielding power like old-fashioned imperial pro-consuls. The new global super-rich have now got themselves into a position where they not only have a 'free market' at their disposal, and not only is this market now global, but they can also command the support of the world's major governments. . . .

ONWARD AND UPWARD

The new super-rich global overclass seems to be possessed of one crucial attribute: a sense of ultimate triumph. As globalisation has proceeded all the bulwarks of social democracy that stood in their way, the cultures that acted as a balancing force and succeeded in civilising, and to some extent domesticating, raw capitalism have fallen. The primary casualty has been the nation-state and its associated public sector and regulated markets. The global economy has also helped to remove that other crucial balancing power available in the Western world—trade unions—which for the most part acted to check unbridled business power and ensure some basic rights to employees, often at the expense of rises in short-term money incomes.

Finally, the end of the Cold War was a seminal moment and played a fateful role as midwife. At one fell swoop the end of command communism (in Eastern Europe, in Russia and, in the economic field, in China) made footloose capital both possible and highly attractive by adding a large number of low-cost production and service centres and new markets to the economy. It also removed the need for the Western super-rich to be 'patriotic' (or pro-Western). It also made redundant the instinct of social appeasement held by many Western capitalists and induced by the need, in the age of Soviet communism, to keep Western publics from flirting with an alternative economic model.

The stark truth is that not one of these obstacles—not the public sector, not the trade unions, not an alternative economic and social model—is ever likely to be reerected. In the short to medium term, without a change in the political climate of the Western world there is nothing to stop further globalisation, higher and higher profits, more and more millionaires. For the new overclass it is onward and upward.

NOTES

1. The median income of US families was about $37 000 in 1993. US Census Bureau, Income and Poverty, CD-ROM, table 3F (1993). The median income of UK households was about $16 500 (The exchange rate used here is $1.6 to the pound) in 1990 at 1993 prices. See John Hills, *Income and Wealth*, vol. 2 (Joseph Rowntree Foundation, Feb. 1995).

2. Thomas J. Stanley, and William D. Danko, *The Millionaire Next Door* (Atlanta, GA: 1997). Some scholars have suggested defining 'the rich' not in terms of millions but rather as those with a family income over nine times the poverty line—in US terms about $95 000 a year in 1987. See S. Danziger, P. Gottschalk and E. Smolensky, 'How The Rich Have Fared, 1973-87', *American Economic Review*, vol. 72, no. 2 (May, 1989), p. 312.

3. The US Finance House Merrill Lynch in conjunction with Gemini Consulting, 'World Wealth Report 1997' (London: Merrill Lynch, 1997).

4. Stanley, and Danko, *The Millionaire Next Door*, op. cit., p. 12.

5. US figures for 1995 from Arthur B. Kennickell, (board of governors of the Federal Reserve System) and R. Louise Woodburn, 'Consistent Weight Design for the 1989, 1992 and 1995 SCF's and the Distribution of Wealth', revised July, 1997 unpublished. The UK figures are for 1993-4. For the UK figures, which include pensions, see Hills, *Income and Wealth*, op. cit., ch. 7.

6. Merrill Lynch, 'World Wealth Report, 1997', op. cit.

7. The 'World Wealth Report, 1997' (Merrill Lynch, op. cit,) projected, before the late 1997 Asian economic decline, that in 2000 the division of high net worth assets by source region would be Europe 7.1, North America 5.8, Asia 6.1, Latin America 3.8, Middle East 1.2 and Africa 0.4.

8. Of these multimillionaire Americans, families of British (that is English, Scottish, Welsh and Irish) and German descent account for 41.3 per cent of the total.

9. *Newsweek*, 4 Aug. 1997 (source IRS).

10. The UN *Human Development Report* (1966) put the figure at 358, and *Forbes* magazine's 1997 wealth list put the figure at 447, up from 274 in 1991.

11. *Newsweek* 4 Aug. 1997 (source *Forbes*, op. cit.).

12. See also, Phillip Hall, *Royal Fortune: Tax, Money and The Monarchy* (London: 1992) for a systematic account of the mysteries of the royal finances. One fact about the Queen's money remains: since 1998 she has remained above the law as far as taxation is concerned as she is not treated in exactly the same way—with all tax laws applying to her—as every other British person.

13. *Newsweek*, 'The New Rich', 4 Aug. 1997.

14. Hills, *Income and Wealth*, op. cit., p. 9. Hills suggests that 'If Britain's richest man, Soho millionaire Paul Raymond, receives a modest 3 per cent net real return on his reported £1.65 billion fortune' his income would be £1 million a week.

15. Figures from *Newsweek*, 4 Aug. 1997, reporting *Forbes* in June 1997. The figures for the Queen were for 1992 (as published in *The Sunday Times*' 'Rich List', 1997), and were subsequently revised downwards following a complaint to the Press Complaints Commission.

16. John Gray, 'Bill Rules the World—And I Don't Mean Clinton', *Daily Express*, 11 Sep. 1998.

17. UN, *Human Development Report*, (1966). Comparing wealth with income is highly problematic, but nonetheless serves to display the enormity of the comparison. These comparisons—between asset net worth and gross national product (GNP) are not of course comparing like with like, but are used in order to show the extent of the egregious financial and economic power of the high net worth individuals. The most reasonable method of comparison would be to compare the net worth of super-rich individuals and groups of super-rich individuals with the total net worth of each country (That is, each individual/family in the country). These figures are not available for more than a handful of countries.

18. 'Billion' here and throughout the book is used in the US sense that is, nine noughts.

19. Wealth figures from *The Sunday Times*', 1997 'Rich List', op. cit., population figures for 1995 from *World Development Report* (Washington, DC: World Bank, 1997).

20. The US figure is from *Forbes*, June 1977, and the European and British from *The Sunday Times*, 6 April, 1977. For GNP figures see World Bank, *The World Atlas*, op. cit.

21. *Guardian*, 23 Sep. 1997.

22. These estimates are based upon the net worth estimates cited in *Forbes* magazine, June 1977, and in 'The Wealth Register', compiled by Dr Richard Beresford for *The Sunday Times* (extracts published in *The Sunday Times*, 6 April 1997), who also cites *Forbes* magazine. *The World Atlas*, op. cit.

23. As I argue throughout this book, the super-rich are in reality global; but they all need a passport, and we are talking here about US passport holders.

24. The percentage of net worth of the total marketable net worth of all British passport holders.

25. *The Sunday Times*, 6 April 1997. This is the British billion, that is, 12 noughts as opposed to the US nine noughts.

26. Andrew Lycett, 'Who Really Owns London?', *The Times*, 17 Sep. 1997.

27. Alan Blinder, former vice chairman of the US Federal Reserve, quoted in *Newsweek*, 23 June 1997.

28. Figures from 'Consistent Weight Design for the 1989, 1992 and 1995 SCF's and the Distribution of Wealth' by Arthur Kennickell (Federal Reserve System) and R. Louise Woodburn (Ernst and Young), revised July 1997 (unpublished). Figures derived from the Survey of Consumer Finances sponsored by the US Federal Reserve System and the Statistics of Income Division of the IRS.

29. See Charles Feinstein, 'The Equalising of Wealth In Britain Since The Second World War', *Oxford Review of Economic Policy*, vol. 12, no. 1 (Spring 1996), p. 96 ff. In British estimates distinctions tend to be made between marketable wealth and total wealth—marketable wealth excludes state pensions, occupational pensions and tenancy rights.

30. *New York Times*, 18 April, 1997.

31. The US Alternative Minimum Tax is levied on those who have substantial incomes but, because of their use of tax shelters and exemptions, submit a zero tax return.

Relocating Law: Making Corporate Crime Disappear

Laureen Snider

What is the "Official Version of Law?" Who and what are subject to this Law? In this article Laureen Snider argues that justice is not blind, and the law is not equally binding on all. Corporations, in particular, are less and less subject to the law. While corporate crime has not ceased to be a problem, in her words, it is "disappearing." Where, then, has it gone?

A cursory content analysis of most newspapers and magazines appears to support her argument. While there are some very high-profile cases of corporate and white-collar crime in the news, much of what is reported is framed in the moral language of "scandals" rather than the judicial language of "crimes," portraying the wrongdoing as a breach of ethics more than law, even when laws have been violated.

Internationally, corporations have the legal status of "persons." Should institutions such as corporations be as accountable and liable as people?

The study of corporate crime provides a dramatic illustration of the influence hegemonic class interests exert over law. Many ... essays ... show how difficult it is to apply the promises of Western legal systems embodied in what is called "the Official Version of Law"—its claims of universalism and equality, the notion that justice is "blind" to class, race and gender; in other words, to workers, people of colour and women, groups with little ideological and/or economic power. But the most cursory attempt to use the Official Version of Law to explain the passage or enforcement of laws governing the antisocial, acquisitive acts of business quickly illustrates the overwhelming, and overt, failure of law to impose meaningful limits on the offences of corporations. State law has consistently, though not exclusively or simplistically, accommodated the interests of Canada's economic elites, particularly where they spoke with

Source: Laureen Snider, "Relocating Law: Making Corporate Crime Disappear," from Elizabeth Cormack (editor), *Locating Law: Race, Class and Gender Connections.* Fernwood Publishing, 1999. Reprinted with permission.

unanimity on issues they saw as important. In the last two decades, this has meant that the few laws and weak enforcement mechanisms which were enacted to control corporate crime—themselves the product of a century of struggle by employees, consumers, feminists, environmentalists and others—have been systematically repealed and dismantled. In direct contrast to state law in every other jurisdiction, which has become ever more intrusive and increasingly punitive (with rates of incarceration spiralling for all traditional offences despite falling crime rates (Rothman 1995; Snider 1998)), laws governing corporate crime have become more lax and lenient, and many types of corporate crime have entirely disappeared from the law books.

This disappearance has not happened because corporate crime—defined as "white-collar crimes of omission or commission by an individual or group of individuals in a legitimate formal organization—which have a serious physical or economic impact on employees, consumers or the general public" (Box, 1983:20)—has ceased to be a problem. The toll of lives lost, injuries sustained, species obliterated, watercourses decimated, savings and pensions destroyed and life chances ruined by the various types of corporate crimes has dramatically increased with the advent of the global marketplace, the spread of capitalist workplaces and production to Third World countries, the new-found dominance of finance capital, and the decline in the power of the nation-state. It has happened, rather, because a successful corporate *counter-revolution* has succeeded in reversing progress towards a more egalitarian society, by taking back "such gains as the working classes have made" (Glasbeek, 1995: 112). In the latter half of the 1970s, the margin of profitability—that is, the surplus value accruing to capital determined by the difference between the all-in costs of production and the all-in profits of production (per unit)—began to decline. From roughly 1980 on, signalled by the election of right-wing governments under Ronald Reagan in the United States and Margaret Thatcher in the United Kingdom, the owners and controllers of capital and their allies in political, media and knowledge elites have waged a highly successful campaign to increase the margin of exploitation, the absolute and relative profitability of capitalist enterprises. This has meant decreasing the incomes and life chances of the bottom 75 to 80 percent of the population to benefit the top one to two percent.

Thus we have observed that, over the last twenty years, rates of inequality have dramatically increased. In the bellwether United States from 1977–89, the top one percent of American families received 60 percent of after-tax income gains while the incomes of the bottom 40 percent went down, in real (absolute) as well as relative dollars (Miyoshi, 1993: 738). In Washington, D.C., for example, the top 20 percent of families today have incomes 28 times larger than the poorest 20 percent. The bottom quintile take home an average of $5290 per year, nearly $2000 less than they received 20 years ago (*Globe and Mail,* December 23, 1997, A13). In Canada, between 1977 and 1991, the average total family income for the bottom fifth of families stood at $17,334 in 1996, a 6.1 percent share of the national income, the lowest in two decades, while the average income of the top fifth rose to $114,874, a 40.6 percent share, the highest in two decades (Statistics Canada, 1997). Meanwhile the size, power and profitability of corporations has dramatically increased—the total wealth of the world's 385 billionaires equals the combined incomes of 45 percent of the world's population or 2.3 billion people (United Nations, 1996). Corporate salaries, perks and bonuses totalling $3 million a year for Chief Executive Officers are not uncommon in Canada; in the United States CEOs routinely get double and triple this (*Globe and Mail,* May 5, 1998: B16; June 22, 1998: A1). World-wide, the top 20 percent of the world's population increased their share of total global wealth from 70 percent in 1960 to 85 percent in 1991, and the share "enjoyed" by the poorest 20 percent actually *declined* from 1960 to 1991, falling from 2.3 percent to 1.4 percent (United Nations 1996). "The richest 200 largest corporations [now] have more economic clout than the poorest four-fifths of humanity" (Dobbin, 1998: 74-5).

Getting rid of laws to control the acquisitive and antisocial acts of business, in general, and corporations, in particular, has been an essential component of this counter-revolutionary reversal. Historically, one of the main tasks of the nation-state has been protecting citizens from the harm caused by corporations. This was, however, a duty that capitalist states undertook with the utmost reluctance. Typically such laws were passed only after major environmental or industrial disasters made some sort of state response imperative (Snider, 1991 and 1993). Thus, it is not surprising that national governments in the 1980s and 90s, facing deficits and declining revenues (caused, in large part, by another successful corporate initiative, this one reducing the tax rates of corporations and the rich), have been quick to repeal laws which restrict and criminalize potentially profitable acts and thereby annoy the corporate sector. For corporate crimes—ignoring costly regulations on mine safety, not paying overtime wages, marketing drugs with harmful side-effects, conspiring to increase the price of necessary goods—are always committed in order to increase profits or prevent losses (and they usually do). By abandoning efforts to prevent, monitor or sanction such acts, the nation-state signals its acquiescence to corporate (and corporatist) agendas and acknowledges its inability (or disinclination) to protect its own citizens from the predations of the global marketplace. In the race to achieve maximal profitability, destroying profit threatening measures such as environmental restrictions and the minimum wage shows that a country is, in the immortal words of former Prime Minister Brian Mulroney, "open for business." It signals that a century of struggle to use the laws of the nation-state to force capital to meet certain standards of behaviour and impose limits on the exploitation of human and natural resources has been abandoned.

. . .

THE DISAPPEARANCE OF CORPORATE CRIME

Item:

Bolar Pharmaceutical Company admitted to selling adulterated and mislabelled drugs, and lying to investigators from the federal (U.S.) Food and Drug Administration about the quality and origin of the medicines (*New York Times,* March 24, 1991: 26; *Orlando Sentinel,* February 28: A17).

Mer/29 (triparanol), a drug developed by Richardson Merrell to reduce cholesterol levels, went on sale in [the] United States in 1960. When skin damage, cataracts and changes in reproductive organs were reported in patients, investigation revealed that Richardson Merrell knew about these side effects (through animal tests done in its own labs) but suppressed the damaging evidence (Clarke, 1990: 205).

Every year thousands of Canadians suffer adverse drug reactions; ineffective, impure and unsafe drugs cause much human anguish and cost millions of dollars. From the Dalkon Shield (an intrauterine device that caused miscarriages, sterility, pelvic infection and several deaths) to the Meme breast implant (a silicon gel coated with poly-urethane foam that decomposed under certain conditions to produce a dangerous chemical, finally banned by Health and Welfare Canada in 1992), to defective heart valves and surgical gloves (more than half of the latex medical and surgical gloves failed quality tests), un-monitored, unassessed medical devices have "killed, mutilated, electrocuted, blinded, burned and injured hundreds, if not thousands, of Canadians["] (Regush, 1991: 9). Every year hundreds of new medical devices, products ranging from heart valves to incubators, are brought into Canada, a business worth more than $2 billion with 300,000 medical devices produced by some 6595 manufacturing companies, the vast majority in Third World countries (Regush, 1991: 16).

However, in the summer of 1997 Health and Welfare Canada closed down its research laboratories, known as the Bureau of Drug Research, eliminating 68 jobs and saving the federal government $2 million (*Globe and Mail,* July 11, 1997: A1). Laboratories and programs in the food directorate, which sets standards for acceptable levels of chemical residues or growth hormones in food, have also been cut. The Bureau of Drug Research had been the agency responsible for investigating the safety and effectiveness of drugs sold in Canada, and for monitoring problems after medications were approved. Drug safety will henceforth depend on the validity of research conducted by pharmaceutical companies, and their truthfulness in reporting adverse findings and risks. This research is either conducted by scientists employed by these companies, or on contract to them.

. . .

Item:

Owners and operators of nearly half of all underground coal mines in the United States have systematically tampered with coal dust samples sent to federal safety inspectors and monitored to control black lung disease. More than 5000 incidents of sampling fraud have been discovered thus far (*Washington Post,* April 4, 1991: 1).

Approximately 4000 coal miners die every year from black lung disease, 4.5 percent of the workforce have contracted it (Cullen et al., 1987: 69).

In 1992, the Westray Mine in Pictou County, Nova Scotia exploded, killing 26 miners. Subsequent investigations showed that the owners of Westray routinely violated safety laws and failed to make essential repairs. Inspectors for the province of Nova Scotia provided advance notice of impending visits, routinely overlooked minor and major law-breaking, and generally adopting the perspective of management, whose goal was to minimize the costs of production. Workers who reported unsafe conditions were seen, by both government and in-dustry (and sometimes by their peers as well) as malcontents and rabble-rousers (Richard, 1997; Comish, 1993). By July of 1998, all of the 52 non-criminal and three remaining criminal charges laid against the owners and managers of the mine and the government inspectors had been dropped (*Globe and Mail,* July 1, 1998: A1; Jobb 1998).

Nova Scotia, and virtually every other province, has consistently cut the budgets and workforce of regulatory agencies charged with protecting the health and safety of employees. Self-regulation, a system which asks workers and managers to regulate themselves, with minimal government oversight, has become the norm in virtually every industry (Tucker, 1995; Walters et al., 1995; Noble, 1995). Between 1976 and 1993, almost 16,700 people died from work-related causes, an average of more than two deaths per day. The average annual rate of seven per 100,000 (1988–93 data, all industries) is four times greater than the average homicide rate (which hovers around 2.2 per 100,000). Two hundred and eighty-one of every 100,000 miners died as a result of their work in the 1988–93 period (Statistics Canada, 1996; *Globe and Mail,* August 10, 1996: A6).

. . .

Item:

Taking Stock, a report released by the NAFTA Commission for Environmental Cooperation, identifies Ontario as the third-biggest polluting jurisdiction in North America, behind only Tennessee and Texas in contributing to air, water and land pollution. Ontario Hydro dumped at least 1800 tonnes of heavy metals into Lake Ontario in the last 25 years. Smog alone kills 1800 people per year in Ontario (*Globe and Mail,* July 30, 1997; also January 13, 1997: A15; July 30, 1997: A3; August 19, 1997: A1; June 22, 1998: A1).

However, fines against polluters in Ontario declined to $955,000 in 1997, the lowest in more than a decade, less than one third of the 1995 amount. Since 1995 the province's Environment Ministry has lost 45 percent of its budget and 32 percent of its staff; the allied department of Natural Resources has dropped 19 and 30 percent respectively (*Globe and Mail,* June 22, 1998: A1). The number of charges laid against polluters has been cut in half since the Progressive Conservatives took office in Ontario, dropping from 1640 in 1994 to 724 in 1996. The average fines on polluting companies dropped from a high of $3,633,095 in 1992 to $1,204,034 in 1997. According to Premier Mike Harris, many of Ontario's rules frighten away investment because they are "extreme" (*Globe and Mail,* November 14, 1996).

. . .

Item:

With the advent of global capitalism and the new-found dominance of financial capital, trillions of dollars circulate daily throughout the stock exchanges and currency markets of the world. Individual traders in these markets buy and sell thousands of shares each day, seeking to maximize their own profits and increase their personal fortunes. Traders, essentially unregulated actors, have the power to bankrupt individuals, companies and countries—and have done so, most recently by driving the value of the Russian ruble down 90 percent in less than a week. In the last five years, Indonesia, Japan and many Asian countries, and Mexico have suffered similar fates. The decision to "set capital free" and facilitate millions of untaxed, unregulated trades has destroyed the standard of living of hundreds of thousands of people in many parts of the world, producing starvation and revolution in some countries, the destruction of personal savings, destitution and penury in others.

Much of the harm done by such actors is not defined as criminal. However, scams such as insider trading, falsifying records, failure to disclose and outright fraud (as in the alleged "salting" of gold samples taken from the Busang site at Bre-X, inflating the value of Bre-X stock by millions of dollars and fleecing countless of innocent investors) have been criminalized in many countries. While most stock market traders are closely allied with multinational capital, "rogue traders" have the power to destroy centuries-old institutions, as Nick Leeson did at Barings Bank in the United Kingdom or Yasuo Hamanaka at Sumitomo Incorporated in Japan. Financial corporate crime has gone global: the BCCI bank, which imploded causing an estimated $15 billion in losses, was organised to be off-shore everywhere. Corporations have always attempted to incorporate in jurisdictions with the lowest taxes and the fewest regulations, but the BCCI shows that it is possible to avoid all nation-state restrictions.

The systematic removal of regulatory controls over financial institutions and stock exchanges has been a cherished component of those leading the corporate counter-revolution. Thus, when the Reagan government took office in the United States in 1980, it repealed regulations and fired regulators responsible for overseeing stock exchanges, such as the federal Security and Exchange Commission (SEC). Staff at the SEC fell by 300, while the number of securities requiring monitoring doubled from 1981–86. In the financial arena, a series of restrictions on institutions known as Savings and Loans companies were removed, leading to the most costly series of corporate crimes ever, with estimated losses topping $500 billion. . . . This caused some momentary re-regulation; however in 1994 the U.S. House of Representatives again loosened controls on banks, and in 1995 it tabled legislation to gut the SEC once again (Calavita et al., 1997).

. . .

Item:

Competition/Combines Offences are anti-competitive practices designed to inflate profits through such deceptive practices as conspiracy to restrict trade, mergers and monopolies, predatory pricing, price discrimination, resale price maintenance and refusal to supply. False Advertising, an allied offence, refers to deceptive trade practices such as inflated claims about a product's effectiveness, or misleading consumers by misrepresenting a product's regular price as a sales price.

Combines, mergers and monopolies were outlawed in Canada's first piece of corporate crime legislation, the Combines Investigation Act, passed in 1889. It was regularly amended and enlarged (though never effectively enforced) from that time until 1976.

. . .

In 1986, new legislation was passed in the House of Commons, abolishing the Combines Investigation Act. Its replacement, the Competition Act, had a very different mission: "to improve and facilitate corporate operations" *not* to control or sanction conspiracies to restrict trade, drive up prices or engage in monopolistic, predatory practices. To this end:

- Criminal sanctions were *removed* from merger/monopoly sections.
- "The Public Interest" was *removed* as a criterion or directive to be used in evaluating a proposed merger.
- A "compliance-centred" approach was adopted to deal with "clients" (who were no longer "offenders" suspected of "crimes").[1] The new goal of law was to provide a stable and predictable climate for business, with the promotion of business prosperity made key.
- Prosecutions for conspiracy, discriminatory and predatory pricing, misleading or deceptive practices and price maintenance dropped from 37 in 1982–84 and 36 in 1984–86 to 23 in 1986–88. By 1995–96, all regional offices of the Competition Bureau were terminated; the number of inquiries commenced dropped from 82 to eight in the four year period from 1991–92 to 1995–96, the number of cases referred for prosecution dropped from 55 to seven, prosecutions declined from 44 to seven, and convictions dropped from 43 to fourteen (Canada 1997: 36). In the spring of 1998, a bill was introduced to further decriminalize misleading advertising and deceptive marketing practices (*Globe and Mail*, May 5, 1998: B3; Canada 1998: 3).
- From 1986–89, 402 merger files were opened, 26 "monitored," seven abandoned, nine mergers were restructured, five went to Competition Tribunal and two were under appeal. By 1996-97, the Competition Bureau reviewed a total of 319 mergers, but only 23 were deemed problematic enough to require follow-up. There are 369 slated for review in 1997–98 (*Globe and Mail*, March 30, 1998: B4), a record the aforementioned *Globe* decries as "Cracking Down" by "Competition Cops" (Milner, 1998) (*Sources*: Snider, 1993 and 1978; Stanbury, 1977, 1986–7 and 1988; Canada, 1989; Varrette, 1985; Goldman, 1989).

Monopolistic markets have become a fixture of the developed world. They mean that the power of capital is increasingly concentrated in the hands of a small number of transnational corporations with monopolistic control over the life-style and life-chances of most of the world's citizens. No representative citizenry elect the owner-controllers and, increasingly, no public body regulates their actions or behaviour. Fifty-one of the largest economies in today's world are not countries but corporations

(Dobbin, 1998; McQuaig, 1997). In Canada in 1991, a mere ten corporations (excluding banks) made up more than one-fifth of the Gross National Product. (Many are transnationals such as General Motors of Canada, Chrysler, Ford and Imperial Oil; others such as Bell Canada, Noranda, George Weston and Thomson are normally controlled in Canada.) Monopolistic, mammoth companies do not provide a commensurate number of jobs—the top 200 transnational corporations, with sales accounting for 28.3 percent of the world's GDP, employ less than one percent of its workforce (Dobbin, 1998: 76). Nor do they pay a commensurate share of taxe[s]—Nortel, for example, received "at least" $880 million in federal research and development tax credits, but paid a mere 0.4 percent of its total revenue back in income tax; as of 1996 it owed $213 million in deferred taxes (Dobbin, 1998: 78).

WHY DOES IT MATTER?

Thus far, this essay has documented the harm caused by corporate crime and the virtual abandonment of attempts to proscribe or sanction it. The nation-state has, in effect, given up the struggle to control corporate criminals through law. The disappearance of corporate crime matters—but not because state law was ever particularly successful in punishing it. Because of the power of the perpetrators and the collective weakness of the victims, laws governing corporate crime have always been full of loopholes, regulating authorities starved for funds, convictions few and far between, and sanctions totally incommensurate with the damage inflicted.[2] It matters because this retreat is part of an ideological retrenchment whereby the ability to censure, monitor or signal disapproval of the antisocial acts of capital is being lost. So is legal recognition of the fact that corporate acts cause harm. It matters because "the growing incapacity of sovereign states to control the behaviour of corporations" (Reiss, 1993: 190) is a created event, not an inevitable, unalterable fact (McQuaig, 1997).

And it matters because the data to counter heavily promoted business claims that corporate crimes are no more than one-time accidents, committed by "good citizens" rather than "criminals," disappear. This argument is both beside the point and incorrect. Beside the point because many traditional offences are equally unintentional, but intention is no defence *except* where the crimes of corporations are at issue. . . .

CONCLUSION

This essay has argued that the disappearance of corporate crime signalled by the abandonment of state law is an event with massive political, legal and ideological consequences. State law has not "kept pace" with the growth and development of corporate wealth and power (Wells, 1993); indeed, it has retreated, fled in ignominious defeat, ceded victory to corporate counter-revolutionary forces. For those who seek to create more equitable societies, this development represents a giant step backwards. It is one thing to argue that state law was inefficient—that governments seldom took effective action against corporations—and quite another to argue that they should jettison the capacity, and the legal obligation, to do so. As we have seen, abandoning state sanctions has far-reaching symbolic and practical consequences. State laws are public statements which convey important public messages about the obligations of the employer classes (Ayres and Braithwaite, 1992).[3] The situation is paradoxical indeed: while crimes of the powerful were never effectively sanctioned by state law, such laws are nonetheless essential to the operation of democratic societies.

NOTES

1. The most recent discourse switch has been the transformation of corporate criminals into "stakeholders" (Canada, 1998: 3).

2. In this context, one could cite virtually the entire corpus of literature on corporate crime. In the American literature this record has been documented in studies from Sutherland's (1949) classic *White Collar Crime* to Clinard and Yeager (1980), Shapiro (1984), Coleman (1985) and Green (1994); more international accounts are found in Box (1983), Clarke (1990), and Punch (1996). Recent studies go beyond description to explanation and reform (see, for example: Braithwaite 1995; Yeager 1991; Tombs 1996; and articles in Pearce and Snider 1995).

3. It is surprising that the citizens in so many democratic countries have allowed this to happen. Abandoning regulation and de-criminalizing harmful acts was not achieved by state dictatorship in Canada. It was a public process, negotiated into existence. This does not mean everyone was consulted. It was largely a process of changing the minds of elites and of those citizens who vote. But, with the partial exception of environmental regulation (where active social movements were and are present), the disappearance of corporate crime never became a major news story, or election issue.

References

Ayres, I. and J. Braithwaite, 1992, *Responsive Regulation: Transcending the Deregulation Debate*, New York: Oxford.

Box, S., 1983, *Power, Crime and Mystification*, London: Tavistock Publications Ltd.

Braithwaite, J., 1995, "Corporate Crime and Republican Criminological Praxis," in F. Pearce and L. Snider (eds.), *Corporate Crime: Contemporary Debates,* Toronto: University of Toronto Press.

Calavita, K., H. Pontell and R. Tillman, 1997, *Big Money Crime*, Berkeley: University of California Press.

Canada, Bureau of Competition Policy, 1989, *Competition Policy in Canada: The First Hundred Years,* Ottawa: Consumer and Corporate Affairs.

Canada, Industry Canada, 1997, Annual Report of the Director of Investigation and Research, Competition Act (for the year ending March 31, 1996), Ottawa.

Canada, Industry Canada, 1998, Annual Report of the Director of Investigation and Research, Competition Act (for the year ending March 31, 1997), Ottawa.

Clarke, M., 1990, *Business Crime,* Cambridge: Polity Press.

Clinard, M. B. and P. Yeager, 1980, *Corporate Crime,* New York: Free Press.

Coleman, J. W., 1985, *The Criminal Elite: The Sociology of White Collar Crime,* New York: St. Martin's Press (2nd ed., 1989).

Comish, Shaun, 1993, *The Westray Tragedy: A Miner's Story,* Halifax: Fernwood Publishing.

Cullen, F., W. Maakestadt and G. Cavender, 1987, *Corporate Crime Under Attack: The Ford Pinto Case and Beyond,* Cincinnati: Anderson.

Dobbin, M., 1998, "Unfriendly Giants," *Report on Business Magazine,* 15 (July): 73–80.

Glasbeek, H., 1995, "Preliminary Observations on Strains of, and Strains in, Corporate Law Scholarship," in F. Pearce and L. Snider (eds.), *Corporate Crime: Contemporary Debates,* Toronto: University of Toronto Press.

Goldman, C., 1989, "The Impact of the Competition Act of 1986," Address given to the National Conference on Competition Law and Policy in Canada, Toronto, October 24–25.

Green, G., 1994, *Occupational Crime,* Chicago: Nelson-Hall.

Jobb, Dean, 1998, "Westray: A Deadly Misuse of Power," Paper presented at (Ab) Using Power: The Canadian Experience, Conference held at Simon Fraser University, Vancouver, B.C., May 7–9.

McQuaig, L., 1997, *The Cult of Impotence: Selling the Myth of Powerlessness in the Global Economy,* Toronto: Viking Press.

Milner, B., 1998, "Competition Cops Flex Muscle," *Globe and Mail Report on Business,* March 30: B1, 3

Miyoshi, M., 1993, "A Borderless World? From Colonialism to Transnationalism and the Decline of the Nation-State," *Critical Inquiry,* University of Chicago: 726–51.

Noble, C. 1995, "Regulating Work in a Capitalist Society," in F. Pearce and L. Snider(eds.), *Corporate Crime: Contemporary Debates,* Toronto: University of Toronto Press.

Pearce, F. and L. Snider, 1995, "Regulating Capitalism," in F. Pearce and L. Snider (eds.), *Corporate Crime: Contemporary Debates,* Toronto: University of Toronto Press.

Punch, M., 1996, Dirty Business: Exploring Corporate Misconduct, London: Sage.

Regush, N., 1991, "Health and Welfare's National Disgrace," *Saturday Night* (April): 9–18; 62–3.

Reiss, A., 1992, "The Institutionalization of Risk," in J. Short and L. Clarke (eds.), *Organizations, Uncertainties and Risk,* Boulder: Westview Press.

Richard, Justice K. Peter (Commissioner), 1997, *The Westray Story: A Predictable Path to Disaster,* Report of the Westray Mine Public Inquiry, Province of Nova Scotia.

Rothman, D., 1995, "More of the Same: American Criminal Justice Policies in the 1990s," in T. Blomberg and S. Cohen (eds.), *Punishment and Social Control,* New York: Aldine de Gruyter.

Shapiro, S., 1984, *Wayward Capitalists,* New Haven: Yale University Press.

Snider, L., 1993, *Bad Business: Corporate Crime in Canada,* Scarborough: ITP Nelson.

Snider, L., 1978, "Corporate Crime in Canada: A Preliminary Report," *Canadian Journal of Criminology* 20: 142–68.

Snider, L., 1991, "The Regulatory Dance: Understanding Reform Processes in Corporate Crime," *International Journal of Sociology of Law* 19: 209–36.

Snider, L., 1998, "Towards Safer Societies," *British Journal of Criminology* 38 (1): 1–38.

Stanbury, W., 1977, Business Interests and the Reform of Canadian Competition Policy 1971–75, Toronto: Carswell/Methuen.

Stanbury, W., 1986–87, "The New Competition Act and Competition Tribunal Act: Not with a Bang but a Whimper?" *Canadian Business Law Journal* 12: 2–42.

Stanbury, W., 1988, "A Review of Conspiracy Cases in Canada, 1965–66 to 1987–88," *Canadian Competition Policy Record* 10 (1): 33–49.

Statistics Canada, 1996, *Death and Injury Rates on the Job: 75-001-XPE,* Ottawa: Ministry of Supply and Services, Summer.

Statistics Canada, 1997, *Income Distributions by Size in Canada,* Ottawa: Ministry of Supply and Services.

Statistics Canada, 1998, National Census, Ottawa: Supply and Services.

Sutherland, E., 1949, *White Collar Crime,* New York: Dryden.

Tombs, S., 1996, "Injury, Death and the Deregulation Fetish: The Politics of Occupational Safety Regulation in United Kingdom Manufacturing Industries," *International Journal of Health Services* 26 (2): 309–29.

Tucker, E., 1995, "And Defeat Goes On: An Assessment of Third Wave Health and Safety Regulation," in F. Pearce and L. Snider (eds.), *Corporate Crime: Contemporary Debates,* Toronto: University of Toronto Press.

United Nations, 1996, *Human Development Report,* New York: United Nations.

Varrette, S. E., C. Meredith, P. Robinson and D. Huffman, ABT Association of Canada, 1985, *White Collar Crime: Exploring the Issues,* Ottawa: Ministry of Justice.

Walters, V., W. Lewchuk, J. Richardson, L. Moran, T. Haines and D. Verma, 1995, "Judgments of Legitimacy regarding Occupational Health and Safety," in F. Pearce and L. Snider (eds.), *Corporate Crime: Contemporary Debates,* Toronto: University of Toronto Press.

Wells, C., 1993, *Corporations and Criminal Responsibility,* Oxford: Oxford University Press.

Yeager, P., 1991, *The Limits of Law: The Public Regulation of Private Pollution,* Cambridge: Cambridge University Press.

On the Assimilation of Racial Stereotypes among Black Canadian Young Offenders*

*John F. Manzo and
Monetta M. Bailey*

What is the stereotype of the Black Canadian? How does the media contribute to this stereotype? In what ways do Black Canadians—and especially Black Canadian young offenders—identify with and embrace these stereotypes? What are some of the consequences of this identification?

This sociological study investigates the influence of racial stereotypes of Blacks in the media on their self-concepts and identity. Specifically, did the interviewees' views of themselves incorporate "criminal" as an aspect of Black identity as a result of the "gangsta" stereotype frequently portrayed in the mass media?

The study's findings address not only if, but also how, stereotypical views may influence an audience. While the interviewees recognized and related to the "gangsta" stereotype, they also assessed other stereotypical views of Blacks in deciding which to relate to. So why did these individuals choose to adopt the "gangsta" image to the extent that they did?

This paper investigates the assimilation and iteration of racial stereotypes among Black[1] Canadians by inspecting open-ended interviews with eight Black or mixed-race respondents who are adjudicated young offenders. The focus of this investigation is on whether, and to what extent, assimilation can be observed in interviewees' discourse and, moreover, whether the speakers' self-concepts entail their incorporation of "criminal" as an aspect of Black identity.

An association between race and criminal justice processing in Canada has been documented, particularly with respect to Black and

Source: John F. Manzo and Monetta M. Bailey, "On the Assimilation of Racial Sterotypes Among Black Canadian Young Offenders." *Canadian Review of Sociology and Anthropology* Vol. 42. No. 3 (2005): 283–300.

Native persons. Wortley (1999) notes that, in 1997, Native persons represented about four percent of the population but constituted fourteen percent of federal prison inmates. Black persons accounted for roughly two percent of the population while representing over six percent of those in federal correctional institutions. Native persons had an incarceration rate of 184.85 per 100,000 persons, while that of Black Canadians was 146.37; non-Native, non-Black Canadians were incarcerated at a rate of about 100 per 100,000 (Wortley, 1999: 261).

This evident association between race and crime (or incarceration), among other factors, has led many in society to develop negative stereotypes of persons based on their racial identities. In Canada these negative impressions stem not only from actual experiences of prisoners in the criminal justice system, but also from images in North American culture and media. Despite the relatively small Black population in Canada, Canadians are almost certainly familiar with the image of the Black "gangsta" from media imagery imported from the U.S., a nation with more than six times the population of Black persons, per capita, and embracing a Black population with a history, culture, and level of social segregation different from that in Canada.

Mass-cultural images of Black Canadians, it would seem, not only motivate stereotyping on the part of those who are not Black: they should also influence racial identities and related self-concepts among Black persons themselves. This paper considers results of a study that investigated the association between crime and the formation of a racial identity among Black young offenders. The study entailed open-ended interviews on topics including police-minority relations, the racialization and criminalization of their racial groups, the connection between their lifestyle and cultural influences such as rap music, and the relationships between their racial group and the dominant (White) culture.

The focus of this paper is on the responses given by respondents with regard to the social depiction of their race, the possible impact of this depiction in their racial identity formation, and the relationship between this depiction and their criminal actions. First, we consider how the youth believe their racial group is portrayed in society. We then ascertain if they believe these images comprise "criminal" elements. Finally, we investigate whether and how this portrayal has been internalized by these youth to inform or influence their criminal actions.

THEORETICAL PERSPECTIVES

We consider "race" to be a socially constructed, malleable, interpersonally relevant and, thus, a "micro"-level phenomenon; we also recognize that "race" has an historical and otherwise "macro" social resonance and meaning that exists over and above individuals' perception of and claims to it. For these reasons, the theoretical perspectives of this paper adopt views that partake of both historical and social-interactional construal of race. This paper deploys social construction perspectives as developed by Berger and Luckmann (1966) with notions of the historically embedded construction of race derived from post-colonial theory (Fanon, 1967; Said, 1978), and, at the level of lived and lively social experience, we rely on the notion of cultural transmission that is based on the contributions of C. Wright Mills (1963) with respect to what he termed "vocabularies of motive." The first two of these theories account for "race" as socially and historically defined and embedded, as aspects of a cultural endowment that is given and, more clearly for post-colonial theory, imposed, on persons; the last theoretical theme considers *how*, through what concrete discursive means, the *content* of racial typifications is "taught" to occupants of those historically and socially constructed racial categories.

Social construction theory (cf. Berger and Luckmann, 1966) maintains that individuals define themselves based on social conceptions of the group to which they claim membership. Social construction theory holds that the basis for "subjective" reality is in fact the social world: the self is created through a dialectical, reflexive relationship between the individuals and their social milieux. Social construction theory thus argues that persons see themselves in the same terms that society views them.

With respect to racial identity and criminal propensities, social constructionists such as Blakey (1999), Holdaway (1997), Rodkin (1993) and Schiele (1998) argue that most social theories about race and crime tend to reify race and ignore the social process that is involved in the creation of racial categories. Such theories do so by treating race as endogenous to the person and assigned as any other biological feature. We, on the other hand, side with constructionists who view "race" not as a static, ascribed quality of persons but as a process achieved and learned through social interaction and as a consequence of the receipt of cultural definitions of race. This construction of race is known as racialization, a process through which meanings and definitions become associated with what become socially defined as different racial categories. It is the way in which race is constructed in everyday life and becomes, in effect, "real" in society and to the individual.

Colonial theory adds to social-construction approaches by accounting for social conceptualizations of race based on historical relations among different racial and ethnic groups. In his seminal *Black Skin, White Masks,* Fanon (1967) proposes that, in a former colonial society, socially accepted modes of thought are based on the views of the dominant, "colonizing" group. The culture, language and customs of the colonizers come to be normative and to be considered superior to both local indigenous cultures and to those who were part of subsequent non-White diasporas to post-colonial societies. Thus, in North American society, definitions and stereotypes of races, among other topics, are created and organized by persons of Northern and Western European origin. Colonial theorists see present-day society as evincing the racial relationships and subjugation that characterized colonial times. Negative images associated with minority groups therefore derive from a colonial history, and the self-concepts that minority persons adopt owe to their place vis-à-vis White persons historically.

. . .

Our research is ... finally informed by the notion of "cultural transmission," which suggests concrete ways though which negative self-concepts and stereotypes are adopted through social praxis. Cultural transmission theories concern how popular culture, among other discursive forms, can influence individual action. Cultural transmission theory also proposes how individuals justify their seemingly deviant actions by referring to socially accepted accounts for doing so, such as "I am owed this money, so I am not really stealing it," or "I am defending my family's honour by harming this person; I am not committing a crime." These statements are known as "vocabularies of motive" (Mills, 1963) and were most famously explicated in "drift theory," as developed by Sykes and Matza (1957). Vocabularies of motive are cognitive and linguistic concepts that furnish motives (before the fact) and accounts (after the fact; cf. Lyman and Scott, 1968) for committing certain classes of behaviours. The concept has seen greatest use as an explanation for the tendency for persons to drift in and out of criminal or otherwise deviant behaviour, as with studies that encompass deviant acts from non-criminal activities like cheating on tests (McCabe, 1992), to suicide (Stephens, 1984), rape and murder (Scully and Marolla, 1984). The concept has also informed understandings relating to the motivations of some victims of domestic violence to remain with their abusers (Ferraro and Johnson, 1983); "vocabulary of motive" thus need not only be a resource to permit the forming of motive to commit deviant or criminal acts. With respect to the topics under investigation here, cultural transmission is important in understanding how Black persons may internalize societal depictions of themselves as resources to "motivate" them to behave in manners consistent with those stereotypes. These cultural messages must, moreover, come from concrete sources of communication, and our research suggests that these can and do emerge from discourses in popular media. As such, we argue that media not only "teach" racial stereotypes (in positive as well as pejorative senses), but also that these stereotypes themselves can facilitate criminal and otherwise deviant motives and rationalizations.

. . .

METHODOLOGY

The Sample

This paper examines interviews with respondents from a study entailing interviews of eight Black or mulatto young offenders between 14 and 18 years old in Alberta, Canada. Three were in open-custody residential "group homes," and five were in secure custody at a youth detention facility. The ages and placements for each interviewee are indicated the first time each is cited in this report.

The second author, who is herself a Black Canadian originally from Barbados, conducted the interviews. As part of our protocol for the protection of human subjects and following the insistence of our gate-keeping agencies, she was not permitted to inquire about our subjects' crimes (although they were, of course, permitted to discuss or allude to them themselves). . . .

FINDINGS

Respondent's Perspectives on Societal Views of Black Persons

Overall, respondents expressed, unsurprisingly, the view that stereotypical ideas of Black persons did exist socially, and that these images owed largely to what the youths construed as representations depicted in media and in the larger culture. The respondents seemed, moreover, to identify with these stereotypes, some more than others.

Stereotypes of Blacks

Respondents suggested that stereotypes of Blacks were of two categories. First, people saw Blacks as being "dangerous," as possessing at best defiant attitudes and at worst criminal tendencies, in line with what might be called a "gangsta" image. Although the respondents demonstrated a partial acceptance of this stereotype in that they also stated that style of dress influences how they judge other Black persons, they expressed some anger that non-Blacks did this. They were quite vocal in their objection to others placing Blacks into categories, and determining membership in a category according to style of dress. This objection was largely founded on the fact that the respondents believed that when others viewed them, based on their style of dress, as in the "gangsta" category, they associated criminal behaviour with it.

The second stereotype that respondents noted saw Black persons as entertainers, that is, as athletes, actors, musical performers and so on. This view of Blacks is not mutually exclusive with respect to the "gangsta" image; indeed, the essence of "gangsta" is demonstrated by a look adopted by rap stars and other Black celebrities, including some athletes. Consequently, respondents stated that they believe that many people in society assumed that they were criminals. Carl (16, secure custody) said,

> ... because I'm Black, everybody looks at you like ... a gangsta, playa, baller right. I don't look at myself like that but everyone else calls me that, like "What's up thug?", you know, like a gangsta ... because you are Black you gotta be a thug, you gotta be a gangsta, that's how they all think. They look at you and if you're not that then you are not popular, you are not really Black. I think that's all wrong though.
>
> ...

Black respondents generally believed that social evaluations of them were more negative than positive. The question of how they believed they were seen by members of society drew responses such as: "Probably bad things, like you are, I don't know, in gangs ... like we are going to steal or something," or " ... some people think that Black people are always bad, they're always in gangs, this and that, right." ... The issue arises in accordance with our research questions as to where these stereotypes come from. In every interview, the interviewees' overwhelming response was that media played an important part in promoting views of Black persons and Black culture. The respondents' utterances on these matters follow.

Portrayals of Blacks in the Media

The respondents saw the overall portrayal of Blacks in the media as mixed with respect to the relative amount of positive and negative imagery that are portrayed, but saw these images as conforming to stereotypes regardless. For example, respondents reflected on how Black persons are shown possessing special talents, as athletes in particular sports or as entertainers in a very delimited range of arts (as rappers or comedians, for example).

The other way they are seen in the media, from our respondents' experiences, was as criminals. The interesting thing about the criminal portrayal was the judgment associated with it depended on the media outlet. Some respondents noted that, usually in the news and mainstream media, the criminal image that Blacks had was seen as negative, while in media intended for Black (almost always African-American) audiences, this image, even when the Black persons in them were in fact depicted as criminals, had more positive connotations.

...

While Daniel (17, secure custody) seemed to recognize that the portrayal of Blacks in the media was mixed in the ways expressed by the other respondents, he implied that the negative representations were more common than positive ones:

> They look like—like criminals and stuff like that. Only some, only some 'cause some Black people are talented and positive people, you know, sometimes. But sometimes they just, I don't know, sometimes they look bad. Like I know like when you're sitting watching TV and stuff, they make them look like people from the ghetto and stuff like that all the time. Like every Black person's from the ghetto and stuff, and do a lot of crime and stuff like that.

Daniel further expressed that he believed that the negative views of Blacks in society could be attributed to the depictions of Blacks in the media.

> ... lots of people think of Black people as thugs and robbing people and stuff, you know .... Like some people think that a Black person's not normal. They just think—like how there's a lot of crime and stuff because of movies and stuff, you know, and how Black people that are in the movies, they all live in ghettos and stuff and all do crime and stuff like that. I just think that that's how people see us.

Given this reporting of negative media imagery, the question emerges as to how respondents themselves adopted or rejected those images as constituting their views of themselves.

Respondents' Perspectives on the "Gangsta" Image

It became clear that the majority of respondents felt some connection to a "gangsta" image. The respondents defined this image as specific to Blacks, and as opposed to the "normal" social depiction of Whites as not "gangsta." While the youths under study here allowed that some Blacks were "normal," respondents also saw "gangstas" or "thugs" as uniquely Black constructs. Moreover, respondents expressed no desire to assume a conventional, socially acceptable image. These respondents all suggested that their refusal to conform to "normal" social types, to which some outrightly referred to as "looking and acting White," meant that they were seen as that particular other known as "gangstas." They chose to embrace this image....

This gangsta image encompasses at once dress, language and demeanour. The style of dress entailed an African-American "street" sensibility, comprising labels such as Sean John; however, despite the evident importance that appearance should play in being a "gangsta," respondents were more likely to address the demeanour or "attitude" that "gangsta" encompasses. One aspect of demeanour entails the linguistic performance of gangstas, and indeed the gangsta argot is marked by profanity and a specialized vocabulary, as Ricky [(14, secure custody)] articulates:

> ... [others] don't swear as much as we do, they don't use profanity or whatever as much as we do. They don't use slang as much as us, they're like, "Hey, how are you?" we are like "Wha's up man?" or something like that. I don't know, to [others] our language or our slang, our gangsta language to them is hard.

Gangsta demeanour encompassed a self-presentation that would best be described as carefree but threatening, because all acknowledged that when challenged or "dissed" they could then marshal violence. Kobe [(16, secure custody)] interpreted the difference between the Black and White attitudes this way:

> White people are more snotty. Black people are I think, are more relaxed, you know. 'Cause I have—some of my friends are Black friends and they're just, like, into chillin' and all that.... Basically. I think White people are more snotty and more rude.

This attitude was one that the respondents thought afforded Blacks more fun than other groups. However, the carefree attitude was juxtaposed by the tough attitude that "gangstas" had when they were upset. When referring to a friend, Carl said,

> ... he's like "hard," like he doesn't care what he does, like me. I can do something, I don't care what you do, but if you cross me you wouldn't expect me to hit you or something like that, but I'll pop you without even caring.
>
> ...

Finally, the gangsta image was marked by the prevalence of smoking marijuana, which most respondents admitted to, and associated with their definition of a gangsta. Some respondents commented on the crimes they committed while smoking "weed"; others noted that they were often approached as a source of it. It was also an aspect of the "carefree" attitude described earlier....

It was clear, then, that the majority of respondents identified to some extent with the "gangsta" type. This identification is seen not only in the style of dress that the respondents embraced, but also in the attitude that they held towards others. This was apparent in Ricky, who stated that he wished others to think the following of him.

Like if you mess with me, like I'll kill you ... like I don't care, that's my attitude, I don't care .... In my mind what I think of it is if you do something to me, I think of it as this: a nigga never forgets.

...

The youths deployed the word "nigga" recurrently. Several respondents, explicitly or implicitly, made the distinction between a "nigga" and a "gangsta." Ricky suggested that a "gangsta" just "goes around and causes trouble, or jacks somebody for no reason," on the other hand, a "nigga" "beats up somebody, but there is a reason." For these youth, there was more pride and, one may conclude, social acceptability in being a "nigga" than in a being a "gangsta." Thus, it is fair to say that some of the youth identified more closely with the image of a "nigga" rather than that of a "gangsta." It should also be noted that most of the youth did not make this distinction, and that it is fair to say that most respondents could be said to identify with both typologies.

History and Black Identity

Post-colonial theory recommends that Black persons' views of themselves will partake of the historical position of Black peoples vis-à-vis their oppressors. Questions emerge, however, concerning which Black peoples' history, which oppressors, and where, exactly, these images would come from. When respondents reported knowledge of the history of Black persons in North America, this knowledge was gleaned from televised imagery and the unavoidable American influence that it contains. This is important because, like most Black persons in Canada, all but one of these respondents' Black ancestors emigrated to Canada from the West Indies, mostly Jamaica, rather than from the United States; an additional respondent's family had immigrated from Africa. In addition, all of the respondents were Canadian citizens, yet not one demonstrated any knowledge of the history of Blacks in Canada. Moreover, most admitted little knowledge of the history of Jamaica (or any other Caribbean nation) or its race politics.

...

[W]ith some exceptions, respondents were largely unaware of the history that attached to Black identity and, in particular, the history relevant to their own ancestries. However, post-colonial theory suggests that the effect of the past provides for an unseen shaping of one's contemporary worldview. We would not have expected that respondents would state the precepts of post-colonial theory explicitly, or that they would have to know about their own ancestors' experiences with colonialism in order to experience its latent effects. They should, however, appear to be aware of those posited effects when they speak of their oppression; however, even this discussion does not completely support the tenets of post-colonial theory, because respondents' views did not clearly suggest that they saw themselves as oppressed. The following excerpts demonstrate this finding.

Respondents' Views on "Oppression"

Most respondents suggested that there were disadvantages associated with being Black in Canada, but they expressed pride in being Black. The interviewees suggested, in general, that while there was racism in society, they did not believe themselves to be seriously affected by it, even while admitting to being influenced by and sometimes judged unfairly due to the currency of a "gangsta" image of Blacks. The reason behind this may be that Black respondents seemed to take pride in some of the stereotypes of their race, both positive and negative. These depictions included stereotypes that Blacks are especially talented in certain sports, and as entertainers, but some also took pride in evincing images such as "gangsta" and "thugs." Many also claimed that they simply did not care what others' views of them were in the first place; indeed, we note that respondents even deployed a fatalistic view of societal opinions of them as a variety of vocabulary of motive, one that said, to paraphrase, "if this is how society sees me, I might as well act in this way."

With respect to perceptions of "oppression," respondents were in fact able to identify some areas in which they believed that their race put them at a relative disadvantage. Some indicated that they believed that stereotypical views of Blacks would make it more difficult for them to acquire certain jobs, and that Blacks were often under the surveillance of police and private security. None, however, indicated that these complications made it impossible for them to achieve their goals.

The youth evidenced internalization of negative stereotypes in a process consistent with the claims of "vocabulary of motive" and related approaches in criminology, which address how persons justify deviant behaviours by marshalling cognitive and linguistic scripts that serve as motives, rationalizations and/or excuses for those behaviours. Dale [(15, group home)] exemplified this tendency when he said, "A lot of people think I'm bad, so why would I care?" . . .

Youth assented to other stereotypes as well. For example, Desmond, Daniel and Carl all embraced the stereotype that Blacks are skilled athletically. Desmond [(17, secure custody)], when asked what characteristics distinguished Blacks from other racial groups, stated:

> Some physical characteristics, 'cause when—like we're more built than other people and we're more physical, can do things better, some things. Like we're quicker in running, like athletes. We're more athletic and stuff like that.

. . .

Carl, however, appeared to be the only respondent who saw this stereotype as misleading, albeit one that always permits him to "automatically play" despite his very surprising and counter-stereotypical admission that he "can't play ball, right?"

> People look at you like because you are Black you can do that. Like in here, at this place, they are like, "Good, you can play ball right?" But I can't play ball. Right? But they just think that all Black people can play ball. They always want me on their teams . . . not for hockey though or for volleyball—but for basketball or baseball. 'Cause I'm Black I automatically play . . . just 'cause I'm Black I should be athletic, not for, like, the joy of the game.

. . .

The interviewees consistently demonstrated beliefs that society saw them in negative terms, but they also showed evidence of internalizing and even coveting those characteristics. . . . [T]he youth interviewed all identified with "gangsta." In addition, they saw this image as a generally desirable one for them, and many of them admitted that they have engaged in behaviour that supports such a representation. Their claimed knowledge of society's opinion of them permitted these youth to relinquish responsibility for their criminality; they merely acted in accordance with how, they believed, they were seen by other people.

CONCLUSIONS

The participants in this study articulate a view of themselves and of Black persons in general that is consistent with certain stereotypes. This finding supports the claims of social construction and post-colonial theories, both of which anticipate that a member of any racial minority—particularly one whose history entailed overt oppression, discrimination or slavery—would adopt self-images in line with those prescribed and maintained by the larger society. However, it is vital that we emphasize that our findings do not support these theories *tout court;* in particular, the implication of victimization tacit in social construction and, especially, post-colonial theory is not clearly present in our respondents' discourses. Yes, their self-concepts appear riddled, in one sense, with the typifications of "Black" provided them in their cultures. Paradoxically, in accepting, to varying degrees, the existence of these social stereotypes, our respondents were also distancing themselves from mainstream (especially "White") society by embracing them. Not only do the interviewees admire and in some cases aspire to "gangsta," but they also take pride in certain other stereotypes, for example that Blacks are good athletes, actors, rappers and singers. While these youth sometimes expressed frustration with the stereotypes, they also supported them in their responses.

Of course, any social status comprises behavioural, performative aspects—"roles," in other words—and so we must ask to what extent these youths' acceptance and internalization of certain racial typologies (such as "gangsta" or "nigga," which, despite the positive claims about this status, would appear to have certain deviant associations as well) had consequences for their behaviours, especially their criminality. We argue that this association is evident. The respondents here had all been adjudicated as young offenders, and although we did not overtly discuss specific crimes, we note that these youth took pride in their own "gangsta" aspect, using it as a way to differentiate them from the dominant culture. Such a finding clarifies the importance of considering the content of communication and the ways in which it is received and deployed by its audience. In our study, the cultural imagery of the "gangsta" furnished not only a stereotype for the youth here to lay claim to, it also furnished a resource for motivating and justifying criminal activity. To reiterate a point made earlier in this paper, the perspectives of social construction/post-colonialism might elucidate *that* Black persons adopt stereotypes of themselves as their own identities; these theories do not, however, elucidate *how* this is done or how the stereotypes themselves might impel deviant behaviours. In incorporating all of these theoretical views, we have here accounted for both sides of the problem.

We believe that pride in their racial group demonstrated for these youth a form of resilience that had both positive and negative connotations. In a culture where they are aware of negative racial associations, it is helpful and hopeful for these youth to identify aspects of their race as positive and to take pride in them. However, a problem arises when the aspect that they choose to embrace is inherently criminal and, quite possibly, criminogenic, such as the gangsta image that these youth have adopted and that they extol. We believe that a deeper problem is a lack of viable alternatives for these youth, which is demonstrated in limited role models, a lack of knowledge of the history of their culture, and a society that advertises a limited view of an entire, diverse racial group. To innovate on cultural transmission approaches, the "culture" in this case "transmits" precious little in terms of an identity that Black youth might adopt. In a world in which the only viable alternatives are sport (for males), entertaining, or crime, it is unsurprising that Black youth, in Canada and elsewhere, ally with the option that requires the least specialized talent.

NOTE

1. Our use of the term "Black," and not "African-Canadian" or "West-Indian Canadian" is preferable since it comprises all those of African heritage. We use this term to refer to both Black and mixed race or "mulatto" speakers, since both groups in this study identify as "Black."

References

Berger, P. and T. Luckmann. 1966. *The Social Construction of Reality: A Treatise in the Sociology of Knowledge.* New York: Anchor Books.

Blakey, M. 1999. "Scientific Racism and the Biological Concept of Race." *Literature and Psychology*, Vol. 13, No. 1, pp. 1–29.

Fanon, F. 1967. *Black Skin, White Masks.* New York: Grove Weidenfeld.

Ferraro, K. and J. Johnson. 1983. "How Women Experience Battering: The Process of Victimization." *Social Problems*, Vol. 39, No. 3, pp. 325–35.

Holdaway, S. 1997. "Some Recent Approaches to the Study of Race in Criminological Research: Race as a Social Process." *British Journal of Criminology*, Vol. 37, No. 3, pp. 383–401.

Lyman, S.M. and M.B. Scott. 1968. "Accounts." *American Sociological Review*, Vol. 33, No. 1, pp. 46–62.

McCabe, D. 1992. "Influence of Situational Ethics on Cheating Among College Students." *Sociological Inquiry*, Vol. 82, no. 3, pp. 365–74.

Mills, C.W. 1963. "Situated Actions and Vocabularies of Motive." In *Power, Politics and People*. I.L. Horowitz (ed.). New York: Oxford University Press, pp. 439–52.

Rodkin, P. 1993. "The Psychological Reality of Social Constructions." *Ethnic and Racial Studies*, Vol. 16, No. 4, pp. 633–56.

Said, E. 1978. *Orientalism.* Harmondsworth, U.K.: Penguin.

Schiele, J. 1998. "Cultural Alignment, African American Male Youths and Violent Crime." *Journal of Behavior in the Social Environment*, Vol. 1, No. 2, pp. 165–81.

Scully, D. and J. Marolla. 1984. "Convicted Rapists' Vocabulary of Motives: Excuses and Justifications." *Social Problems*, Vol. 31, No. 4, pp. 530–44.

Stephens, B. 1984. "Vocabularies of Motive and Suicide." *Suicide and Life-Threatening Behavior*, Vol. 14, No. 2, pp. 243–53.

Sykes, G. and D. Matza. 1957. "Techniques of Neutralization: A Theory of Delinquency." *American Sociological Review*, Vol. 22, No. 4, pp. 664–70.

Tatum, B. 2000. *Crime, Violence and Minority Groups.* Aldershot, U.K.: Ashgate Publishing.

Wortley, S. 1999. "A Northern Taboo: Research on Race, Crime and Criminal Justice in Canada." *Canadian Journal of Criminology*, Vol. 41, No. 1, pp. 261–74.

Talking Past Each Other: The Black and White Languages of Race

Bob Blauner

In this article Blauner uses the race riots that occurred after the trial of Rodney King to examine the different worldviews of blacks and whites in America. Focusing on the general understanding of race, racism, and racial equality for both sectors, he argues that there are two "languages" operating within America. What does the author mean when he says that there are "two languages of race"? Do you think that there are two (or more) languages of race in Canada and the U.S.?

Blauner also topicalizes the deeper question of what race means, especially in comparison to ethnicity. How is confusion between race and ethnicity echoed in the popular discourse about race? Given that physical anthropologists and other social scientists have shown the biological notion of race to be a social and political fiction, does it any longer make sense to dignify and promote the concept even though the fiction has been consequential historically and currently?

For many African-Americans who came of age in the 1960s, the assassination of Martin Luther King, Jr. in 1968 was a defining moment in the development of their personal racial consciousness. For a slightly older group, the 1955 lynching of the fourteen-year-old Chicagoan Emmitt Till in Mississippi had been a similar awakening. Now we have the protest and violence in Los Angeles and other cities in late April and early May of 1992, spurred by the jury acquittal of four policemen who beat motorist Rodney King.

The aftermath of the Rodney King verdict, unlike any other recent racial violence, will be seared into the memories of Americans of *all* colors, changing the way they see each other and their society. Spring 1992 marked the first time since the 1960s that incidents of racial injustice against an

Source: Reprinted with permission from Bob Blauner, "Talking Past Each Other: Black and White Languages of Race," *The American Prospect*, Volume 3, Number 10: June 01, 1992. The American Prospect, 11 Beacon Street, Suite 1120, Boston, MA 02108. All rights reserved.

African-America[n]—and by extension the black community—have seized the entire nation's imagination.... The response to the Rodney King verdict is thus a long-overdue reminder that whites still have the capacity to feel deeply about white racism—when they can see it in unambiguous terms.

The videotaped beating by four Los Angeles police officers provided this concreteness. To be sure, many whites focused their response on the subsequent black rioting, while the anger of blacks tended to remain fixed on the verdict itself. However, whites initially were almost as upset as blacks: An early poll reported that 86 percent of European-Americans disagreed with the jury's decision. The absence of any black from the jury and the trial's venue, Simi Valley, a lily-white suburban community, enabled mainstream whites to see the parallels with the Jim Crow justice of the old South. When we add to this mixture the widespread disaffection, especially of young people, with the nation's political and economic conditions, it is easier to explain the scale of white emotional involvement, unprecedented in a matter of racial protest since the 1960s....

While many whites saw the precipitating events as expressions of racist conduct, they were much less likely than blacks to see them as part of some larger pattern of racism. Thus two separate polls found that only half as many whites as blacks believe that the legal system treats whites better than blacks. (In each poll, 43 percent of whites saw such a generalized double standard, in contrast to 84 percent of blacks in one survey, 89 percent in the other.)

This gap is not surprising. For twenty years European-Americans have tended to feel that systematic racial inequities marked an earlier era, not our own. Psychological denial and a kind of post-1960s exhaustion may both be factors in producing the sense among mainstream whites that civil rights laws and other changes resolved blacks' racial grievances, if not the economic basis of urban problems. But the gap in perceptions of racism also reflects a deeper difference. Whites and blacks see racial issues through different lenses and use different scales to weigh and assess injustice.

I am not saying that blacks and whites have totally disparate value systems and worldviews. I think we were more polarized in the late 1960s.... By 1979 blacks and whites had come closer together on many issues than they had been in 1968. In the late 1970s and again in the mid-to-late 1980s, both groups were feeling quite pessimistic about the nation's direction. They agreed that America had become a more violent nation and that people were more individualistic and less bound by such traditional values as hard work, personal responsibility, and respect for age and authority. But with this and other convergences, there remained a striking gap in the way European-Americans and African-Americans evaluated *racial* change. Whites were impressed by the scale of integration, the size of the black middle class, and the extent of demonstrable progress. Blacks were disillusioned with integration, concerned about the people who had been left behind, and much more negative in their overall assessment of change. . . .

I want to advance the proposition that there are two languages of race in America. I am not talking about black English and standard English, which refer to different structures of grammar and dialect. "Language" here signifies a system of implicit understandings about social reality, and a racial language encompasses a worldview.

Blacks and whites differ on their interpretations of social change from the 1960s through the 1990s because their racial languages define the central terms, especially "racism," differently. Their racial languages incorporate different views of American society itself, especially the question of how central race and racism are to America's very existence, past and present. Blacks believe in this centrality, while most whites, except for the more race-conscious extremists, see race as a peripheral reality. Even successful, middle-class black professionals experience slights and humiliations—incidents when they are stopped by police, regarded suspiciously by clerks while shopping, or mistaken for messenger, drivers, or aides at work—that remind them they have not escaped racism's reach. For whites, race becomes central on exceptional occasions: collective, public moments such as the recent events, and private ones, such as a family's decision to escape urban problems with a move to the suburbs. But most of the time European-Americans are able to view racial issues as aberrations in American life. . . .

Because of these differences in language and worldview, blacks and whites often talk past one another, just as men and women sometimes do. . . . Whites locate racism in color consciousness and its absence in color blindness. They regard it as a kind of racism when students of color insistently underscore their sense of difference, their affirmation of ethnic and racial membership, which minority students have increasingly asserted. Many black, and increasingly also Latino and Asian, students cannot understand this reaction. It seems to them misinformed, even ignorant. They in turn sense a kind of racism in the whites' assumption that minorities must assimilate to mainstream values and styles. Then African-Americans will posit an idea that many whites find preposterous: Black people, they argue, cannot be racist, because racism is a system of power, and black people as a group do not have power.

In this and many other arenas, a contest rages over the meaning of racism. . . .

THE WIDENING CONCEPTION OF RACISM

The term "racism" was not commonly used in social science or American public life until the 1960s. "Racism" does not appear, for example, in the

Swedish economist Gunnar Myrdal's classic 1944 study of American race relations, *An American Dilemma*. But even when the term was not directly used, it is still possible to determine the prevailing understandings of racial oppression.

In the 1940s racism referred to an ideology, an explicit system of beliefs postulating the superiority of whites based on the inherent, biological inferiority of the colored races. Ideological racism was particularly associated with the belief systems of the Deep South and was originally devised as a rationale for slavery. Theories of white supremacy, particularly in their biological versions, lost much of their legitimacy after the Second World War due to their association with Nazism. In recent years, cultural explanations of "inferiority" are heard more commonly than biological ones. . . .

By the 1950s and early 1960s, with ideological racism discredited, the focus shifted to a more discrete approach to racially invidious attitudes and behavior, expressed in the model of prejudice and discrimination. "Prejudice" referred (and still does) to hostile feelings and beliefs about racial minorities and the web or stereotypes justifying such negative attitudes. "Discrimination" referred to actions meant to harm the members of a racial minority group. The logic of this model was that racism implied a double standard, that is, treating a person of color differently—in mind or action—than one would a member of the majority group.

By the mid-1960s the terms "prejudice" and "discrimination" and the implicit model of racial causation implied by them were seen as too weak to explain the sweep of racial conflict and change, too limited in their analytical power, and for some critics too individualistic in their assumptions. Their original meanings tended to be absorbed by a new, more encompassing idea of racism. During the 1960s the referents of racial oppression moved from individual actions and beliefs to group and institutional processes, from subjective ideas to "objective" structures or results. . . .

The most notable of these new definitions was "institutional racism." In their 1967 book *Black Power,* Stokely Carmichael and Charles Hamilton stressed how institutional racism was different and more fundamental than individual racism. Racism, in this view, was built into society and scarcely required prejudicial attitudes to maintain racial oppression.

This understanding of racism as pervasive and institutionalized spread from relatively narrow "movement" and academic circles to the larger public with the appearance in 1968 of the report of the commission on the urban riots appointed by President Lyndon Johnson and chaired by Illinois Governor Otto Kerner. The Kerner Commission identified "white racism" as a prime reality of American society and the major underlying cause of ghetto unrest. America, in this view, was moving toward two societies, one white and one black (it is not clear where other racial minorities fit in)....

Another definition of racism, which I would call "racism as atmosphere," also emerged in the 1960s and 1970s. This is the idea that an organization or an environment might be racist because its implicit, unconscious structures were devised for the use and comfort of white people, with the result that people of other races will not feel at home in such settings. Acting on this understanding of racism, many schools and universities, corporations, and other institutions, have changed their teaching practices or work environments to encourage a greater diversity in their clientele, students, or work force.

Perhaps the most radical definition of all was the concept of "racism as result." In this sense, an institution or an occupation is racist simply because racial minorities are underrepresented in numbers or in positions of prestige and authority.

Seizing on different conceptions of racism, the blacks and whites I talked to in the late 1970s had come to different conclusions about how far America had moved toward racial justice. Whites tended to adhere to earlier, more limited notions of racism. Blacks for the most part saw the newer meanings as more basic. Thus African-Americans did not think racism had been put to rest by civil rights laws, even by the dramatic changes in the South....

Whites saw racism largely as a thing of the past. They defined it in terms of segregation and lynching, explicit white supremacist beliefs, or double standards in hiring, promotion, and admissions to colleges or other institutions. Except for affirmative action, which seemed the most blatant expression of such double standards, they were positively impressed by racial change. Many saw the relaxed and comfortable relations between whites and blacks as the heart of the matter....

The newer, expanded definitions of racism just do not make much sense to most whites. I have experienced their frustrations directly when I try to explain the concept of institutional racism to white students and popular audiences. The idea of racism as an "impersonal force" loses all but the most theoretically inclined. Whites are more likely than blacks to view racism as a personal issue....

The new meanings make sense to blacks, who live such experiences in their bones. But by 1979 many of the African Americans in my study, particularly the older activists, were critical of the use of racism as a blanket explanation for all manifestations of racial inequality. Long before similar ideas were voiced by the black conservatives, many blacks sensed that too heavy emphasis on racism led to the false conclusions that blacks could only progress through a conventional civil rights strategy of fighting prejudice and discrimination.... Overemphasizing racism, they feared, was interfering with the black community's ability to achieve greater self-determination through the politics of self-help. In addition, they told me that the prevailing rhetoric of the 1960s had affected many young blacks. Rather than taking responsibility for their own difficulties, they were now using racism as a "cop-out."

. . .

BACK TO BASICS

The question then becomes what to do about these multiple and confusing meanings of racism and their extraordinary personal and political change. I would begin by honouring both the black and white readings of the term....

... While the black understanding of racism is, in some sense, the deeper one, the white views of racism (ideology, double standard) refer to more specific and recognizable beliefs and practices. Since there is also a cross-racial consensus on the immorality of racist ideology and racial discrimination, it makes sense whenever possible to use such a concrete referent as discrimination, rather than the more global concept of racism.... And when feasible, we need to try to bridge the gap by shifting from the language of race to that of ethnicity and class.

RACE OR ETHNICITY?

In the American consciousness the imagery of race—especially along the black-white dimension—tends to be more powerful than that of class or ethnicity. As a result, legitimate ethnic affiliations are often misunderstood to be racial and legitimate.

Race itself is a confusing concept because of the variance between scientific and common sense definitions of the term. Physical anthropologists who study the distribution of those characteristics we use to classify "races" teach us that race is a fiction because all peoples are mixed to various degrees. Sociologists counter that this biological fiction unfortunately remains a sociological reality. People define one another racially, and thus divide society into racial groups. The "fiction" of race affects every aspect of peoples' lives, from living standards to landing in jail.

The consciousness of color differences, and the invidious distinctions based on them, have existed since antiquity and are not limited to any one corner of the world. And yet the peculiarly modern division of the world into a discrete number of hierarchically ranked races is a historic product of Western colonialism. In precolonial Africa the relevant group identities were national, tribal, or linguistic. There was no concept of an African or black people until this category was created by the combined effects of slavery, imperialism, and the anticolonial and Pan-African movements.... Thus race is an essentially political construct, one that translates our tendency to see people in terms of their color or other physical attributes into structures that make it likely that people will act for or against them on such a basis.

The dynamic of ethnicity is different, even though the results at times may be similar. An ethnic group is a group that shares a belief in its common past. Members of an ethnic group hold a set of common memories that make them feel that their customs, culture, and outlook are distinctive. In short, they have a sense of peoplehood. Sharing critical experiences and sometimes a belief in their common fate, they feel an affinity for one another, a "comfort zone" that leads to congregating together, even when this is not forced by exclusionary barriers. Thus if race is associated with biology and nature, ethnicity is associated with culture. Like races, ethnic groups arise historically, transform themselves, and sometimes die out.

Much of the popular discourse about race in America today goes awry because ethnic realities get lost under the racial umbrella.... Thus white students, disturbed when blacks associate with each other, justify their objections through their commitment to *racial* integration. They do not appreciate the ethnic affinities that bring this about, or see the parallels to Jewish students meeting at the campus Hillel Foundation or Italian-Americans eating lunch at the Italian house on the Berkeley campus.

When blacks are "being ethnic," whites see them as "being racial." Thus they view the identity politics of students who want to celebrate their blackness, their *chicanismo,* their Asian heritages, and their American Indian roots as racially offensive. Part of this reaction comes from a sincere desire, almost a yearning, of white students for a color-blind society. But because the ethnicity of darker people so often gets lost in our overracialized perceptions, the white students misread the situation. When I point out to my class that whites are talking about race and its dynamics and the students of color are talking about ethnicity and its differing meaning, they can begin to appreciate each other's agendas.

Confounding race and ethnicity is not just limited to the young. The general public, including journalists and other opinion makers, does this regularly, with serious consequences for the clarity of public dialogue and sociological analysis. A clear example comes from the Chicago mayoral election of 1983. The establishment press, including leading liberal columnists, regularly chastised the black electorate for giving virtually all its votes to Harold Washington. Such racial voting was as "racist" as whites voting for the other candidate because they did not want a black mayor. Yet African-Americans were voting for ethnic representation just as Irish-Americans, Jews, and Italians have always done. Such ethnic politics is considered the American way. What is discriminatory is the double standard that does not confer the same rights on blacks....

Such confusions between race and ethnicity are exacerbated by the ambiguous sociological status of African-Americans. Black Americans are *both* a race and an ethnic group. Unfortunately, part of our heritage of racism has been to deny the ethnicity, the cultural heritage of black Americans.... Until the 1960s few believed that black culture was a real ethnic culture.

Because our racial language is so deep-seated, the terminology of black and white just seems more "natural" and common-sensical than more ethnic labels like African-American or European-American. But the shift to the term African-American has been a conscious attempt to move the discourse from a language of race to a language of ethnicity. "African-American," as Jesse Jackson and others have pointed out, connects the group to its history and culture in a way that the racial designation, black, does not. The new usage parallels terms for other ethnic groups....

The issue is further complicated by the fact that African-Americans are not a homogeneous group. They comprise a variety of distinct ethnicities. There are the West Indians with their long history in the U.S., the darker Puerto Ricans (some of whom identify themselves as black), the more recently arrived Dominicans, Haitians, and immigrants from various African countries, as well as the native-born African-Americans, among whom regional distinctions can also take on a quasi-ethnic flavor....

For white Americans, race does not overwhelm ethnicity. Whites see the ethnicity of other whites; it is their own whiteness they tend to overlook. But even when race is recognized, it is not conflated with ethnicity. Jews, for example, clearly distinguish their Jewishness from their whiteness. Yet the long-term dynamic still favors the development of a dominant white racial identity. Except for recent immigrants, the various European ethnic identifies have been rapidly weakening. Vital ethnic communities persist in some cities, particularly on the East Coast. But many whites, especially the young, have such diverse ethnic heritages that they have no meaningful ethnic affiliation....

Instead of dampening the ethnic enthusiasms of the racial minorities, perhaps it would be better to encourage the revitalization of whites' European heritages. But a problem with this approach is that the relationship between race and ethnicity is more ambiguous for whites than for people of color.

. . .

Out of this vacuum the emerging identity of "European-American" has come into vogue. I interpret the European-American idea as part of a yearning for a usable past. Europe is associated with history and culture.

"America" and "American" can no longer be used to connote white people. "White" itself is a racial term and thereby inevitably associated with our nation's legacy of social injustice.

. . .

It is unrealistic to expect that the racial groupings of American society can be totally "deconstructed," as a number of scholars now are advocating. After all, African-Americans and native Americans, who are not immigrants, can never be exactly like other ethnic groups. Yet a shift in this direction would begin to move our society from a divisive biracialism to a more inclusive multiculturalism.

To return to the events of spring 1992, I ask what was different about these civil disturbances. Considering the malign neglect of twelve Reagan–Bush years, the almost two decades of economic stagnation, and the retreat of the public from issues of race and poverty, the violent intensity should hardly be astonishing.

More striking was the multi-racial character of the response. In the San Francisco Bay area, rioters were as likely to be white as non-white. In Los Angeles, Latinos were prominent among both the protesters and the victims. South Central Los Angeles is now more Hispanic than black, and this group suffered perhaps 60 percent of the property damage. The media have focused on the specific grievances of African-Americans toward Koreans. But I would guess that those who trashed Korean stores were protesting something larger than even the murder of a fifteen-year-old black girl. Koreans, along with other immigrants, continue to enter the country and in a relatively short time surpass the economic and social position of the black poor. The immigrant advantage is real and deeply resented by African-Americans, who see that the two most downtrodden minorities are those that did not enter the country voluntarily. . . .

Will this widened conflict finally lead Americans toward a recognition of our common stake in the health of the inner cities and their citizens, or toward increased fear and division? The Emmitt Till lynching in 1955 set the stage for the first mass mobilization of the civil rights movement, the Montgomery bus boycott later that year. Martin Luther King's assassination provided the impetus for the institution of affirmative action and other social programs. The Rodney King verdict and its aftermath must also become not just a psychologically defining moment but an impetus to a new mobilization of political resolve.

READING THIRTY-EIGHT

Anti-Semite and Jew

Jean Paul Sartre

What is racism? How does it develop and what sustains it? How can we formulate it as a distinctive course of social action that represents the self and orients to another? In this excerpt from his larger work, Sartre answers these questions by examining the anti-Semite as one particular expression of racism. Does his characterization of the anti-Semite accurately depict other forms of racism with which you may be familiar?

If a man attributes all or part of his own misfortunes and those of his country to the presence of Jewish elements in the community, if he proposes to remedy this state of affairs by depriving the Jews of certain of their rights, by keeping them out of certain economic and social activities, by expelling them from the country, by exterminating all of them, we say that he has anti-Semitic *opinions.*

This word *opinion* makes us stop and think. It is the word a hostess uses to bring to an end a discussion that threatens to become acrimonious. It suggests that all points of view are equal; it reassures us, for it gives an inoffensive appearance to ideas by reducing them to the level of tastes. All tastes are natural; all opinions are permitted. Tastes, colors, and opinions are not open to discussion. In the name of democratic institutions, in the name of freedom of opinion, the anti-Semite asserts the right to preach the anti-Jewish crusade everywhere.

. . .

But I refuse to characterize as opinion a doctrine that is aimed directly at particular persons and that seeks to suppress their rights or to exterminate them. The Jew whom the anti-Semite wishes to lay hands upon is not a schematic being defined solely by his function, as under administrative law; or by his status or his acts, as under the Code. He is a Jew, the son of Jews, recognizable by his physique, by the color of his

Source: From ANTI-SEMITE AND JEW by Jean-Paul Sartre, translated by George Becker, copyright 1948 and renewed 1976 by Schocken Books, a division of Random House, Inc. Preface Copyright © 1995 by Michael Walzer. Used by permission of Schocken Books, a division of Random House, Inc.

hair, by his clothing perhaps, and, so they say, by his character. Anti-Semitism does not fall within the category of ideas protected by the right of free opinion.

Indeed, it is something quite other than an idea. It is first of all a *passion*. No doubt it can be set forth in the form of a theoretical proposition. The "moderate" anti-Semite is a courteous man who will tell you quietly: "Personally, I do not detest the Jews. I simply find it preferable, for various reasons, that they should play a lesser part in the activity of the nation." But a moment later, if you have gained his confidence, he will add with more abandon: "You see, there must be *something* about the Jews; they upset me physically."

This argument, which I have heard a hundred times, is worth examining. First of all, it derives from the logic of passion. For, really now, can we imagine anyone's saying seriously: "There must be something about tomatoes, for I have a horror of eating them"? In addition, it shows us that anti-Semitism in its most temperate and most evolved forms remains a syncretic whole which may be expressed by statements of reasonable tenor, but which can involve even bodily modifications. Some men are suddenly struck with impotence if they learn from the woman with whom they are making love that she is a Jewess. There is a disgust for the Jew, just as there is a disgust for the Chinese or the Negro among certain people. Thus it is not from the body that the sense of repulsion arises, since one may love a Jewess very well if one does not know what her race is; rather it is something that enters the body from the mind. It is an involvement of the mind, but one so deep-seated and complete that it extends to the physiological realm, as happens in cases of hysteria.

This involvement is not caused by experience. I have questioned a hundred people on the reasons for their anti-Semitism. Most of them have confined themselves to enumerating the defects with which tradition has endowed the Jews. "I detest them because they are selfish, intriguing, persistent, oily, tactless, etc."—"But, at any rate, you associate with some of them?"—"Not if I can help it!" A painter said to me: "I am hostile to the Jews because, with their critical habits, they encourage our servants to insubordination." Here are examples a little more precise. A young actor without talent insisted that the Jews had kept him from a successful career in the theater by confining him to subordinate roles. A young woman said to me: "I have had the most horrible experiences with furriers; they robbed me, they burned the fur I entrusted to them. Well, they were all Jews." But why did she choose to hate Jews rather than furriers? Why Jews or furriers rather than such and such a Jew or such and such a furrier? Because she had in her a predisposition toward anti-Semitism.

. . .

People speak to us also of "social facts," but if we look at this more closely we shall find the same vicious circle. There are too many Jewish lawyers, someone says. But is there any complaint that there are too many Norman lawyers? Even if all the Bretons were doctors would we say anything more than that "Brittany provides doctors for the whole of France"? Oh, someone will answer, it is not all the same thing. No doubt, but that is precisely because we consider Normans as Normans and Jews as Jews. Thus wherever we turn it is the *idea of the Jew* which seems to be the essential thing.

It has become evident that no external factor can induce anti-Semitism in the anti-Semite. Anti-Semitism is a free and total choice of oneself, a comprehensive attitude that one adopts not only toward Jews but toward men in general, toward history and society; it is at one and the same time a passion and a conception of the world. No doubt in the case of a given anti-Semite certain characteristics will be more marked than in another. But they are always all present at the same time, and they influence each other. It is this syncretic totality which we must now attempt to describe.

I noted earlier that anti-Semitism is a passion. Everybody understands that emotions of hate or anger are involved. But ordinarily hate and anger have a *provocation:* I have someone who has made me suffer, someone who contemns or insults me. We have just seen that anti-Semitic passion could not have such a character. It precedes the facts that are supposed to call it forth; it seeks them out to nourish itself upon them; it must even interpret them in a special way so that they may become truly offensive. Indeed, if you so much as mention a Jew to an anti-Semite, he will show all the signs of a lively irritation. If we recall that we must always *consent* to anger before it can manifest itself and that, as is indicated so accurately by the French idiom, we "put ourselves" into anger, we shall have to agree that the anti-Semite has *chosen* to live on the plane of passion. It is not unusual for people to elect to live a life of passion rather than one of reason. But ordinarily they love the *objects* of passion: women, glory, power, money. Since the anti-Semite has chosen hate, we are forced to conclude that it is the *state* of passion that he loves. Ordinarily this type of emotion is not very pleasant: a man who passionately desires a woman is impassioned because of the woman and in spite of his passion. We are wary of reasoning based on passion, seeking to support by all possible means opinions which love or jealousy or hate have dictated. We are wary of the aberrations of passion and of what is called monoideism. But that is just what the anti-Semite chooses right off.

How can one choose to reason falsely? It is because of a longing for impenetrability. The rational man groans as he gropes for the truth; he knows that his reasoning is no more than tentative, that other considerations may supervene to cast doubt on it. He never sees very clearly where he is going; he is "open"; he may even appear to be hesitant. But there are people who are attracted by the durability of a stone. They wish to be massive and impenetrable; they wish not to change. Where, indeed, would change take them? We have here a basic fear of oneself and of truth. What frightens them is not the content of truth, of which they have no conception, but the form itself of truth, that thing of indefinite approximation. It is as if their own existence were in continual suspension. But they wish to exist all at once and right away. They do not want any acquired opinions; they want them to be innate. Since they are afraid of reasoning, they wish to lead the kind of life wherein reasoning and research play only a subordinate role, wherein one seeks only what he has already found, wherein one becomes only what he already was. This is nothing but passion. Only a strong emotional bias can give a lightninglike certainty; it alone can hold reason in leash; it alone can remain impervious to experience and last for a whole lifetime.

The anti-Semite has chosen hate because hate is a faith; at the outset he has chosen to devaluate words and reasons. How entirely at ease he feels as a result. How futile and frivolous discussions about the rights of the Jew appear to him. He has placed himself on other ground from the beginning. If out of courtesy he consents for a moment to defend his point of view, he lends himself but does not give himself. He tries simply to project his intuitive certainty onto the plane of discourse. I mentioned awhile back some remarks by anti-Semites, all of them absurd: "I hate Jews because they make servants insubordinate, because a Jewish furrier robbed me, etc." Never believe that anti-Semites are completely unaware of the absurdity of their replies. They know that their remarks are frivolous, open to challenge. But they are amusing themselves, for it is their adversary who is obliged to use words responsibly, since he believes in words. The anti-Semites have the *right* to play. They even like to play with discourse for, by giving ridiculous reasons, they discredit the seriousness of their interlocutors. They delight in acting in bad faith, since they seek not to persuade by sound argument but to intimidate and disconcert. If you press them too closely, they will abruptly fall silent, loftily indicating by some phrase that the time for argument is past. It is not that they are afraid of being convinced. They fear only to appear ridiculous or to prejudice by their embarrassment their hope of winning over some third person to their side.

If then, as we have been able to observe, the anti-Semite is impervious to reason and to experience, it is not because his conviction is strong. Rather his conviction is strong because he has chosen first of all to be impervious.

He has chosen also to be terrifying. People are afraid of irritating him. No one knows to what lengths the aberrations of his passion will carry him—but he knows, for this passion is not provoked by something external. He has it well in hand; it is obedient to his will: now he lets go the reins and now he pulls back on them. He is not afraid of himself, but he sees in the eyes of others a disquieting image—his own—and he makes his words and gestures conform to it. Having this external model, he is under no necessity to look for his personality within himself. He has chosen to find his being entirely outside himself, never to look within, to be nothing save the fear he inspires in others. What he flees even more than Reason is his intimate awareness of himself. But someone will object: What if he is like that only with regard to the Jews? What if he otherwise conducts himself with good sense? I reply that that is impossible. There is the case of a fishmonger who, in 1942, annoyed by the competition of two Jewish fishmongers who were concealing their race, one fine day took pen in hand and denounced them. I have been assured that this fishmonger was in other respects a mild and jovial man, the best of sons. But I don't believe it. A man who finds it entirely natural to denounce other men cannot have our conception of humanity; he does not see even those whom he aids in the same light as we do. His generosity, his kindness are not like our kindness, our generosity. You cannot confine passion to one sphere.

The anti-Semite readily admits that the Jew is intelligent and hardworking; he will even confess himself inferior in these respects. This concession costs him nothing, for he has, as it were, put those qualities in parentheses. Or rather they derive their value from the one who possesses them: the more virtues the Jew has the more dangerous he will be. The anti-Semite has no illusions about what he is. He considers himself an average man, modestly average, basically mediocre. There is no example of an anti-Semite's claiming individual superiority over the Jews. But you must not think that he is ashamed of his mediocrity; he takes pleasure in it; I will even assert that he has chosen it. This man fears every kind of solitariness, that of the genius as much as that of the murderer; he is the man of the crowd. However small his stature, he takes every precaution to make it smaller, lest he stand out from the herd and find himself face to face with himself. He has made himself an anti-Semite because that is something one cannot be alone. The phrase, "I hate the Jews," is one that is uttered in chorus; in pronouncing it, one attaches himself to a tradition and to a community—the tradition and community of the mediocre.

We must remember that a man is not necessarily humble or even modest because he has consented to mediocrity. On the contrary, there is a passionate pride among the mediocre, and anti-Semitism is an attempt to give value to mediocrity as such, to create an elite of the ordinary. To the anti-Semite, intelligence is Jewish; he can thus disdain it in all tranquility, like all the other virtues which the Jew possesses. They are so many ersatz attributes that the Jew cultivates in place of that balanced mediocrity which he will never have. The true Frenchman, rooted in his province, in his country, borne along by a tradition twenty centuries old, benefiting from ancestral wisdom, guided by tried customs, does not *need* intelligence. His virtue depends upon the assimilation of the qualities which the work of a hundred generations has lent to the objects which surround him; it depends on property. It goes without saying that this is a matter of inherited property, not property one buys. The anti-Semite has a fundamental incomprehension of the various forms of modern property: money, securities, etc. These are abstractions, entities of reason related to the abstract intelligence of the Semite. A security belongs to no one because it can belong to everyone; moreover, it is a sign of wealth, not a concrete possession. The anti-Semite can conceive only of a type of primitive ownership of land based on a veritable magical rapport, in which the thing possessed and its possessor are united by a bond of mystical participation; he is the poet of real property. It transfigures the proprietor and endows him with a special and concrete sensibility. To be sure, this sensibility ignores eternal truths or universal values: the universal is Jewish, since it is an object of intelligence. What his subtle sense seizes upon is precisely that which the intelligence cannot perceive. To put it another way, the principle underlying anti-Semitism is that the concrete possession of a particular object gives as if by magic the meaning of that object. Maurras said the same thing when he declared a Jew to be forever incapable of understanding this line of Racine:

*Dans l'Orient désert, quel devint mon ennui.**

But the way is open to me, mediocre me, to understand what the most subtle, the most cultivated intelligence has been unable to grasp. Why? Because I possess Racine—Racine and my country and my soil. Perhaps the Jew speaks a purer French than I do, perhaps he knows syntax and grammar better, perhaps he is even a writer. No matter; he has spoken this language for only twenty years, and I for a thousand years. The correctness of his style is abstract, acquired; my faults of French are in conformity with the genius of the language. We recognize here the reasoning that Barrès used against the holders of scholarships. There is no occasion for surprise. Don't the Jews have all the scholarships? All that intelligence, all that money can acquire one leaves to them, but it is as empty as the wind. The only things that count are irrational values, and it is just these things which are denied the Jews forever. Thus the anti-Semite takes his stand from the start on the ground of irrationalism. He is opposed to the Jew, just as sentiment is to intelligence, the particular to the universal, the past to the present, the concrete to the abstract, the owner of real property to the possessor of negotiable securities.

* *Bérénice.*

Besides this, many anti-Semites—the majority, perhaps—belong to the lower middle class of the towns; they are functionaries, office workers, small businessmen, who possess nothing. It is in opposing themselves to the Jew that they suddenly become conscious of being proprietors: in representing the Jew as a robber, they put themselves in the enviable position of people who could be robbed. Since the Jew wishes to take France from them, it follows that France must belong to them. Thus they have chosen anti-Semitism as a means of establishing their status as possessors. . . .

We begin to perceive the meaning of the anti-Semite's choice of himself. He chooses the irremediable out of fear of being free; he chooses mediocrity out of fear of being alone, and out of pride he makes of this irremediable mediocrity a rigid aristocracy. To this end he finds the existence of the Jew absolutely necessary. Otherwise to whom would he be superior? Indeed, it is vis-à-vis the Jew and the Jew alone that the anti-Semite realizes that he has rights. If by some miracle all the Jews were exterminated as he wishes, he would find himself nothing but a concierge or a shopkeeper in a strongly hierarchical society in which the quality of "true Frenchman" would be at a low valuation, because everyone would possess it. He would lose his sense of rights over the country because no one would any longer contest them, and that profound equality which brings him close to the nobleman and the man of wealth would disappear all of a sudden, for it is primarily negative. His frustrations, which he has attributed to the disloyal competition of the Jew, would have to be imputed to some other cause, lest he be forced to look within himself. He would run the risk of falling into bitterness, into a melancholy hatred of the privileged classes. Thus the anti-Semite is in the unhappy position of having a vital need for the very enemy he wishes to destroy.

The equalitarianism that the anti-Semite seeks with so much ardor has nothing in common with that equality inscribed in the creed of the democracies. The latter is to be realized in a society that is economically hierarchical, and is to remain compatible with a diversity of functions. But it is in protest *against* the hierarchy of functions that the anti-Semite asserts the equality of Aryans. He does not understand anything about the division of labor and doesn't care about it. From his point of view each citizen can claim the title of Frenchman, not because he co-operates, in his place or in his occupation, with others in the economic, social, and cultural life of the nation, but because he has, in the same way as everybody else, an imprescriptible and inborn right to the indivisible totality of the country. Thus the society that the anti-Semite conceives of is a society of juxtaposition, as one can very well imagine, since his ideal of property is that of real and basic property. Since, in point of fact, anti-Semites are numerous, each of them does his part in constituting a community based on mechanical solidarity in the heart of organized society.

. . .

Any anti-Semite is therefore, in varying degree, the enemy of constituted authority. He wishes to be the disciplined member of an undisciplined group; he adores order, but a *social* order. We might say that he wishes to provoke political disorder in order to restore social order, the social order in his eyes being a society that, by virtue of juxtaposition, is egalitarian and primitive, one with a heightened temperature, one from which Jews are excluded. These principles enable him to enjoy a strange sort of independence, which I shall call an inverted liberty. Authentic liberty assumes responsibilities, and the liberty of the anti-Semite comes from the fact that he escapes all of his. Floating between an authoritarian society which has not yet come into existence and an official and tolerant society which he disavows, he can do anything he pleases without appearing to be an anarchist, which would horrify him. The profound seriousness of his aims—which no word, no statement, no act can express—permits him a certain frivolity. He is a hooligan, he beats people up, he purges, he robs; it is all in a good cause. If the government is strong, anti-Semitism withers, unless it be a part of the program of the government itself, in which case it changes its nature. Enemy of the Jews, the anti-Semite has need of them. Anti-democratic, he is a natural product of democracies and can only manifest himself within the framework of the Republic.

We begin to understand that anti-Semitism is more than a mere "opinion" about the Jews and that it involves the entire personality of the anti-Semite. But we have not yet finished with him, for he does not confine himself to furnishing moral and political directives: he has a method of thought and a conception of the world all his own. In fact, we cannot state what he affirms without implicit reference to certain intellectual principles.

The Jew, he says, is completely bad, completely a Jew. His virtues, if he has any, turn to vices by reason of the fact that they are his; work coming from his hands necessarily bears his stigma. If he builds a bridge, that bridge, being Jewish, is bad from the first to the last span. The same action carried out by a Jew and by a Christian does not have the same meaning in the two cases, for the Jew contaminates all that he touches with an I-know-not-what execrable quality. The first thing the Germans did was to forbid Jews access to swimming pools; it seemed to them that if the body of an Israelite were to plunge into that confined body of water, the water would be completely befouled. Strictly speaking, the Jew contaminates even the air he breathes.

If we attempt to formulate in abstract terms the principle to which the anti-Semite appeals, it would come to this: A whole is more and other than the sum of its parts; a whole determines the meaning and underlying character of the parts that make it up. There is not *one* virtue of courage which enters indifferently into a Jewish character or a Christian character in the way that oxygen indifferently combines with nitrogen and argon to form air and with hydrogen to form water. Each person is an indivisible totality that has its own courage, its own generosity, its own way of thinking, laughing, drinking, and eating. What is there to say except that the anti-Semite has chosen to fall back on the spirit of synthesis in order to understand the world. It is the spirit of synthesis which permits him to conceive of himself as forming an indissoluble unity with all France. It is in the name of this spirit that he denounces the purely analytical and critical intelligence of the Jews. But we must be more precise. For some time, on the Right and on the Left, among the traditionalists and among the socialists, it has been the fashion to make appeal to synthetic principles as against the spirit of analysis which presided over the foundation of bourgeois democracy....

We are now in a position to understand the anti-Semite. He is a man who is afraid. Not of the Jews, to be sure, but of himself, of his own consciousness, of his liberty, of his instincts, of his responsibilities, of solitariness, of change, of society, and of the world—of everything except the Jews. He is a coward who does not want to admit his cowardice to himself; a murderer who represses and censures his tendency to murder without being able to hold it back, yet who dares to kill only in effigy or protected by the anonymity of the mob; a malcontent who dares not revolt from fear of the consequences of his rebellion. In espousing anti-Semitism, he does not simply adopt an opinion, he chooses himself as a person. He chooses the permanence and impenetrability of stone, the total irresponsibility of the warrior who obeys his leaders—and he has no leader. He chooses to acquire nothing, to deserve nothing; he assumes that everything is given him as his birthright—and he is not noble. He chooses finally a Good that is fixed once and for all, beyond question, out of reach; he dares not examine it for fear of being led to challenge it and having to seek it in another form. The Jew only serves him as a pretext; elsewhere his counterpart will make use of the Negro or the man of yellow skin. The existence of the Jew merely permits the anti-Semite to stifle his anxieties at their inception by persuading himself that his place in the world has been marked out in advance, that it awaits him, and that tradition gives him the right to occupy it. Anti-Semitism, in short, is fear of the human condition. The anti-Semite is a man who wishes to be pitiless stone, a furious torrent, a devastating thunderbolt—anything except a man.

Realms of Freedom, Realms of Necessity

Ian McKay

Poverty, disease, global warming, war. What is to be done? In this inspiring call to action Ian McKay reinvigorates the idea of the political left as the "possibility of living otherwise." What might we do differently, and how might we think differently, about persistent social problems? McKay says that we begin by locating the patterns of problems and the system underlying them. He urges us to see the connections between struggles, formulate a vision of an "otherwise," and then act in ways that exemplify this vision. One central problem he raises is that, despite all the political talk of freedom in Western societies, most of us live trapped in the realm of necessity. Why must this be so?

In 1998 the planet's two hundred wealthiest residents had a net worth equal to about 41 per cent of the total world population. A very few favoured individuals—Bill Gates, the principal owners of Wal-Mart, and the Sultan of Brunei—together enjoyed accumulations of wealth equal to the national incomes of thirty-six of the world's most impoverished countries.

Meanwhile about 1.3 billion people around the world were making do on the equivalent of about one U.S. dollar a day. In Canada, one of the wealthiest countries in the world, poverty increased dramatically from 1990 to 1995, particularly in metropolitan centres. In large cities the general population grew by 6.9 per cent between 1900 and 1995; those with living standards below the Statistics Canada poverty line increased in numbers by 24.5 per cent. Women in Canada are still the poorest of the poor: their pre-tax incomes amount to 62 per cent of men's incomes; they make up a disproportionate share of the population with low incomes— 2.4 million in 2001 compared to 1.9 million men. At the turn of the twenty-first century, more than a decade after the House of Commons

Source: Ian McKay, "Realms of Freedom, Realms of Necessity," from *Rebels, Reds and Radicals: Rethinking Canada's Left History,* Between the Lines, 2005, 1–14, 16–21. Reprinted with permission.

unanimously passed a dramatic resolution to "seek to eliminate child poverty by the year 2000," about one in every six Canadian children was—according to the state's own statistics—impoverished. At least four out of every ten renter households were paying more than 30 per cent of their monthly incomes on shelter, leaving them little left over for food, transportation, or other basic necessities.[1]

In the first decade of the new century, global warming proceeds at a faster pace than at any time during the past four hundred to six hundred years. Since the beginning of the twentieth century the mean surface temperature of the Earth has increased by about 0.6 degrees Celsius; about half of that warming has taken place in the past forty years. The impact on the Arctic has been striking. Scholars report a 40 per cent reduction in the thickness of the ice pack and new ailments such as lungworms in muskoxen. The global sea level rose faster in the last century than it did in the previous three thousand years. With continued global warming, billions of people face unimaginable calamities. The glacier-fed rivers of the Himalayas, which supply water to one-third of the world's population, are likely to flood. Latin Americans confront the prospect of a severe water shortage.[2]

The forty-two million people who have contracted HIV/AIDS confront, as did the twenty-two million already killed by the disease, crumbling health-care systems and a profit-oriented pharmaceutical industry. In 2003, one country—the United States—voted against a United Nations resolution calling for open access to drugs to meet this "global health emergency."[3]

Globally a vast engine of accumulation transforms almost every human activity into a dollars-and-cents proposition. Across North America cities are penned in by look-alike malls, full of commodities designed to slake recently invented consumer desires. In multiplex movie theatres and supersized grocery stores, consumers are enveloped by a system of goods and services that doubles as a system of meaning and transcendence. Yet all too often there is seemingly no clear purpose of direction to everyday life: activities seem geared to *means* and not to *ends*, to *fragmented* rather than *integrated* experiences, to an eternal "present" and not to any history or future. Struggling for *something* beyond the shopping mall, North Americans grasp at the occult, countless schemes of self-improvement, new diets, "nature"—all of which require further trips to the shopping mall.[4]

This general state of affairs, we are told again and again, every day, by a hundred voices and in a hundred ways, is the only way things can possibly be: all of these massive patterns are beyond human control; you might try to change one or two details, you cannot change the big picture; to imagine a radically different world that does not generate patterns like the ones we are now seeing is to succumb to a delusion.

This "delusion"—that another world is possible—is traditionally called the left.

THE REAL UTOPIA

To be a leftist—a.k.a. socialist, anarchist, radical, global justice activist, communist, socialist-feminist, Marxist, Green, revolutionary—means believing, at a gut level, "It doesn't have to be this way." *Vivre autrement*—"live otherwise! Live in another way!"—was a slogan used by one Quebec radical group in the 1970s. *Reasoning Otherwise* was the slogan of William Irvine, the legendary Prairie socialist. Words like these are inscribed on the heart of every leftist.

Of course, every one of the social problems of the day—from growing inequality to global warming—has its own story. It is properly addressed by its own experts. Such problems cannot simply be lumped together. Each demands its own response. So why not just do what is pragmatically possible, and tackle one issue at a time?

Just so. Living otherwise means engaging with the life-and-death, down-to-earth issues as they present themselves. Living and reasoning otherwise mean the mobilization of resources to handle the emergencies of everyday life.

Yet many people engaged in these emergencies are forced to the conclusion that living otherwise demands more than pragmatic, one-issue-at-a-time responses. Consider the HIV/AIDS pandemic. Suppose, instead of some grandiose scheme of ridding the planet of the disease, you just settle for a more modest objective: reducing the projected death toll over the next few decades, say, from forty-two million to ten million. You come up with the most practical, common-sense ways of doing so: making drugs as effective as possible, promoting the use of safe sex, attacking the other ailments that facilitate the spread of AIDS, and fighting the stigma often attached to people living with the disease. Quite soon you will find yourself up against people who are actively working against you. The Catholic Church will fight you on "moral" grounds about the human rights of gays and the legitimacy of contraception. Pharmaceutical companies will fight you economically on producing free and effective medicines. The U.S. government, the mightiest in the world, will fight you on both fronts. How are you going to make an effective difference, if your struggle necessarily means working in a world dominated by these forces?

Or suppose, instead of some revolutionary vision of humanity living in a harmonious balance with the rest of nature, you settle for a more modest objective—say, a 50 per cent reduction in carbon-based air pollution over the next ten years. You come up with the most practical, common-sense proposals for doing so: reducing emissions of carbon dioxide, switching 35 per cent of the power grid to alternative energy sources, cutting back on coal-burning generating plants, exploring new energy sources such as wind or solar power. Even though you can argue that every human has a long-term interest in the success of these modest proposals, you will quite quickly find yourself up against people who are actively working against you. Automobile manufacturers will fight your demand that they make only less-polluting cars. Powerful oil companies will hire advertising firms and scientific consultants to discredit you. And, once again, the U.S. government will oppose even the most pragmatic, down-to-earth measures—even if many of its own scientific experts are convinced that a capitalist system reliant on fossil fuels is one that is riding for a fall.

Do what is possible, one issue at a time? Of course—there's no realistic alternative. But you will most likely soon reach conclusions about the patterns of opposition and support that shape each and every one of these issues and connect them together. You may well decide that the persistent general relations *behind* that specific pattern also need to be understood and changed. You will start to see not just a random pattern of problems, but a system underlying them.

Every leftist, at some level, believes and acts on this insight: there are ways of explaining not just the individual problems but the connections between them. Once grasped in thought, these connections have to be transformed in reality. To tackle even one problem—eliminating HIV/AIDS, preventing global environmental meltdown—means struggling to puzzle out why that problem arose in the first place. As soon as you start pursuing the process of figuring each problem out, and connecting it with other problems, you have started down the road to leftism. You will be led, step by step, to a recovery of the down-to-earth historical explanations of why such patterns emerged and why large groups of people respond to them in such different ways.

To struggle against each of these problems means that you think alternatives are possible. War, mass starvation, death from disease, global environmental devastation—maybe these are aspects of life that have always been and always will be with us. Maybe they reflect unchangeable human nature. Maybe they reflect the Will of God. Maybe they are part of an unstoppable process of evolution. Once you start trying to change these patterns, even in the most direct and down-to-earth ways, you are acting on a different conviction. You are saying, in your own way, that humanity's future is not completely predetermined. Collectively, human beings have the ability to shape different destinies for themselves.

You are also saying that some futures are better than others. We humans face strategic choices. A world without hunger, disease, poverty, war, environmental degradation, the subordination of women and gays, and wars fought in the name of nations and religions would be better than our present-day world. To be a leftist means thinking that human beings could organize themselves in such a way that these evils would be at least diminished if not ultimately eliminated. To be a leftist means throwing oneself into the problems of the present in the gamble that these problems are not just eternal aspects of the human condition.

The sociologist Zygmunt Bauman has developed these simple insights into a brilliant distillation of the project of the left, which in most recent human history has gone under the name of "socialism." (I'll get back to that word, which I use in its broadest possible sense, later.) Bauman sees socialism as a kind of "utopianism." As soon as many leftists hear that word, their backs go up. Isn't that just what their enemies have always said—that the left is full of idealistic daydreamers, people clinging to a childish dream of "heaven on earth?" But Bauman doesn't mean that kind of "utopianism." What he means is that leftists typically put forward visions of the future that are radically different from the conditions of present-day reality. "Utopias" in this sense are aspects of culture in which "possible extrapolations of the present are explored." They are, in a sense, thought-experiments in living otherwise.

In general, Bauman says, leftists are more inclined to realism than to romanticism. When they draw upon their experiences in solving particular issues, they have been surprisingly down-to-earth. When leftists use utopias, they are doing so as a technique to "help to lay bare and make conspicuous the major divisions of interest within a society." Their utopias are present-day expressions of the "other world" that human collective action might make possible. Although utopias generally address society as a whole—"here's a future that would be good for all of us"—they actually work to reveal that society is made up of very different groups with radically different interests.

"In other words," writes Bauman, "utopias relativise the future into a bundle of class-committed solutions, and dispel the conservative illusion that one and only one thread leads on from the present." Against the many people who say, of a given social problem, "Well, that's just human nature" or "That's just the way things have always been" or "The poor ye have always with you," concrete utopias suggest that things that seem to be just "natural" parts of life are actually the outcome of history and politics—of the forces and choices people made, perhaps many generations ago, that still shape our world today. Utopias "portray the future as a set of competing projects, and thereby reveal the role of human volition and concerted effort in shaping and bringing it about."[5]

No law in the universe lays down that some Torontonians and Montrealers live in cardboard boxes and others in 5,000-square-foot houses equipped with plasma-screen television sets and hot tubs. No inescapable logic rules that many Aboriginal Canadians in the north are required to have a life expectancy far lower than that of Euro-Canadians in the south. These are matters of *history* and *politics*. Consequently, they are within the limits that every one of us inherits from human choices made in the past.

In considering the HIV/AIDS pandemic, for example, critics might argue that it is "utopian" to think that the population of the world could mobilize its resources to save a majority of the estimated forty-two million people living with the disease. After all, to date, heavy evidence indicates that the most powerful, rich, and priestly people in the world are against making that happen. Yet other conflicting indications also exist. There is Stephen Lewis, for example, the one-man left-wing crusade for justice for Africa. There is the 1980s legacy of brave struggle on the part of the gay communities—historically among the most despised and outcast minorities in North American society—fighting for dear life against historic patterns of indifference and prejudice, much of it found in the conventional left. Most impressively, there is emergent grassroots activism in Africa itself, with some notable victories in some states. In North America's gay communities in the 1980s huge victories were won when an oppressed community took up a life-and-death struggle and linked it to a more general vision of freedom. We can project from the reality of corporate greed, indifference to the poor, and religious and official prejudice and inhumanity; or we can project from a reality of successful grassroots activism that has already changed lives from San Francisco to South Africa.

With global warming, for example, it is certainly possible to project into the future the continuance of current practices, which might quite possibly spell the end of human life on the planet. These practices are deeply rooted in how most Westerners, living in the world's dominant capitalist economies, make their living. A realistic "utopian" projection of a more balanced, long-term approach begins with a scientific under-standing of biology, physics, and chemistry, with explorations of the Earth's atmosphere, with an understanding that human beings, as animals, confront real limits to what they can or should do if they want to survive on this planet. Neither "projection" is unscientific, but the second, Bauman would say, is an example of a "concrete utopia." Just to point out that human beings have a collective interest in survival that global capitalism may be placing at risk is to "portray the future as a set of competing projects."

To be a leftist, then, means an immersion in urgent day-by-day struggles and a willingness to see the connections linking them together. But it also means introducing into the world a vision of the future and producing a logical program for its realization. It means defending that vision against constant hostility. Projecting a "utopia" into the present means understanding all the forces—such as those organized by class, gender, race, sexual orientation, nationality—that are likely to fight against it, even if the utopia in question is just a modest proposal for cleaner air that would bring even the left's enemies healthier and longer lives.

When Karl Marx, in the posthumously published third volume of *Capital,* considered this concept of the "real utopia," he used the term "realm of freedom." Marx was an ardent democrat, back when democracy was a far-fetched and disreputable revolutionary idea. He despised the world of privilege and elitism and scorned liberals who talked non-stop about the rights of individuals without realizing that none of these individuals and none of their "inalienable" rights could exist apart from society. Today Marx and the Marxists are often depicted as crackpots urging their followers into a mad "lovers' leap" into an unknown future. (And many of the twentieth-century regimes supposedly based on Marx's ideas were guilty as charged.) But when you actually read Marx, you will find the opposite message. Marx spent a lifetime reasoning otherwise. His message to people who needed to believe without evidence and without doing the hard work of analysis was in accord with his own personal motto: "Doubt everything."[6]

Marx fully described his "real utopia" in his masterpiece, *Capital,* and—if you can get past some of his dated Victorian expressions—his words resonate today.

> The realm of freedom really begins only where labour determined by necessity and external expediency ends; it lies by its very nature beyond the sphere of material production proper. Just as the savage must wrestle with nature to satisfy his needs, to maintain and reproduce his life, so must civilized man, and he must do so in all forms of society and under all modes of production. This realm of natural necessity expands with his development, because his needs do too; but the productive forces to satisfy these expand at the same time. Freedom, in this sphere, can consist only in this, that socialized man, the associated producers, govern the human metabolism with nature in a rational way, bringing it under their collective control instead of being dominated by it as a blind power; accomplishing it with the least expenditure of energy and in conditions most worthy and appropriate for their human nature. But this always remains a realm of necessity. The true realm of freedom, the development of human powers as an end in itself, begins beyond it, though it can only flourish with this realm of necessity as its basis.[7]

Like so much in Marx, it is a passage that you could spend a lifetime pondering. Those who have written about it disagree with each other. I see in it an approach very similar to the one described by Bauman: it is possible to live otherwise, but keep your feet on the ground. In any imaginable future there will be fields to be ploughed, dishes to be washed, diapers to be changed, folks to look after. Yet even as we carry out all the mundane tasks that keep body and soul together, we can still live otherwise. Even as we do the things we need to do to survive, we can manage things collectively more rationally than we now do. We can invest the everyday world with meaning and purpose. But *alongside* that realm of necessity—notice, Marx says "beyond" but not "above"—there begins another realm, the "true realm of freedom." In that realm the development of human creativity—building relationships, making music or drawing pictures, doing philosophy, birding, quilting, playing hockey— is an end in itself.

Marx is not talking about a "dualistic vision," in the same way that some Christians view heaven as the complete opposite of a sinful and troubled Earth. He is talking about an Earth transformed. The realm of freedom is not removed from the realm of necessity—it couldn't exist without it—but it is also not the same thing. In this realm of freedom, there is a capacity to create that can only emerge when people feel prosperous and comfortable and at home in providing for themselves and those they love. It is the vision of a utopian, but a utopian with dirt under the fingernails.

This passage has had a long underground life. In Latin America, for example, it inspired a whole tradition of reading Christ's gospel as a call to social revolution. It has also appealed to environmental activists trying to transcend long-standing exploitive approaches to nature, because the "rational management" of a metabolism is a far cry from a one-way strategy of corporate dominance. Instead of seeing the "realm of freedom" as a socialist version of the bland Christian heaven—with white-coated scientists and a gleaming Starship Enterprise taking the place of sexless celestial choirs and pearly gates—Marx sees it more as an unfolding here-and-now process, a freeing of human possibilities stifled prematurely or destroyed altogether in our modern world. This is a Marx, for example, that I can imagine celebrating both the necessity to struggle against HIV/AIDS and the project of free sexual expression that lay his-torically at the heart of the major community that began this struggle in North America; and savouring the prospect of the hours of hard labour, and the human solidarities, that are entailed in the struggle against global warming.[8]

In the everyday world of the corporate twenty-first century, "necessity" and "freedom" confront each other as opposites. Many of us work so hard for those weekends off, when we can finally be ourselves, listen to music, get outdoors, or simply enjoy each other's company— when we can find the time to do so after completing the other mundane tasks of life that more and more seem to fill up the weekends. But weekends would not exist without our working weeks; and times at work can also be filled with pockets of freedom and spontaneity. The hard collective work of wiring a hospital building, for instance, can be a time of fun, solidarity, excitement, and creativity, and if it isn't so, it could be made so. In fact, "freedom and necessity" aren't neatly separated opposites. You can't have one without the other. Human beings collectively can expand the "realm of freedom" through a more rational management of the "realm of necessity." Instead of working for the weekend, we can experience both our workweeks and our leisurely weekends with a sense of participating in an important human project, as part of a caring and generous human community. Marx's projection of a human potential for creativity makes us look twice at the everyday world. Most of all, he makes us realize that this everyday world is not just a "thing" but a process and a problem, which can be addressed only by working to understand it.

Marx was the most brilliant and influential of all the "concrete utopians."

. . .

Entire generations of both socialists and anti-socialists have thought they had found in Marx an "economic determinist" who provided iron-clad laws of social development. Others found in the young Marx a radically voluntarist "revolutionary humanist" with a vision of a true human essence imprisoned in the material world. Conflicts between essence and construction, agency and structure, freedom and fate— conflicts that in other contexts are identified with radically different schools and thinkers—are in Marx fought out in the same texts, often in the same paragraphs and even within the same sentence. Marx's most influential single text, *Capital: A Critique of Political Economy,* is a "scientific" text that "critiques"—on political, logical, and ethical grounds—the very object it constructs. (A parallel from natural science might be something like *Symbiosis: A Critique of Biological Science.*) A tension between "structure" and "agency" is built right into the very title of this famous book. Marx wanted to found the realm of freedom on the most rigorous understanding of the massive network of relations that make us unfree so that our critique of those relations can be all the more effective.[9] As the Sardinian thinker and activist Antonio Gramsci so nicely put it:

> Karl Marx is not, for us, the infant whimpering in the cradle or the bearded man who frightens priests. . . . He is an individual moment in the anxious search that humanity has been conducting for centuries to acquire consciousness of its being and its becoming, to grasp the mysterious rhythm of history and disperse the mystery, to be stronger in its thinking and to act better. He is a necessary and integral part of our spirit, which would not be what it is if he had not lived, had not

thought, had not sent sparks of light flying from the collision with his passions and his ideas, his sufferings and his ideals.[10]

In the conversation with "Marx" that means most to me, the "realm of freedom" is not just a way of talking about some distant utopia. It is a way of understanding and extending the democratic spaces that we are able to experience in the daily world. In the everyday world of capitalism, small realms of freedom can be carved out of working lives—spaces and times in which individuals and groups can find time to think about the world and their places in it. At other times everyday realms of freedom become much larger and the daily rhythms of life are suspended. Mind-transforming questions are asked. Why should workers not receive the full value of the commodities they produce? Why should people starve on Third World streets when there is enough food in the world to feed everyone? Why should Aboriginal peoples be treated like second-class citizens in their own country? Why should parliament be dominated by men? Why should so many women have to flee their homes and find refuge in shelters? Why should gay men be bashed in back alleys and pilloried from the pulpit? Why should the new regimes of world trade tyrannize and impoverish much of the world?

Most of us have our moments of freedom when we feel as though we are more truly at home with the world, and at those moments large groups of people also share this feeling. A core part of these realms of freedom is the freedom to criticize the everyday world and to project alternative worlds into the future. In these and other anticipations of the realm of freedom, each specific to its time and all of them an anticipation of an authentic liberation, radicals and socialists have found their own social spaces and their own spaces in time. They have drawn upon the past to draw an imagined future closer. They have created social and intellectual spaces in which isolated individuals can see themselves as part of a much bigger story.

Socialist realms of freedom do not refer to science-fiction realms of abstract possibility. As Gramsci dryly observes in the classic *Prison Notebooks*—and here the echoes of Marx's earlier position are clear:

> Possibility is not reality: but it is in itself a [kind of] reality. Whether a man can or cannot do a thing has its importance in evaluating what is done in reality. Possibility means "freedom." [And that] measure of freedom enters into the concept of man. That objective possibilities exist for people not to die of hunger, and that people do die of hunger has its importance, or so one would have thought. But the existence of 'objective' conditions, of possibilities or of freedom is not yet enough: it is necessary to 'know' them, and know how to use them. And to want to use them.[11]

Like Bauman, Gramsci is talking about using a set of *objective possibilities* as a way of awakening people in the present and distinguishing friends from foes. To be a leftist is to use the possibility, the objective possibility, of living otherwise. It means building a sense of possibility into the very concept we hold of ourselves—of "knowing" it at a deep level—and understanding our context within a historical process bigger than we are. Knowing what this living otherwise entails means struggling to make the possibility a reality. We can abstractly imagine a world in which the lines of power and wealth are drawn very differently. Over time, if our projection is in fact something possible within a feasible world and not simply a "candy mountain" fantasy, it can become a kind of reality, as more and more people mobilize around it. People can only come to really "know" this possibility, in the full sense, by trying to make it happen; and making it happen means finding others who share the conviction that the possibility is there.

Take, for example, the sense of weary alienation experienced by many North Americans in travelling along congested stretches of fast-food restaurants, muffler shops, malls, and monster big box stores that lock in our cities. We might simply gripe or make snide jokes about them. We might vow to frequent only family-owned eateries or shop at small independent stores in the downtown core—if we can find any. We might resolve only to buy from our local organic farmer and boycott all branded merchandise. These kinds of personal decisions capture an authentic, resistant vision of an "otherwise." They create small spaces of personal critique and freedom, outside that manipulative and ugly commercialism that we might want to avoid. No discerning leftist should ridicule such small-scale acts of resistance, but neither should he or she be content with them. Isolated and dispersed across the social landscape, such little acts of freedom are vulnerable and short-lived. The "freedom" is confined to one person, one family, one moment, and it is often purchased through the "unfreedom" of others. The "realm of freedom" that the left, and the left alone, can act upon is one open to the vast majority of humankind. It might begin with small collective acts—such as "no shopping" days or local campaigns to stop the spread of Wal-Marts—but to be truly "of the left," it must connect these acts with a larger strategy, a more inclusive storyline. It must see every such struggle as a partial answer to a much bigger question, to which it contributes part of the answer: "How can we live differently?" What possibilities can we use to turn the little "measures" of freedom available to a few into much bigger realms of freedom open to the many? How can our small acts of resistance snowball into a system-changing social movement?

With this kind of question in mind, we can see in the grey shadowy details of the history of the Canadian left since 1890 the coherent shapes of the past possibilities that people tried to take up in order to live otherwise. Living in an often hostile individualistic social and political order—one that since the 1840s was increasingly well-fortified against dissenting opinions and that treated "democracy" first as a term of abuse and then as an "optional extra"—socialists were able to carve out limited realms of freedom, where they could, for a time, develop their alternatives. In some places and in some times, these realms of freedom—freedom from commodities, from necessity, freedom to self-expression, to enjoyment—attained a physical presence in parties and neighbourhoods and movements.

In these little realms of freedom it was possible to speak and act with a freedom generally reserved for the elite. With creativity and ingenuity, in the time remaining after work, left-wing men and women created spaces and traditions and even specialized bodies of concepts. They created a succession of realms of freedom in the interstices of everyday necessity. At the core of the set of objective possibilities that the Canadian left has characteristically defended is an ideal of a rational, just democracy.

NOTES

1. The Canadian poverty rates are determined by Statistics Canada's "Low Income Cut-off." See *Campaign 2000, Child Poverty in Canada* (Toronto: The Campaign, 2000), citing Survey of Consumer Finances, Statistics Canada, microdata files. For the world poverty figures, see Alex Callinicos, *Equality* (Cambridge: Polity, 2000), citing United Nations World Development Report statistics. A portion of those 1.3 billion people around the world were not technically "impoverished" because they were not caught up in world market relations. Under conditions of globalization, their autonomy from capitalist social relations is not likely to last indefinitely. For the Canadian figures, see Kevin K. Lee, *Urban Poverty in Canada: A Statistical Profile* (Ottawa: Canadian Council on Social Development, 2000); and "Women's Income and Poverty," *The CCPA Monitor*, II, 2 (June 2004), 29.

2. See www.ucsusa.org/global_environment/global_warming/index.cfm (30 Oct. 2004) for these and other statistics, as well as "Ecological Responses to Recent Climate Change," *Nature,* March 28, 2002; Susan Klutz, Eric Hoberg, and Lydden Polley, "Global Warming and Host-Parasite Systems in the Arctic: Should We Be Concerned?" wildlife.usask.ca/SatellitePages/iwap/abstracts/kutzb (14 Nov. 2003); www.commondreams.org/headlines03/1128-04 (29 Nov. 2003). For more general jargon-free information, see Dinyar Godrej, *The No-Nonsense Guide to Climate Change* (Toronto: Between the Lines, 2001).

3. See Shereen Usdin, *The No-Nonsense Guide to HIV/AIDS* (Toronto: Between the Lines, 2004).

4. I am drawing this part of the discussion from Henri Lefebvre, *Everyday Life in the Modern World*, trans. S. Rabinowich (New Brunswick, N.J.: Transaction Publishers, 1984), 89.

5. Zygmunt Bauman, *Socialism: The Active Utopia* (London: George Allen & Unwin, 1976), especially 14–15.

6. See Francis Wheen, *Karl Marx* (London: Fourth Estate, 1999), for an engaging biography—one that whets the appetite for other more scholarly studies.

7. Karl Marx, *Capital,* trans. David Fernbach, vol. 3 (New York: Vintage Books, 1981), 959.

8. See, for instance, Franz J. Hinkelammert, *The Ideological Weapons of Death: A Theological Critique of Capitalism*, trans. Phillip Berryman (Maryknoll, N.Y.: Orbis Books, 1986), 52–57.

9. These points about structure and agency owe much to Lucio Colletti, whose works on Marx and Hegel were unfortunately not much heeded when they first came out in English in the 1970s. See Lucio Colletti, *From Rousseau to Lenin: Studies in Ideology and Society* (London: New Left Books, 1972).

10. Antonio Gramsci, *Il Grido del Popolo,* 4 May 1918, as cited in David Forgacs, ed., *The Antonio Gramsci Reader* (New York: New York University Press, 2000), 39.

11. Antonio Gramsci, *Selections from the Prison Notebooks*, ed. and trans. Quinton Hoare and Geoffrey Nowell-Smith (London: Lawrence and Wishart, 1971), 360.

Jihad vs. McWorld

Benjamin R. Barber

Barber argues that underlying many current changes and conflicts in the world are two opposing historical forces—globalism and tribalism—both of which are a threat to democracy. Barber's article raises several questions. First, of the more than 200 nation-states that make up the international system today, few have credible liberal democracies. Is democracy necessarily the best form of political organization for everyone, as Barber assumes?

Second, as a solution Barber recommends "confederal democracy" where powers are divided between a central government with coordinative functions and smaller units, like provinces or states, which also hold major political power. While he suggests we examine the United States' Articles of Confederation as an early example of such a system, Canada's experiment with Confederation, particularly as it has been unfolding in recent years, offers a more mature example. What lessons does Canada offer for understanding and managing the conflicting forces of globalism and tribalism?

Just beyond the horizon of current events lie two possible political figures—both bleak, neither democratic. The first is a retribalization of large swaths of humankind by war and bloodshed: a threatened Lebanonization of national states in which culture is pitted against culture, people against people, tribe against tribe—a Jihad in the name of a hundred narrowly conceived faiths against every kind of interdependence, every kind of artificial social cooperation and civic mutuality. The second is being borne in on us by the onrush of economic and ecological forces that demand integration and uniformity and that mesmerize the world with fast music, fast computers, and fast food—with MTV, Macintosh, and

Source: Published originally in *The Atlantic Monthly* March 1992 as an introduction to the *Jihad vs. McWorld* (Ballantine paperback, 1996), a volume that discusses and extends the themes of the original article. Benjamin R. Barber is Kekst Professor of Civil Society at the University of Maryland; Distinguished Senior Fellow, Demos; Director, CivWorld and the author of many books including the classic *Strong Democracy* (1984), international bestseller *Jihad vs. McWorld* (Times Books, 1995), and the forthcoming *Consumed: The Fate of Citizens Under Capitalism Triumphant.*

McDonald's, pressing nations into one commercially homogenous global network: one McWorld tied together by technology, ecology, communications, and commerce. The planet is falling precipitantly apart and coming reluctantly together at the very same moment.

These two tendencies are sometimes visible in the same countries at the same instant: thus Yugoslavia, clamoring just recently to join the New Europe, is exploding into fragments; India is trying to live up to its reputation as the world's largest integral democracy while powerful new fundamentalist parties like the Hindu nationalist Bharatiya Janata Party, along with nationalist assassins, are imperilling its hard-won unity. States are breaking up or joining up: the Soviet Union has disappeared almost overnight, its parts forming new unions with one another or with like-minded nationalities in neighboring states. The old interwar national state based on territory and political sovereignty looks to be a mere transitional development.

The tendencies of what I am here calling the forces of Jihad and the forces of McWorld operate with equal strength in opposite directions, the one driven by parochial hatreds, the other by universalizing markets, the one re-creating ancient subnational and ethnic borders from within, the other making national borders porous from without. They have one thing in common: neither offers much hope to citizens looking for practical ways to govern themselves democratically. If the global future is to put Jihad's centrifugal whirlwind against McWorld's centripetal black hole, the outcome is unlikely to be democratic—or so I will argue.

McWORLD, OR THE GLOBALIZATION OF POLITICS

Four imperatives make up the dynamic of McWorld: a market imperative, a resource imperative, an information-technology imperative, and an ecological imperative. By shrinking the world and diminishing the salience of national borders, these imperatives have in combination achieved a considerable victory over factiousness and particularism, and not least of all over their most virulent form—nationalism. It is the realists who are now Europeans, the utopians who dream nostalgically of a resurgent England or Germany, perhaps even a resurgent Wales or Saxony. Yesterday's wishful cry for one world has yielded to the reality of McWorld.

The market imperative. Marxist and Leninist theories of imperialism assumed that the question for ever-expanding markets would in time compel nation-based capitalist economies to push against national boundaries in search of an international economic imperium. Whatever else has happened to the scientific predictions of Marxism, in this domain they have proved farsighted. All national economies are now vulnerable to the inroads of larger, transnational markets within which trade is free, currencies are convertible, access to banking is open, and contracts are enforceable under law. In Europe, Asia, Africa, the South Pacific, and the Americas such markets are eroding national sovereignty and giving rise to entities—international banks, trade associations, transnational lobbies like OPEC and Greenpeace, world news services like CNN and the BBC, and multinational corporations that increasingly lack a meaningful national identity—that neither reflect nor respect nationhood as an organizing or regulative principle.

The market imperative has also reinforced the quest for international peace and stability, requisites of an efficient international economy. Markets are enemies of parochialism, isolation, fractiousness, war. Market psychology attenuates the psychology of ideological and religious cleavages and assumes a concord among producers and consumers—categories that ill fit narrowly conceived national or religious cultures. Shopping has little tolerance for blue laws, whether dictated by pub-closing British paternalism, Sabbath-observing Jewish Orthodox fundamentalism, or no-Sunday-liquor-sales Massachusetts puritanism. In the context of common markets, international law ceases to be a vision of justice and becomes a workday framework for getting things done—enforcing contracts, ensuring that governments abide by deals, regulating trade and currency relations, and so forth.

Common markets demand a common language, as well as a common currency, and they produce common behaviors of the kind bred by cosmopolitan city life everywhere. Commercial pilots, computer programmers, international bankers, media specialists, oil riggers, entertainment celebrities, ecology experts, demographers, accountants, professors, athletes—these compose a new breed of men and women for whom religion, culture, and nationality can seem only marginal elements in a working identity. Although sociologists of everyday life will no doubt continue to distinguish a Japanese from an American mode, shopping has a common signature throughout the world. Cynics might even say that some of the recent revolutions in Eastern Europe have had as their true goal not liberty and the right to vote but well-paying jobs and the right to shop (although the vote is proving easier to acquire than consumer goods). The market imperative is, then, plenty powerful; but, notwithstanding some of the claims made for "democratic capitalism," it is not identical with the democratic imperative.

The resource imperative. Democrats once dreamed of societies whose political autonomy rested firmly on economic independence. The Athenians idealized what they called autarky, and tried for a while to create a way of life simple and austere enough to make the polis genuinely self-sufficient. To be free meant to be independent of any other community or polis. Not even the Athenians were able to achieve autarky, however: human nature, it turns out, is dependency. By the time of Pericles, Athenian politics was inextricably bound up with a flowering empire held together by naval power and commerce—an empire that, even as it appeared to enhance Athenian might, ate away at Athenian independence and autarky. Master and slave, it turned out, were bound together by mutual insufficiency.

The dream of autarky briefly engrossed nineteenth-century America as well, for the underpopulated, endlessly bountiful land, the cornucopia of natural resources, and the natural barriers of a continent walled in by two great seas led many to believe that America could be a world unto itself. Given this past, it has been harder for Americans than for most to accept the inevitability of interdependence. But the rapid depletion of resources even in a country like ours, where they once seemed inexhaustible, and the maldistribution of arable soil and mineral resources on the planet, leave even the wealthiest societies ever more resource-dependent and many other nations in permanently desperate straits.

Every nation, it turns out, needs something another nation has; some nations have almost nothing they need.

The information-technology imperative. Enlightenment science and the technologies derived from it are inherently universalizing. They entail a quest for descriptive principles of general application, a search for universal solutions to particular problems, and an unswerving embrace of objectivity and impartiality.

Scientific progress embodies and depends on open communication, a common discourse rooted in rationality, collaboration, and an easy and regular flow and exchange of information. Such ideals can be hypocritical covers for power-mongering by elites, and they may be shown to be wanting in many other ways, but they are entailed by the very idea of science and they make science and globalization practical allies.

Business, banking, and commerce all depend on information flow and are facilitated by new communication technologies. The hardware of these technologies tends to be systemic and integrated—computer, television, cable, satellite, laser, fiber-optic, and microchip technologies combining to create a vast interactive communications and information network that can potentially give every person on earth access to every other person, and make every datum, every byte, available to every set of eyes. If the automobile was, as George Ball once said (when he gave his blessing to a Fiat factory in the Soviet Union during the Cold War), "an ideology on four wheels," then electronic telecommunication and information systems are an ideology at 186,000 miles per second—which makes for a very small planet in a very big hurry. Individual cultures speak particular languages; commerce and science increasingly speak English; the whole world speaks logarithms and binary mathematics.

Moreover, the pursuit of science and technology asks for, even compels, open societies. Satellite footprints do not respect national borders; telephone wires penetrate the most closed societies. With photocopying and then fax machines having infiltrated Soviet universities and *samizdat* literary circles in the eighties, and computer modems having multiplied like rabbits in communism's bureaucratic warrens thereafter, *glasnost* could not be far behind. In their social requisites, secrecy and science are enemies.

The new technology's software is perhaps even more globalizing than its hardware. The information arm of international commerce's sprawling body reaches out and touches distinct nations and parochial cultures, and gives them a common face chiseled in Hollywood, on Madison Avenue, and in Silicon Valley. Throughout the 1980s one of the most-watched television programs in South Africa was *The Cosby Show*. The demise of apartheid was already in production. Exhibitors at the 1991 Cannes film festival expressed growing anxiety over the "homogenization" and "Americanization" of the global film industry when, for the third year running, American films dominated the awards ceremonies. America has dominated the world's popular culture for much longer, and much more decisively. In November of 1991 Switzerland's once insular culture boasted bestseller lists featuring *Terminator 2* as the No. 1 movie, *Scarlett* as the No. 1 book, and Prince's *Diamonds and Pearls* as the No. 1 record album. No wonder the Japanese are buying Hollywood film studio sets even faster than Americans are buying Japanese television sets. The kind of software supremacy may in the long term be far more important than hardware superiority, because culture has become more potent than armaments. What is the power of the Pentagon compared with Disneyland? Can the Sixth Fleet keep up with CNN? McDonald's in Moscow and Coke in China will do more to create a global culture than military colonization ever could. It is less the goods than the brand names that do the work, for they convey life-style images that alter perception and challenge behavior. They make up the seductive software of McWorld's common (at times much too common) soul.

Yet in all this high-tech commercial world there is nothing that looks particularly democratic. It lends itself to surveillance as well as liberty, to new forms of manipulation and covert control as well as new kinds of participation, to skewed, unjust market outcomes as well as greater productivity. The consumer society and the open society are not quite synonymous. Capitalism and democracy have a relationship, but it is something less than a marriage. An efficient free market after all requires that consumers be free to vote their dollars on competing goods, not that citizens be free to vote their values and beliefs on competing political candidates and programs. The free market flourished in junta-run Chile, in military-governed Taiwan and Korea, and, earlier, in a variety of autocratic European empires as well as their colonial possessions.

The ecological imperative. The impact of globalization on ecology is a cliché even to world leaders who ignore it. We know well enough that the German forests can be destroyed by Swiss and Italians driving gas-guzzlers fueled by leaded gas. We also know that the planet can be asphyxiated by greenhouse gases because Brazilian farmers want to be part of the twentieth century and are burning down tropical rain forests to clear a little land to plough, and because Indonesians make a living out of converting their lush jungle into toothpicks for fastidious Japanese diners, upsetting the delicate oxygen balance and in effect puncturing our global lungs. Yet this ecological consciousness has meant not only greater awareness but also greater inequality, as modernized nations try to slam the door behind them, saying to developing nations, "The world cannot afford *your* modernization; ours has wrung it dry!"

Each of the four imperatives just cited is transnational, transideological, and transcultural. Each applies impartially to Catholics, Jews, Muslims, Hindus, and Buddhists; to democrats and totalitarians; to capitalists and socialists. The Enlightenment dream of a universal rational society has to a remarkable degree been realized—but in a form that is commercialized, homogenized, depoliticized, bureaucratized, and, of course, radically incomplete, for the movement toward McWorld is in competition with forces of global breakdown, national dissolution, and centrifugal corruption. These forces, working in the opposite direction, are the essence of what I call Jihad.

JIHAD, OR THE LEBANONIZATION OF THE WORLD

OPEC, the World Bank, the United Nations, the International Red Cross, the multinational corporation...there are scores of institutions that reflect globalization. But they often appear as ineffective reactors to the world's real actors: national states and, to an ever greater degree, subnational factions in permanent rebellion against uniformity and integration—even the kind represented by universal law and justice. The headlines feature these players regularly: they are cultures, not countries; parts, not wholes; sects, not religions; rebellious factions and dissenting minorities at war not just with globalism but with the traditional nation-state. Kurds, Basques, Puerto Ricans, Ossetians, East Timoreans, Quebecois, the Catholics of Northern Ireland, Abkhasians, Kurile Islander Japanese, the Zulus of Inkatha, Catalonians, Tamils, and, of course, Palestinians—people without countries, inhabiting nations not their own, seeking smaller worlds within borders that will seal them off from modernity.

A powerful irony is at work here. Nationalism was once a force of integration and unification, a movement aimed at bringing together disparate clans, tribes, and cultural fragments under new, assimilationist flags. But as Ortega y Gasset noted more than sixty years ago, having won its victories, nationalism changed its strategy. In the 1920s, and again today, it is more often a reactionary and divisive force, pulverizing the very nations it once helped cement together. The force that creates nations is "inclusive," Ortega wrote in *The Revolt of the Masses.* "In periods of consolidation, nationalism has a positive value, and is a lofty standard. But in Europe everything is more than consolidated, and nationalism is nothing but a mania...."

This mania has left the post-Cold War world smoldering with hot wars; the international scene is little more unified that it was at the end of the Great War, in Ortega's own time. There were more than thirty wars in progress last year, most of them ethnic, racial, tribal, or religious in character, and the list of unsafe regions doesn't seem to be getting any shorter. Some new world order!

The aim of many of these small-scale wars is to redraw boundaries, to implode states and resecure parochial identities: to escape McWorld's dully insistent imperatives. The mood is that of Jihad: war not as an instrument of policy but as an emblem of identity, an expression of community, an end in itself. Even where there is no shooting war, there is fractiousness, secession, and the quest for ever smaller communities. Add to the list of dangerous countries those at risk: In Switzerland and Spain, Jurassian and Basque separatists still argue the virtues of ancient identities, sometimes in the language of bombs. Hyperdisintegration in the former Soviet Union may well continue unabated—not just a Ukraine independent from the Soviet Union but a Bessarabian Ukraine independent from the Ukrainian republic; not just Russia severed from the defunct union but Tatarstan severed from Russia. Yugoslavia makes even the disunited, ex-Soviet, nonsocialist republics that were once the Soviet Union look integrated, its sectarian fatherlands springing up within factional motherlands like weeds within weeds within weeds. Kurdish independence would threaten the territorial integrity of four Middle Eastern nations. Well before the current cataclysm Soviet Georgia made a claim for autonomy from the Soviet Union, only to be faced with its Ossetians (164,000 in a republic a 5.5 million) demanding their own self-determination within Georgia. The Abkhasian minority in Georgia has followed suit. Even the good will established by Canada's once promising Meech Lake protocols is in danger, with francophone Quebec again threatening the dissolution of the federation. In South Africa the emergence from apartheid was hardly achieved when friction between Inkatha's Zulus and the African National Congress's tribally identified members threatened to replace Europeans' racism with an indigenous tribal war[.A]fter thirty years of attempted integration using the colonial language (English) as a unifier, Nigeria is now playing with the idea of linguistic multiculturalism—which could mean the cultural breakup of the nation into hundreds of tribal fragments. Even Saddam Hussein has benefited from the threat of internal Jihad, having used renewed tribal and religious warfare to turn last season's mortal enemies into reluctant allies of an Iraqi nationhood that he nearly destroyed.

The passing of communism has torn away the thin veneer of inter-nationalism (workers of the world unite!) to reveal ethnic prejudices that are not only ugly and deep-seated but increasingly murderous. Europe's old scourge, anti-Semitism, is back with a vengeance, but it is only one of many antagonisms. It appears all too easy to throw the historical gears into reverse and pass from a Communist dictatorship back into a tribal state.

Among the tribes, religion is also a battlefield. ("Jihad" is a rich word whose generic meaning is "struggle"—usually the struggle of the soul to avert evil. Strictly applied to religious war, it is used only in reference to battles where the faith is under assault, or battles against a government that denies the practice of Islam. My use here is rhetorical, but does follow both journalistic practice and history.) Remember the Thirty Years War? Whatever forms of Enlightenment universalism might once have come to grace such historically related forms of monotheism as Judaism, Christianity, and Islam, in many of their modern incarnations they are parochial rather than cosmopolitan, angry rather than loving, proselytizing rather than ecumenical, zealous rather than rationalist, sectarian rather than deistic, ethnocentric rather than universalizing. As a result, like the new forms of hypernationalism, the new expressions of religious fundamentalism are fractious and pulverizing, never integrating. This is religion as the Crusaders knew it: a battle to the death for souls that if not saved will be forever lost.

The atmospherics of Jihad have resulted in a breakdown of civility in the name of identity, of comity in the name of community. International relations have sometimes taken on the aspect of gang war—cultural turf battles featuring tribal factions that were supposed to be sublimated as integral parts of large national, economic, postcolonial, and constitu-tional entities.

THE DARKENING FUTURE OF DEMOCRACY

These rather melodramatic tableaux vivants do not tell the whole story, however. For all their defects, Jihad and McWorld have their attractions. Yet, to repeat and insist, the attractions are unrelated to democracy. Neither McWorld nor Jihad is remotely democratic in impulse. Neither needs democracy; neither promotes democracy.

McWorld does manage to look pretty seductive in a world obsessed with Jihad. It delivers peace, prosperity, and relative unity—if at the cost of independence, community, and identity (which is generally based on difference). The primary political values required by the global market are order and tranquility, and freedom—as in the phrases "free trade," "free press," and "free love." Human rights are needed to a degree, but not citizenship or participation—and no more social justice and equality than are necessary to promote efficient economic production and consumption. Multinational corporations sometimes seem to prefer doing business with local oligarchs, inasmuch as they can take confidence from dealing with the boss on all crucial matters. Despots who slaughter their own populations are no problem, so long as they leave markets in place and refrain from making war on their neighbors (Saddam Hussein's fatal mistake). In trading partners, predictability is of more value than justice.

The Eastern European revolutions that seemed to arise out of concern for global democratic values quickly deteriorated into a stampede in the general direction of free markets and their ubiquitous, television-promoted shopping malls. East Germany's Neues Forum, that courageous gathering of intellectuals, students, and workers which overturned the Stalinist regime in Berlin in 1989, lasted only six months in Germany's mini-version of McWorld. Then it gave way to money and markets and monopolies from the West. By the time of the first all-German elections, it could scarcely manage to secure three percent of the vote. Elsewhere there is growing evidence that *glasnost* will go and *perestroika*—defined as privatization and an opening of markets to Western bidders—will stay. So understandably anxious are the new rulers of Eastern Europe and whatever entities are forged from the residues of the Soviet Union to gain access to credit and markets and technology—McWorld's flourishing new currencies—that they have shown themselves willing to trade away democratic prospects in pursuit of them: not just old totalitarian ideologies and command-economy production models but some possible indigenous experiments with a third way between capitalism and socialism, such as economic cooperatives and employee stock-ownership plans, both of which have their ardent supporters in the East.

Jihad delivers a different set of virtues: a vibrant local identity, a sense of community, solidarity among kinsmen, neighbors, and countrymen, narrowly conceived. But it also guarantees parochialism and is grounded in exclusion. Solidarity is secured through war against outsiders. And solidarity often means obedience to a hierarchy in governance, fanaticism in beliefs, and the obliteration of individual selves in the name of the group. Deference to leaders and intolerance toward outsiders (and toward "enemies within") are hallmarks of tribalism—hardly the attitudes required for the cultivation of new democratic women and men capable of governing themselves. Where new democratic experiments have been conducted in retribalizing societies, in both Europe and the Third World, the result has often been anarchy, repression, persecution, and the coming of new, non-communist forms of very old kinds of despotism. During the past year, Havel's velvet revolution in Czechoslovakia was imperiled by partisans of "Czechland" and of Slovakia as independent entities. India seemed little less rent by Sikh, Hindu, Muslim, and Tamil infighting than it was immediately after the British pulled out, more than forty years ago.

To the extent that either McWorld or Jihad has a *natural* politics, it has turned out to be more of an antipolitics. For McWorld, it is the antipolitics of globalism: bureaucratic, technocratic, and meritocratic, focused (as Marx predicted it would be) on the administration of things—with people, however, among the chief things to be administered. In its politico-economic imperatives McWorld has been guided by laissez-faire market principles that privilege efficiency, productivity, and beneficence at the expense of civic liberty and self-government.

For Jihad, the antipolitics of tribalization has been explicitly anti-democratic: one-party dictatorship, government by military junta, theocratic fundamentalism—often associated with a version of the *Führerprinzip* that empowers an individual to rule on behalf of a people. Even the government of India, struggling for decades to model democracy for a people who will soon number a billion, longs for great leaders; and for every Mahatma Gandhi, Indira Gandhi, or Ranjiv Gandhi taken from them by zealous assassins, the Indians appear to seek a replacement who will deliver them from the lengthy travail of their freedom.

THE CONFEDERAL OPTION

How can democracy be secured and spread in a world whose primary tendencies are at best indifferent to it (McWorld) and at worst deeply antithetical to it (Jihad)? My guess is that globalization will eventually vanquish retribalization. The ethos of material "civilization" has not yet encountered an obstacle it has been unable to thrust aside. Ortega may have grasped in the 1920s a clue to our own future in the coming millennium.

> Everyone sees the need of a new principle of life. But as always happens in similar crises—some people attempt to save the situation by an artificial intensification of the very principle which has led to decay. This is the meaning of the "nationalist" outburst of recent years ... things have always gone that way. The last flare, the longest; the last sigh, the deepest. On the very eve of their disappearance there is an intensification of frontiers—military and economic.

Jihad may be a last deep sigh before the eternal yawn of McWorld. On the other hand, Ortega was not exactly prescient; his prophecy of peace and internationalism came just before blitzkrieg, world war, and the Holocaust tore the old order to bits. Yet democracy is how we remonstrate with reality, the rebuke our aspirations offer to history. And if retribalization is inhospitable to democracy, there is nonetheless a form of democratic government that can accommodate parochialism and communitarianism, one that can even save them from their defects and make them more tolerant and participatory: decentralized participatory democracy. And if McWorld is indifferent to democracy, there is nonetheless a form of democratic government that suits global markets passably well—representative government in its federal or, better still, confederal variation.

With its concern for accountability, the protection of minorities, and the universal rule of law, a confederalized representative system would serve the political needs of McWorld as well as oligarchic bureaucratism or meritocratic elitism is currently doing. As we are already beginning to see, many nations may survive in the long term only as confederations that afford local regions smaller than "nations" extensive jurisdiction. Recommended reading for democrats of the twenty-first century is not the U.S. Constitution or the French Declaration of Rights of Man and Citizen but the Articles of Confederation, that suddenly pertinent document that stitched together the thirteen American colonies into what then seemed a too loose confederation of independent states but now appears a new form of political realism, as veterans of Yeltsin's new Russia and the new Europe created at Maastricht will attest.

By the same token, the participatory and direct form of democracy that engages citizens in civic activity and civic judgment and goes well beyond just voting and accountability—the system I have called "strong democracy"—suits the political needs of decentralized communities as well as theocratic and nationalist party dictatorships have done. Local neighborhoods need not be democratic, but they can be. Real democracy has flourished in diminutive settings: the spirit of liberty, Tocqueville said, is local. Participatory democracy, if not naturally apposite to tribalism, has an undeniable attractiveness under conditions of parochialism.

Democracy in any of these variations will, however, continue to be obstructed by the undemocratic and antidemocratic trends toward uniformitarian globalism and intolerant retribalization which I have portrayed here. For democracy to persist in our brave new McWorld, we will have to commit acts of conscious political will—a possibility, but hardly a probability, under these conditions. Political will requires much more than the quick fix of the transfer of institutions. Like technology transfer, institution transfer rests on foolish assumptions about a uniform world of the kind that once fired the imagination of colonial administrators. Spread English justice to the colonies by exporting wigs. Let an East Indian trading company act as the vanguard to Britain's free parliamentary institutions. Today's well-intentioned quick-fixers in the National Endowment for Democracy and the Kennedy School of Government, in the unions and foundations and universities zealously nurturing contacts in Eastern Europe and the Third World, are hoping to democratize by long distance. Post Bulgaria a parliament by first-class mail. Fed Ex the Bill of Rights to Sri Lanka. Cable Cambodia some common law.

Yet Eastern Europe has already demonstrated that importing free political parties, parliaments, and presses cannot establish a democratic civil society; imposing a free market may even have the opposite effect. Democracy grows from the bottom up and cannot be imposed from the top down. Civil society has to be built from the inside out. The institutional superstructure comes last. Poland may become democratic, but then again it may heed the Pope, and prefer to found its politics on its Catholicism, with uncertain consequences for democracy. Bulgaria may become democratic, but it may prefer tribal war. The former Soviet Union may become a democratic federation, or it may just grow into an anarchic and weak conglomeration of markets for other nations' goods and services.

Democrats need to seek out indigenous democratic impulses. There is always a desire for self-government, always some expression of participation, accountability, consent, and representation, even in traditional hierarchical societies. These need to be identified, tapped, modified, and incorporated into new democratic practices with an indigenous flavor. The tortoises among the democratizers may ultimately outlive or outpace the hares, for they will have the time and patience to explore conditions along the way, and to adapt their gait to changing circumstances. Tragically, democracy in a hurry often looks something like France in 1794 or China in 1989.

It certainly seems possible that the most attractive democratic ideal in the face of the brutal realities of Jihad and the dull realities of McWorld will be a confederal union of semi-autonomous communities smaller than nation-states, tied together into regional economic associations and markets larger than nation-states—participatory and self-determining in local matters at the bottom, representative and accountable at the top. The nation-state would play a diminished role, and sovereignty would lose some of its political potency. The Green movement adage "Think globally, act locally" would actually come to describe the conduct of politics.

This vision reflects only an ideal, however—one that is not terribly likely to be realized. Freedom, Jean-Jacques Rousseau once wrote, is a food easy to eat but hard to digest. Still, democracy has always played itself out against the odds. And democracy remains both a form of coherence as binding as McWorld and a secular faith potentially as inspiriting as Jihad.

PART SIX

Pleasure of Inquiry

Ksenych and Liu

VI. DOING SOCIOLOGY

Throughout the previous five sections of this reader you have been provided with many examples of sociological inquiry and analysis. You have encountered many methods that sociologists use to study the social world and have been exposed to different styles of writing, thinking, and questioning. But, regardless of the particular topic addressed, or the particular approach used to address the topic, the theme throughout this book has been the practice of sociology and pleasure of inquiry. Our aim has been to show you how the practice of questioning the world around you and the relationships you take for granted can be both a worthwhile and pleasurable activity.

Learning takes many forms. Reviewing and critically assessing others' analyses is a crucial step in the development of one's own analytical abilities. Reading about how others have selected topics, and collected and interpreted their data, is part of learning the practice of sociology.

However, in this section of readings the focus is somewhat different. Here, it is students, and their own investigations, that are the center of attention. These selections invite students to begin to do original sociological research. The first two selections outline projects which give students the opportunity to take matters into their own hands, to begin to experience what it is like to gather their own data, and do their own analysis.

All three readings focus on concrete research practices, and the practical and ethical issues which need to be managed in the course of doing so.

The first two articles in this section outline some basic sociology projects which the introductory student can undertake. They can be used in themselves or as a general background to more specific projects that encourage the student to analyze and account for findings in terms of a group's culture and/or social structure. The kinds of projects outlined here have been classroom tested, and have been given the seal of approval by students themselves.

Throughout this book are readings which can work with, or be used as a resource for, the research methods presented in the articles. Readings #3, #5, #13, #15 and #16 employ observation. Readings #1, #2, #11, #15, #17, #31 and #36 are relevant to participant observation. And readings #14, #23, #27, #32 and #33 are applicable to content analysis, either as an example of gathering the data or interpreting and analyzing content.

The value of doing sociology—whether empirical studies, interpretive analyses, critical investigations, or theoretic formulations—is bound up with a concern which Plato posed over 2,000 years ago as the question of the "just" society. The Greek concept we translate as "justice" more correctly gathers together the ideas of justice, morality and goodness. So the question of a just society is more like asking after the nature of the good/moral/just society? To put it differently, in what kind of society can women and men realize the best of themselves together?

Plato insisted that conducting our inquiry into a just/good/moral society needed to proceed in a just and knowledgeable manner. In modern terms this is reformulated as proceeding ethically according to the standards of science. Or, as McIntyre puts it in her article, doing what's right in sociological research. But there was another dimension of Plato's examination which is visible in the way his dialogue proceeds, and which we hope that that you have experienced in the course of your research as well. Such an inquiry can, and should, be a pleasurable one.

Observing

*Ed Ksenych and
David Liu*

In the following article, the authors introduce some basic issues concerning the simple activity of observing as a method for developing our understanding of a social phenomenon, then outline some general guidelines for doing an observational field project. Questions concerning the nature and validity of gathering information are related directly to experiences students encounter in doing observations. The guidelines, which have been classroom tested, can be adapted to both simple observation or more complex participant observation studies.

> *"Reality is that which, when you stop believing in it, doesn't go away."*
> —Philip K. Dick

I. LOOKING AND OBSERVING

Before we begin discussing the activity of observing we'd like you to raise your eyes and look at your surroundings. Perhaps you're in a room, an area of a library, a kitchen, the interior of a bus, or a park. But before reading on, take a look around. . . .

Now that you've returned to this page, without looking again, consider what you just saw. Next, we'd like you to look again. Only this time observe your surroundings more closely and carefully. Take a couple of minutes, even if you're in a familiar place. So before reading on, observe. . . .

Now that you've returned to the page, what did you see? In addition, what was different about your experiences with the two activities? Both exercises were deliberately open-ended. But comparing the two, you likely noted that there's a difference between the activities of casually looking and of observing. Ordinary looking is like driving down the highway: You see everything and very little at the same time. Observing requires a heightened state of awareness. It's a more deliberate and focused mode of seeing.

In the second instance you may have become more alert and generally more attentive. Being more responsive to your environment, you may have noticed things you didn't notice the first time. The fact that you were asked to take a closer look may have made you more sensitive to the activity of looking itself. That is, you became aware of yourself looking. And you might have begun wondering about what you were noticing.

If you did your observing in a public setting, you may have also experienced something else. You likely became aware that others either were, or easily could be, aware you were observing your surroundings and them. Did you try to disguise your observing, perhaps by using long sweeping glances or by pretending to be staring into space?

Observing, unlike casually looking, not only pulls your overall environment and the things within it out of the ordinary flow of everyday life such that you see more. It also pulls you, the observer, out of the ordinary flow of everyday social life. The observer often becomes visible in the setting in a way he/she wasn't before. Observing disturbs unspoken rules regarding how we are expected to visually encounter others and our environment. In many cultures, including ours, there are norms regarding when and how we are to look at our surroundings and those within them. One of the principles behind the norms is to look, but not for too long or too carefully; to see, but not to appear to see too much. Children violate this norm constantly, staring until they've had their fill.

Now we're going to ask you to observe one more time. We're aware you're busy. But that is often another constraint on observing. Norms concerning looking and the ordinary flow of life aren't all that discourage us from careful observation. There's also our own pragmatic focus on the frenetic pace and demands of daily life. Sociological observing takes time and patience as well as an openness to what lies beyond the immediate practical significance of things.

So this time have the following general questions guide your observing. Has the setting been designed for a particular purpose or function? In what ways? If others are present, are there any recurring patterns of behaviour? Do you see any patterns, either general or detailed ones? For instance, if you happen to be sitting near a street corner with stoplights, you'll generally see cars stopping when the light is red. But is there also a pattern to how they stop, wait, and prepare to proceed? Before reading on, observe your surroundings for a couple of minutes with these guiding questions in mind....

Now that you've returned to this page, let's reflect on what you observed. One issue you may have encountered during this last exercise is distinguishing between what's actually occurring and what you expect to see. While we all are guided by expectations we bring to the settings we observe, these expectations shouldn't predetermine what we'll find. The challenge is to remain open to discovering what might not be obvious at first glance. How is this done?

First of all, consider the purpose or function of the setting. Whether you are in a classroom, in a coffee shop, or on a bus, the overt or manifest function of the setting seems quite obvious. But you may have also observed things that seem out of place, or don't seem to relate to that function. Perhaps some features relate to this function but you noted the function could have been met in a different way. For example, why are the seats arranged the way they are in a classroom, in a coffee shop, or on a bus? Could they be arranged differently? These kinds of observations and queries suggest that the setting may be helping to perform a less obvious or latent function.

To stay with our classroom example: In most classrooms the desks and chairs are lined up in rows, all facing the same direction—toward the teacher and blackboard. Certainly this affords all students a good view of the front of the classroom. But how might this particular arrangement serve a further function? If we are observing sociologically, we want to find out how the setting and patterns of interaction are related to social or cultural purposes rather than just practical ones.

We would ask of the classroom setting: How does this promote a certain pattern of interactions between teacher and students? How might this reinforce the authority of the teacher? How might a different pattern of interactions emerge if the teacher and students were all sitting in a circle or around a table?

The idea that a setting may serve more than one function also applies to the many kinds of action and interaction one sees generally. There is, for example, the often observed social interaction ritual of greeting those we know by asking one another how each is and usually answering with a "good" or "fine." While this does serve the practical purpose of acknowledging the presence of a person you know, couldn't it be done otherwise? What is the purpose or social function of the particular ritualistic way we do it, especially if we are not feeling good or fine?

Perhaps there are no covert social functions and/or cultural purposes in your setting beyond the obvious practical one. Observing doesn't always reward the observer with the new and unexpected. Then again, with practice, it often does.

II. OBSERVING AS A RESEARCH METHOD

We engage in research to broaden our shared knowledge, to deepen our understanding of a phenomenon, and to better appreciate the nature of the social world. Of the many research methods such as interviewing, surveying, experimentation, and so on that are available, observation is the most basic. Unlike the open-ended exercises we began with, the formal research process begins with questions, ideas, or theories, and uses a particular research method to systematically arrive at ordered facts about some phenomenon (i.e., research findings). One's analysis begins with interpreting these findings using the initial theory or a newly emerging one, and then trying to understand their larger significance.[1]

Observing is usually associated with the faculty of sight. But it is not limited to seeing. One can, for example, "observe" that fast food is usually very salty and the drinks very sweet. One can observe a professor's voice getting louder when students begin chatting in class. Clearly one does not see these, although they are observations. So while most observation concerns what is visibly present, or noticeably absent, observing—even if it privileges sight—essentially refers to a more intensely aware, focused manner of attending to oneself, others and one's environment which draws on all one's senses.

Of course observing is not restricted to sociological inquiry. All people observe when they want to know more about something of importance to them and share their observations when given the chance. The first day of class most students keenly observe their professor for signs of what he/she is like, and frequently share what they saw with friends afterwards. Yet there is a difference between sociological observation and even the most intensive amateur observing we may ordinarily engage in.

There are five characteristics which distinguish sociological observation: 1) observing systematically; 2) carefully recording one's observations; 3) being alert to details, patterns, and how things occur; 4) being aware of issues regarding observing; and 5) presenting and discussing one's findings in an organized way.

Observing Systematically

Sociological observation is guided by the goals of the particular study the observer has in mind, as well as by the requirement to gather data objectively and to use sociological concepts and theories to organize and analyze the data. To do this effectively we need to consider how we will go about systematically observing and recording the phenomenon under investigation. For instance, suppose the goal is to study how high school students informally interact with one another. We will need to decide on an appropriate setting where such informal interactions will occur often and openly. Settings such as the hallways and grounds at a high school would seem more promising than an event organized and monitored by adults. We will also need to consider how to be unobtrusive while recording the observations. An observer tries not to affect the setting.

Another aspect of systematically observing concerns operationally defining or explicating key terms ahead of time. If we are looking for friendship, rivalry, courtesy, leadership, and the like among high school students, how will we recognize them? Sometimes, we can't be entirely sure how something like friendship or rivalry might appear, or we don't know how to interpret a particular behaviour. So we will need to be ready to clearly explain what we observe and what we are counting as an expression of the phenomenon. Observing requires us to be systematically attentive to what is unexpected as well as to what we are specifically looking for.

Carefully Recording

Ordinarily when we observe, we simply rely on memory and what we remember is what strikes us at the time. However, sociological study requires attending to more detail than we ordinarily do and carrying out a more systematic analysis than we usually engage in. So we need to think about how we will keep written descriptions of what we actually observe.

Moreover, sociological study requires that we should stop, record, and even reflect on what we are seeing as frequently as we can. The observer needs to be attentive to the nature of the participants' actions and reactions, what is sociologically significant, and what sociological concepts, principles, or theories may be at play while observing. The investigator needs to prepare for how he/she will make the time and place to do this. In general, try not to do this too publicly, or you might draw unwanted attention to yourself, attention which which will distract you from further observation.

Being Alert to Details and Patterns

A third characteristic of sociological observation is the need to be alert to details, patterns of behaviour, and how things actually occur. We usually observe long enough to get a take on what's going on and some sense of why. But for sociologists what's going on is often a question and a question with several layers of meaning. As Peter Berger described it, the first wisdom of sociology is that

> things are not what they seem.... Social reality turns out to have different layers of meaning. The discovery of each layer changes the perception of the whole. The fascination of sociology is not the excitement of coming upon the totally unfamiliar, but rather the excitement of finding the familiar transformed in its meaning. The fascination of sociology lies in the fact that its perception makes us see in a new light the very world in which we have lived all our lives.[2]

Careful observation is central to perceiving the sociological significance of what initially seems obvious and/or insignificant.

Part of what we attend to in detail is what is unique or what stands out in the setting. But what may be even more important sociologically are patterns of interaction or recurring events and the relationships they signify. One spot to look for detailed patterns of interaction, for instance, is in how something routinely occurs in a setting. In a coffee shop, we can expect to see the routine of customers ordering and receiving coffee. But finding this shouldn't stop us from further observation. Exactly how does the transaction occur? What do the customer and the server typically say and do, and in what order? Paying close attention to the routine exchange can result in some surprising observations and questions.

Being Aware of Basic Issues in Observing

The example of the coffee shop calls attention to one way an observer brings expectations, values, beliefs, and interests to the situation. And as we've already noted, while our expectations may guide us, they should not predetermine what we find, or prevent our looking into what we expect to see in greater detail, nor limit our interpretation of what we see.

This raises a couple of other issues. First, observing requires that we orient to being objective, but can an investigator ever really provide neutral, disinterested observations?[3] Second, how should the investigator interpret what is observed? What is occurring can have a different meaning for the participants than it ordinarily has for the observer, especially if the observer is unfamiliar with the setting or the group. And it can mean something else yet again to the observer looking through the lens of sociology. So which meaning system is one working with at any point in time while observing?

Regarding the first issue, if we are doing an observational study in the strict sense, we are required to orient to objectively seeing what is actually present, or noticeably absent, in a dispassionate manner, and to note our subjective biases and interpretive decisions. At this point we would suggest that it is possible to orient to the standard of objectivity and that there is a value in doing so for understanding the social world.[4] By "being objective" we mean attending to all that is actually there rather than focusing on what we're used to seeing or want to believe is there. For instance, our friends may objectively notice features about someone we have fallen in love with which we disregard or fail to notice, but eventually will need to attend to. Analogously, it is possible to have fallen in love with the comfort of the routine and familiar.

But even when we orient to the standard of objectivity in an observational study, we want to emphasize that good observing isn't simply a neutral recording of what's present. Video cameras and tape recorders do not observe. They passively and mechanically record. Observing sociologically involves the art of gently challenging what we see by documenting and asking questions about the conventional nature of what's present.

Conventions refer to the unspoken agreements, assumptions, and background expectations that people rely on while engaging in everyday interactions. Recognizing the conventional nature of everyday life is to acknowledge that what usually occurs could be otherwise.[5] There is very little that we regularly do or that routinely goes on that *necessarily* has to be the way it is. A simple example is that, while humans may need to eat in order to survive, a group of people don't really "have to" eat the foods they regularly do in the places they routinely do at the times or in the particular ways they do. It could reasonably be otherwise. So how can we make sense of what we observe them doing?

Presenting and Discussing Findings in an Organized Manner

A final characteristic of sociological observation concerns the organized presentation and discussion of findings. Ordinarily, we report our observations with a short general summary and noteworthy highlights. But sociological observation requires we organize our detailed findings more comprehensively, and discuss them in a clear, systematic way. This involves reproducing what we observed as accurately and comprehensively as possible.[6] But while a sociological description orients to a standard of objectivity, it is nevertheless a description presented and interpreted in terms of concepts or of a theory for understanding social life. As such, we need to call attention to the premises, concepts, and theories we are drawing on to make sense of what we have seen.

The final consideration is how to apply what one has discovered. How much can one generalize from one's findings? Being clear and explicit about the sociological questions or issues that originally animated the research project will help demonstrate how one's findings can be used to shed light on wider concerns.

III. OBSERVATION AND PARTICIPANT OBSERVATION

The activity of observing occurs not only in observation studies, but also in a related research method called participant observation. We are always involved in some way in making sense of what we observe. For instance, if you were to observe a family having dinner in their home dining room from behind a two-way mirror, you would still be involved in selectively framing and interpreting what you see. However, participant observation refers to a research method in which you are actively present in the very setting or group you are observing.

In the preceding example, instead of looking from the outside, you might be involved in the dinner itself. You could be at the dinner in a variety of ways: as a guest, a family member, a waiter, or an interviewer.

> It is probably misleading to regard participant observation as a single method. Rather ... it refers to a characteristic blend or combination of methods and techniques that are employed in studying certain types of subject matter. ... This characteristic blend of techniques ... involves some amount of genuinely social inter-action in the field with the subjects of the study, some direct observation of relevant events, some formal and a great deal of informal interviewing, some systematic counting, some collection of documents and artifacts, and open-endedness in the direction the study takes.[7]

In doing any observing an investigator needs to decide the degree to which his/her role in the setting will be openly defined as a research role or will be concealed beneath the performance of some existing role within the setting or group. The decision is important because it involves what you want to put at risk in gathering and interpreting data:

> It is generally true that with increasingly more observation than participation the chances of "going native" become smaller, although the possibility of ethnocentrism becomes greater. ... A field worker who "goes native" passes the point of field rapport by literally accepting his informant's views as his own. ... Ethnocentrism occurs whenever a field worker cannot, or will not, interact meaningfully with an informant. He then seemingly or actually rejects the informant's views without ever getting to the point of understanding them. [Because] a complete observer remains entirely outside the observed interactions, he faces the greatest danger of misunderstanding the observed.[8]

Participant observation brings to the foreground the issue of which, and whose, meaning system we will be using to interpret what is occurring. The issue is at play in all observing. However, we may not always be as aware of or explicit about what ideas or theories are framing our ordinary observing. For example, what is the difference between these two descriptions? 1) A male person kept rapidly contracting his right eyelid in front of a female person. 2) A guy kept winking at a woman. The first is a flat, clinical description of sheer physical behaviour. But this is no more objective than the second description which takes into account the intersubjective meaning of the action for the observer and the participants.

Sociological observation can help us make explicit the meaning systems we use in making sense of the world in which we live. This is especially important if one is observing a setting or a group with which one is unfamiliar. In cases where one is an outsider, a significant part of observing is determining the social and cultural frameworks in which people are acting.[9] Following the example above, even if everyone agrees winking occurred, what does the winking mean to the people one is observing? Perhaps the observer regards it is a sexual advance, while the two participants regard it as signifying their common understanding about an incident that just occurred. Unless the observer shares in the system of meanings the other participants are employing, then he/she is left to draw on his/her own framework in describing what is going on.

The value of participant observation, especially in settings and groups to which you are an outsider, is that it enables you to gain access to how members routinely interpret and understand what others and you see as going on while maintaining some distance from them. As Aron and Wiseman say, the observer "sees the world, at least part of the time, in the same way other members of the group see it because he is living their kind of life with them. . . . All the contradictions between what people say they believe in doing and what they actually do, their consistencies and inconsistencies, are played out before the observer."[10]

IV. DOING AN OBSERVATIONAL FIELD PROJECT

In this section we offer a brief overview of how to do an observational field project. We've divided it into six steps: selecting your topic of study; focusing your topic and data collection; gathering data; organizing and presenting the data; analyzing the data sociologically; and concluding.

Selecting a topic for an observation or participant observation study is often a difficult task for students. The range of possibilities can be overwhelming: a courtroom trial; a religious ceremony; people interacting in bars; people eating in a cafeteria; the organization of a sports team; encounters in a hotel lobby; the role of a receptionist in an organization; what goes on in coffee shops; the relationship between police and civilians; how parents discipline children.

Whether you have a specific topic in mind that you would like to learn more about, or only a general idea, it is important to consider how much time you have and what resources are available to you. Don't try to take on too much!

Once you have a general topic you will need to narrow it down by formulating a set of guiding questions. This is important not only for creating a manageable study, but also for determining the appropriate site and preparing yourself for logging data. If you are going to observe, say, how parents discipline children, you need to consider where you'll be able to best observe this. Gaining access to parents' homes is possible, but you'll be quite obtrusive even if you disguise the reason for your study. Observing playgrounds, beaches, parks, stores, and other places where parents are supervising children might be preferable given the nature of your topic and the limitations of time and resources. However, note that getting into some settings can raise social, ethical, and even legal issues, especially if you're studying a setting not usually open to the general public.

A second important part of setting up your project is *focusing your topic and data collection sociologically.* Most phenomena can be studied from various perspectives. Consider a fast food restaurant and what occurs within it. We could observe and interpret the phenomenon from a business, a psychological, or a nutritional perspective. But to study it sociologically is to investigate how the setting and what occurs within it are organized in terms of culture and social structure. It involves examining on some level how the setting and what occurs within it relate to the society in which they exist as well as to the kinds of historical changes the society may be undergoing.

Lofland and Lofland offer a useful guide with regard to narrowing your study sociologically. They point out different facets or levels of social organization on which you can focus: meaning systems; recurring practices or activities; the norms which guide behaviour; episodes; encounters; roles (formal, informal, social types, social psychological types); social relationships; and in the case of broader, long-term studies, groups, communities, organizations, social worlds, and lifestyles.[11]

The focus of the earlier example of parents disciplining children is at the level of social practices and activities, namely social control and socialization practices. Related to this level of focus are the roles of parent and child, social types of parents, as well as the parent-child relationship. These could be the level of focus of some other study. But they also may require some consideration here in order to more effectively understand the disciplinary practices being observed.

The third step is *gathering the data.* Once on site, you will need to figure out an approach to logging data by keeping field notes. How you set yourself up will depend on whether you are primarily doing a quantitative or qualitative study. A quantitative approach is appropriate if the research goal is to represent some phenomenon numerically. It involves establishing relevant operational categories ahead of time to codify your anticipated observations.

Continuing with the example of disciplinary practices, you might list the categories of reminding, loudly restating a rule or expectation, warning, scolding, demeaning, lightly slapping, spanking, and explaining. As the behaviours occur, you simply put a mark in the appropriate category in order to keep track of how frequently they occur. But along with these categories, leave plenty of space to comment on the nature of the practices and to jot down and keep track of ones you didn't anticipate, such as ignoring or bribing in this example.

In some studies, tabulating the frequency of certain occurrences along with the character of the occurrences may be the primary research goal. On the other hand, a qualitative approach is best suited if the research goal is studying questions concerning social processes, interactions, or social organization (e.g., how students go about writing tests; interaction rituals in a bar; the social organization of a family dinner). While qualitative approaches don't ignore the frequency of behaviours, they pay more attention to the nuances and details and provide a richer description of social life.

While you should consider ahead of time the sorts of things you will be looking for and how you expect to recognize them, usually the strategy here is to keep your recording more open-ended, jot down as much as you can, and review and organize the observations afterwards in terms of typologies, categories, patterns, and so on that emerged. Since you aren't rigorously establishing operationally defined categories ahead of time, you will need to explicate the key terms you use to describe what you see.

Good observing requires recording what you are observing frequently, accurately, and in reasonable detail. As you observe you may find yourself both describing what you observe and making comments on the observations. Try to keep your field notes and more analytic comments separate.

A participant observation study may also involve informal or formal interviewing. In this case, you will need to be able to document parts of what the interviewees said as accurately as possible, and not rely on simple summaries.

The fourth step is clearly and effectively *organizing and presenting your data*. Your presentation should reflect the sociological focus of your study, research approach, and the data that you've gathered. For instance, if your study is primarily quantitative and you've been tabulating the frequency of some event or behaviour, then your presentation should include a table or chart that neatly summarizes your tabulations. If your research is primarily qualitative, you will probably organize your findings and discussion in terms of patterns, typologies, or stages in a process that emerged from your observations. In either case, presenting one's findings involves providing a systematically organized, accurate, and comprehensive written description of what you found.[12]

The *sociological analysis* is often the most difficult part, especially for students just learning sociology. What does one say that's sociological? What counts as "analyzing" the findings? It may be useful to keep in mind that the aim of analysis is understanding what you've observed by discerning patterns and relationships among parts. Within sociology we're aiming to understand what we've observed in terms of the social structure and the culture of social actors within a setting. Here are some suggested approaches.

1. Show how what you observed, especially the typical or recurring patterns of behaviours and/or events, relates to relevant aspects of the culture and/or social structure of the group, organization, or overall society. For these purposes, culture can be thought of as the prevailing way of life and shared meaning system (i.e., values, beliefs, norms, ideas) within the group. Social structure refers to the social organizational arrangements (e.g., social positions with their role expectations and status rights, social divisions, stratification system(s), broader social institutions) and underlying organizing relationships and principles in terms of which individuals interact.

2. Compare and contrast what you've observed with other research that has been conducted on the topic, and attempt to account for any major differences.

3. Use relevant sociological concepts to explain and account for what you've observed. Make sure to outline your understanding of a concept before using it.

4. Interpret your observations in terms of a general theoretic approach within sociology such as functionalism, conflict theory, feminist theory, or symbolic interactionism, or in terms of the framework developed by a particular theorist you think is relevant. It is always a good idea to provide an overview of the theoretic approach you intend to work with and the particular aspect of it you will emphasize.

The final step is *effectively concluding*. The simplest aspect of concluding is to simply share what you have learned from the study. That can involve a discussion of the difference your study made in relation to views you previously held, or of how your findings relate to prevailing beliefs and opinions. You can also comment on the relation of your study to other research. Finally, it's also useful to highlight questions, problems, or issues you encountered during your study or that emerged for you as a consequence of doing the study.

V. OBSERVING AND KNOWING

Observation and participant observation studies require time and differ in their breadth, depth, and the degree which they can be generalized. They do not always result in solid, astonishing results an inquirer might hope to find. At the same time, they usually draw attention to aspects of a setting, a group, or a phenomenon one simply never noted before. Often, they make available some of the different levels of meaning and organization in which we are immersed in our everyday lives.

But there is an irony to observational studies that those who try them quickly encounter. The seemingly simple activity of observing is fraught with many difficult questions about what we are actually seeing when we observe, and what we can really claim to know given what we see. It is unrealistic to expect newcomers to observational studies to resolve such questions in any conclusive way. But it isn't unrealistic to expect newcomers to be aware of the issues and to be prepared to think about what they claim to have discovered. Indications that they have done so usually signal an excellent effort at observing. And that's an observation we've made regarding students who have tried such projects.

NOTES

1. Matilda White Riley, *Sociological Research: A Case Approach*, New York: Harcourt, Brace and World, 1963: 4. Riley and Nelson also collect excerpts from a number of classic sociological studies which address the key methodological issues involved in observational field work in Matilda White Riley and Edward E. Nelson, *Sociological Observation: A Strategy for New Social Knowledge*, New York: Basic Books, 1974.

2. Peter Berger, *Invitation to Sociology*, Garden City, NJ: Doubleday, 1963.

3. An interesting discussion of the emotional dimension involved in the cognitive activity of observing is taken up by Israel Scheffler in "In Praise of the Cognitive Emotions" from *Inquiries: Philosophical Studies of Language, Science and Learning*, Indianapolis: Hackett Publishing, 1986: 347–362. Scheffler challenges the conventional split between unfeeling knowledge and mindless arousal presented as an indicator of the opposition of cognition and emotion to be unhelpful and inaccurate. Instead, he argues the life of reason "works for a balance in thought, an epistemic justice," and this balance involves rational passions, perceptive feelings, and theoretical imagination.

4. Israel Scheffler provides an extensive discussion of subjectivity in relation to the ideal of objectivity within science in *Science and Subjectivity*, Indianapolis: Bobbs-Merrill, 1967. More recently, Berard revisits the issue of disinterested description as practised, for example, by ethnomethodologists and conversational analysts, in "Evaluative Categories of Action and Identity in Non-Evaluative Human Studies Research: Ethnomethodology," *Qualitative Sociology Review*, Vol. 1, No. 1 (2005). While recognizing the prevalent argument that all descriptions are interested descriptions, he nevertheless argues for its possibility and desirability, even if the subject matter involves evaluations of problematic actions and ideas.

5. Peter McHugh, "A Common-Sense Conception of Deviance," in Jack Douglas (ed.), *Deviance and Respectability: The Social Construction of Moral Meanings*, New York: Basic Books, 1970: 61–88.

6. Although written with regard to understanding different philosophies of composition in fiction, Georg Lukacs offers several important insights into the nature as well as the strengths and weaknesses of uninvolved description, or of being a stranger to an event, and narrating as an active participant in the event in his essay "Narrate or Describe?" in *Writer and Critic*, New York: Grosset and Dunlap, 1970: 110–148.

7. George McCall and J. L. Simmons, "The Nature of Participant Observation," in *Issues in Participant Observation: A Text and Reader*, Don Mills, ON: Addison-Wesley Publishing Co., 1969: 1.

8. Raymond L. Gold, "Roles in Sociological Field Observations," in George J. McCall and J. L. Simmons, 1969: 37.

9. The topic of attending to meaning structures and their implications for reporting what we observe is taken up by Clifford Geertz in his discussion of "thick description" in *The Interpretation of Cultures*, New York: Basic Books, 1973: 3–30. When one is engaged in the thick description of things one is not only attending to the features of the things in and of themselves, but also to the meaningful structures in which "they are produced, perceived, and interpreted, and without which they would not ... in fact exist." The importance of meaning in understanding human conduct also forms the basis of a symbolic interactionist approach to sociology as formulated by Herbert Blumer in *Symbolic Interactionism: Perspectives and Method*, Englewood Cliffs, NJ: Prentice-Hall, 1969. For a contrasting position on the ability to understand the meaning an action has for the actor, see Harold Garfinkle, *Seeing Sociologically: The Routine Grounds of Social Action*, Boulder: Paradigm Publishers, 2006, including Rawls' introductory discussion of Garfinkle's interest in the problem of the scientific description of action. "For Garfinkle, the actor's point of view is not a private matter. ... Intersubjective meaning could only be achieved through some witnessable, public, seeable process. ... Thus, reflection is a feature of the witnessable order of the conversation itself, not something happening in the minds of the speakers." (page 34).

10. Jacqueline P. Wiseman and Marcia S. Aron, *Field Projects for Sociology Students*, Cambridge, Mass.: Schenkman Publishing Co., 1970: 52.

11. John Lofland and Lyn H. Lofland, *Analyzing Social Settings: A Guide to Qualitative Observation and Analysis*, 2nd edition, Belmont, CA: Wadsworth Publishing Co., 1984.

12. Lee Cuba provides a very helpful guide to writing sociology papers based on original research in *A Short Guide to Writing About Social Science*, 4th edition, New York: Longman, 2002: 79–119.

Content Analysis

David Liu and
Ed Ksenych

This article provides detailed guidelines for setting up and conducting a basic content analysis of "texts" drawn from the mass media. The authors describe what a basic quantitative content analysis is, and contrast it with more qualitative approaches to content analysis. The article concludes by discussing how the mass media are involved in the construction of a group's culture, and alerts students to differences between how the mass media represent cultural practices and how they actually come to pass.

I. STUDYING "TEXTS"

Content analysis is used to help us understand a group or institution by systematically examining collections of symbols, artifacts, and/or practices and activities produced and used by its members. This may include written works such as books, newspapers, or magazines; letters, transcripts, or graffiti; or visual images such as films or photographs, television shows or comic strips. This technique requires that we collect, organize, and compare data in order to reveal what a group's norms, values, beliefs, ideas, or concerns are, and how they may have changed over time. In short, content analysis, as defined by Robert Weber, is "a research methodology that utilizes a set of procedures to make valid inferences from text."[1]

The key difference between content analysis and other forms of analysis is that it is the content of *texts*[2] or, more broadly speaking, *media* which is the focus of the research. We are not observing people's behaviour directly but, rather, looking at what George Zito calls "published communication." By this he means that, regardless of the form the communication originally took, it has been made concrete, it has been set down in a way that is publicly available. It has taken a form which is observable, measurable, and analyzable.

II. VARIETIES OF CONTENT ANALYSES

Until recently content analysis was regarded as a special, secondary research method in sociological inquiry.[3] In contrast to observing and interviewing individuals more or less directly about their social behaviour and attitudes, content analysis investigates a group's culture and social structure indirectly by studying the content of the communication and cultural artifacts of group members. However, in an era in which communication technologies have proliferated and communication is influenced by many forms of mass media, content analysis has become an increasingly important research method for studying and understanding modern social life and society.

Paralleling Sigmund Freud's famous psychoanalytic statement that dreams are the royal road to the unconscious, contemporary sociological inquiry has found cultural artifacts and communications to be a royal road to a group's conscious and unconscious beliefs, attitudes, values, prevailing ideas, and norms. This requires studying these texts with the same systematic attention to details and analyzing them with the same level of interpretive skills as a psychoanalyst might decipher dreams. The sociologist's focus, however, is on understanding the texts in relation to a group's cultural meaning system and social organizational arrangements.

Content analyses can cover a wide range of different research practices. They range from highly detailed qualitative analyses of a single case to more quantitatively oriented studies which survey a number of examples from the media to ascertain trends, commonalities, and patterns in the content of the communication. While a textual analysis of the meaning structures in a single newspaper article about an abnormal fetus,[4] a discourse analysis of courtroom proceedings,[5] or a semiotic analysis of a fashion ad[6] can provide rich, rewarding insights into the meaning and social systems in which they occur, these types of content analyses do involve an exposure to theories of communication, culture, and meaning beyond the scope of this paper.[7] They also involve analytic and interpretive techniques that are rather sophisticated for an introductory field project.[8] So we will limit our discussion of content analysis to more basic, quantitatively oriented, descriptive surveys of messages and artifacts. And while content analyses can be used to study interpersonal communications such as letters, diaries, phone conversations, or even family photographs, we will also limit ourselves to content analyses of more public cultural artifacts and communiqués usually associated with the mass media and its different forms.

Basic content analysis can serve a variety of research purposes. It can be used to investigate or test an idea or hypothesis that one has about a group, such as testing the hypothesis that traditional gender stereotypes continue to be dominant in a group.[9] It is a useful research tool if one is doing a comparative study of two or more groups or institutions, or if one is looking at changes within a group or institution over time, such as investigating whether our society is more diverse today than it was four decades ago. But it is primarily used to systematically explore and uncover particular overt and/or underlying representations, images, values, themes, or associations being presented through forms of mass media. For example, what kinds of values are communicated through the national news?[10] What images or metaphors concerning the Canadian nation-state underlie current political cartoons?[11] What are the characteristics of the male hero presented in Harlequin romance novels?[12] How are youth gangs portrayed in newspapers?[13] How is family life portrayed on American television?[14] In all these cases, content analysis always carries with it an accompanying question: What is the social and cultural significance of any trends, patterns, or commonalities that we are able to ascertain using this method?

III. DOING A BASIC CONTENT ANALYSIS

In this section we offer a relatively brief, but more detailed overview of how to do a basic descriptive content analysis. We've divided it into six steps: selecting a topic; setting up the data collection; gathering data; presenting the data; analyzing the data sociologically; and concluding. For the purpose of this overview we'll be working primarily with examples of basic exploratory content analyses in which the investigator sets out to determine and document how some phenomenon is portrayed in a particular media genre.

Selecting a Topic

If your research goal is to systematically investigate how a particular phenomenon appears, or is portrayed, in a specific medium you will need to make two basic decisions: selecting a phenomenon to focus on, and selecting the type of media you will be examining. The following are some examples: crime in TV crime dramas; men and women in newspaper personal ads; children in classical fairy tales; evil in video games; families in TV ads; CEOs in business magazines; mental illness in U.S. films; sin in evangelical shows; black males in hip hop lyrics; images of social class in family sitcoms; politicians in TV newscasts; beauty in fashion magazines; animals in comic strips.

The range of possible topics is almost endless. But for the study to be a workable field project, before beginning you need to practically consider the time involved, the availability of resources, and the level of difficulty of effectively interpreting and documenting content. It is worth noting, for instance, that novels and feature films can be very time consuming to work with. Magazines, newspapers, or TV shows from a different era can be hard to find. And some media genres, such as adult cartoons or satirical TV shows, are more difficult, though not impossible, to examine quantitatively because often two or more levels of meaning are being conveyed through the comedic portrayals of people and situations. Moreover, while it is possible to examine various media, it is best to focus on a single medium or media type.[15]

You will also need to focus your study as finely as you can by selecting a particular aspect of the phenomenon you are interested in, and by narrowing down the media genre. Consider the example of investigating how crime is portrayed in TV crime dramas.[16] Given the time limits for doing a field project within a semester you will need to decide on whether you want to focus on the types of crime, the characteristics of the criminals, or the reasons for crime. All are legitimate topics, but the investigator needs to be prudent about how much he/she is willing to take on. With regard to narrowing the medium, will you be focusing on crime shows from one country or globally? Will they be current TV shows or cover the last couple of decades? In general, set clear, explicit limits to what you will study and the techniques you will use in your field project.

Preparing for the Data Collection

In order to collect the data effectively you will need to spend some time preparing. Specifically, clarify even more precisely what you are going to explore, and identify the specific categories, qualities, characteristics, or features associated with the topic which you will be examining. For example, if you're investigating how crime is portrayed, then you need to decide whether you are going to consider all types of crime or focus on a particular type of crime, say, violent crimes. In either case, you will need to identify the categories of relevant crimes as defined by the law. For instance, the Criminal Code of Canada identifies the following as the major types of violent criminal offences: homicide, assault, sexual assault (unlike the U.S., which has rape), robbery, extortion, and kidnapping/abduction, and other less frequent crimes such as intimidation, inciting hatred, unlawful confinement, and infanticide. So if you are investigating violent crimes in a Canadian context these will be your categories.

If your topic is violent offenders, then you will need to select which characteristics you are looking for. Some possibilities are sex, race/ethnicity, social class (e.g., upper, middle, working, lower), general age category (e.g., juvenile, young adult, middle aged, older adult), psychological condition at time of the crime (e.g., insane, rational, angry, motivated by fear, motivated by greed), committed the crime alone or with another, member of a gang or criminal organization, and so on. Sometimes it's helpful to look at a few examples of the portrayal before deciding on the characteristics you will focus on.

Even when you have established the characteristics, it is a good idea to leave an "other" category in case a characteristic you didn't think would be relevant appears. However, as a rule of thumb, select qualities, characteristics, or categories clearly pertinent to your topic and research goal. This is especially important if you have a specific guiding question or a hypothesis you want to test.

Once you've selected the categories you will need to operationally define them. That is, you need to explain how you will go about determining what counts as an instance within a category. If the topic is violent crimes, the Criminal Code provides legal definitions, and the police in the TV show will usually announce what crime has been committed. However, if the topic is violent offenders, not all the categories may be either formally defined or self-evident. How you are going to determine the general age category, social class, or psychological condition of the criminal? You will also need to be prepared for the possibility that a characteristic may be unknown or indeterminable.

A second part of preparing for the data collection is deciding on a representative sample. Specifically, how many and what kinds of cases of the media depiction are you going to examine? It is difficult to provide more than general guidelines for sample sizes in introductory field projects. One needs to take into consideration the nature of the research project, type of genre, and time available to do the project, as well as whether the project is quantitatively or qualitatively oriented.

As a rule of thumb try to work with about ten examples of the media genre for a field project of this kind. For example, if you are doing an analysis of the kinds of topics covered in evening newscasts, watch ten newscasts. However, you need to be aware that while feasible for an introductory field project, statistically this sample size may not be sufficiently representative to allow you to make valid and reliable generalizations. Another issue concerns the range of newscasts you will watch. Should they be from one broadcasting system or a variety of channels? Unless it is important to your particular research goal to focus on one media organization, TV show, etc., use a variety of examples from the genre.

Once you have decided on the characteristics as well as the number and range of examples, you will need to create a "working form" so you can tabulate how often the qualities, characteristics, or features appear in your survey of examples. If you're examining the kinds of crime committed in TV crime dramas, for instance, you'll need to create a chart or form with a grid that lists the categories of crime down the left side and the various shows across the top. This way you can easily check off the number of times a particular kind of crime occurs in the shows while you're watching them. And it will enable you to easily total the numbers of each kind of crime and calculate their frequencies relative to the total of all the crimes you noted. The working form can be rough; it's for the investigator to work with.

Gathering the Data

If you've prepared well then gathering the data chiefly consists of going through each of the chosen media examples and documenting the frequency of the characteristics, qualities, or categories of the cases which appear in them. Continuing with the example of types of crime in TV crime dramas, in ten examples of crime shows you may witness the occurrence of, say, fifty crimes. You simply tick off which category each crime belongs to in your working form as you go along.

As straightforward as this may seem, issues can arise when recording the content. For instance, if the TV show presents an individual seriously assaulting another person as part of the storyline, but the police in the show are unaware of it, do you count it as a crime or not? In some types of projects you may encounter the issue of multiple appearances of the same case. For instance, if you are analyzing the types of celebrities reported in local newspapers, you may encounter the very same story about, say, a sports celebrity in three different newspapers. Do you count this as one case of a sports celebrity overall or three? And if the story is followed by a newspaper for four days, do you count this as one case of a sports celebrity overall or four? Such issues are part of the research process and are not to be ignored. If you encounter such an issue, you will need to identify it and describe how you handled it either in your discussion of your methodology or in the report of your findings. In this instance, keep in mind what is best for accomplishing your research goal.

There is another aspect to gathering the data. Even in quantitatively oriented content analyses you may notice many things outside the scope of your formal chart, but of interest or relevance to your research goal. It is important to keep written notes as you transcribe your observations on the working form. For example, in analyzing TV crime dramas, you may notice that most of the victims know their offenders, though you weren't setting out to look for this. Or you may realize that the crimes are always solved, but that we are never shown what happens to the criminals after they've been apprehended. You may encounter a significant exception to trends or expected patterns in the TV shows. Observations of these kinds are important and should be documented even if you are not entirely sure what you want to say or do with them.

Presenting Your Findings

Your presentation should provide a final table which clearly and effectively summarizes your data, as well as a written report in which you describe your findings in more detail, including a discussion of any significant or unanticipated aspects of your findings not reflected in your table or chart. The table or chart should indicate the number of examples of the media genre you used and the number of cases (N) of the topic you noted in the examples. The findings for your categories should be presented in terms of percentages. For instance, after investigating types of crime in ten examples of TV crime dramas you observed 108 cases of offences (N=108). Of these, 35 were property crimes; 16 were drunk driving; 24 were homicides; and 33 were assaults. So in this case you'd make a final table which graphically presented that property crimes made up 32% (35/108); drunk driving, 15% (16/108); homicides, 22% (24/108); and assaults, 31% (33/108) of the total offences. If you are reporting on the characteristics of offenders, you will encounter a slightly more complicated situation. Here you need to provide percentages within each of the separate and distinct categories you were working with. For example, if the crimes were committed by 25 offenders, then you would provide percentages totaling 100% for each category. That is, 80% were male, 20% were female, 55% were Caucasian, 25% Black, 10% Hispanic, 10% unknown, and so on.

Your written report is a detailed description of your findings. But why repeat what the chart already indicates? That is a fair question. It enables you to introduce specific, vivid examples to illustrate and elaborate on your findings. The written report also provides an opportunity to share any other observations you made but didn't tabulate in the formal table.

Analyzing Your Findings Sociologically

There are different approaches one can take to doing a sociological analysis in a field project such as this. Which one is most appropriate will depend on the nature of your particular topic, findings, and emerging interests and questions.

The key to an analysis is probing what you found with questions that help you interpret and understand the findings in terms of the nature of our society, its culture, and its social structure, as well as the historical transformations it is undergoing.

The following are some suggested ways of approaching the analysis:

1. Consider the cultural meaning or social significance of what you found in terms of the ramifications for the wider society.
2. Compare the portrayal of the topic you've selected with relevant social scientific research and/or official statistics.
3. Use a particular sociological concept, theory, or theorist to discuss or account for what you found. For example, what is the social function of a particular portrayal? Which groups stand to gain and lose from it? Can we gain any insights into the phenomenon and/or its portrayal in terms of a feminist framework? If you draw on a

particular concept, theory, or theorist, provide a brief overview of it or the aspect of the theorist's work you intend to use.

4. Examine your findings in terms of relevant media research and theory. Such an examination might include the nature of mass media and its effects on content; the effects of commercialization on mass media; and if and how the media are participating in the social construction of the phenomenon.

What about discussing the effects on the audience? Although it is tempting to make claims about the effects a media portrayal will have on an audience, a content analysis is not designed to study the effects or influence on the audience. Moreover, research has found that the audience's interpretation can vary significantly from what one might expect.[17] At best, you can only speculate on the influence (unless you cite other research which has carefully and systematically studied the effects). Be cautious about invoking overly deterministic assumptions in your deliberations. Viewers or readers are social actors who select and interpret what they observe and read. Your analysis should focus more on what the content signifies about the society, its culture, and its social institutions, including how they function and how they may be changing, rather than on its behavioural effects on an audience.

Concluding

The final step is *concluding*. The key aspect of concluding is to share what you have learned from the study. This can involve a discussion of the difference your study made to the views you previously held, or how your findings relate to prevailing beliefs and opinions. Your conclusion can also comment on the relation of your study to other relevant research. Finally, it's also useful to highlight questions, problems, or issues you encountered during your study or that emerged for you as a consequence of doing the study.

IV. COMMERCIALIZED MASS MEDIA AND CULTURE

Since we have generally limited ourselves to analyzing content from the mass media for this field project, it can be useful to call attention to some aspects of the mass media which you might keep in mind when interpreting and analyzing the data you gather.

Although we usually associate mass media with types of communication technologies and industries like newspapers, TV, radio, and film, the mass media also involve a particular form of culture that is produced, communicated, and consumed through them. Specifically, it refers to culture generally shared in a standardized form by large numbers of people.

Culture—that is, the way of life and meaning system of a people—as a feature of being created and transmitted via the mass media becomes standardized. This means in part that images, messages, and ideas will be processed for mass consumption by large numbers of people and, if successfully communicated, become generally shared *as such*. Exploring how this occurs and its implications moves us into the area of media and communication studies, an area which extends beyond the scope of this article.

However, there is one aspect of it we do wish to draw attention to: What happens when the medium is commercialized? Specifically what is the effect on content being transmitted primarily through a commercialized mass medium? As Czerny and Swift have pointed out, there is an emphasis on the intense, the unambiguous, the familiar, and the marketable. By the intense, they mean there is a selective focus on action and the dramatic in selecting and presenting content. The unambiguous means news and stories are generally presented in terms of clear and simple moral constructs, and complex issues and events are simplified. By the familiar, they are alerting us to the ways in which messages in the mass media are constructed to fit in with common notions and broadly shared prejudices. And the marketable means the content is geared to consumer taste through the use of graphic violence and sex, and emphasizes the sensate and the concrete. In short, the commercialized mass media commodifies news, stories, and events and in doing so distorts what is being reported upon. So there may be a significant difference between a phenomenon as it occurs in actual everyday life and how it is portrayed in the mass media.[18]

But all communication involves some form of distortion, and commercialization is not the only distorting factor we need to be aware of. All mass media work with cultural codes which humans create to represent and think about reality and which serve as a basis for participating within it.

Codes, with their systems of signs, symbols, relations among signs, and rules for interpreting signs, are an important part of how we constitute meaning, and create a meaningful world, together. Consider the example of the alphabet. It consists of a series of signs comprised of marks, relations among the marks based on spaces, and rules for interpreting the marks as letters of the alphabet and words. Codes can also include symbols, which are less arbitrary in their representation of some thing. For example, consider the symbol of the cross in Christianity or the maple leaf in Canada. The cross, while representing the Christian religion, is not a perfect reproduction of all that Christianity is. Rather we associate various ideas, beliefs, events, and personages with it.

Cultural codes are used to depict and interpret the world around us. Because they are symbolic in nature they often resemble, but are not exact renderings, of the things they represent. For example, consider the term *prostitute*. This can be defined in fairly neutral terms. In an objective sense a prostitute can be defined as someone selling sexual services indiscriminately to a wide range of clients. But what are the qualities, characteristics, and other attributes generally associated with a prostitute? How do they appear, for instance, on television or in the movies? Who appears? When do they appear? What are they associated with? As members of this society we can offer answers based on our experience and common stock of knowledge. However, in sociology our task is not to guess, but to inquire, observe, and understand the codes we use.

For instance, does the portrayal of the prostitute in the mass media, or in our culture generally, reflect the facts that, in Canada, about one in five are male, about one in five are married, except for tobacco they do not use drugs at significantly higher rates than the general public, or that street prostitutes constitute only about 5 to 20 per cent of prostitution activity?[19]

As we can see, cultural codes may more or less accurately represent, or distort quite significantly, the realities they represent, as part of effectively symbolizing what they are referring to. A code always conventionalizes the thing in terms of some particular form of life or tradition. One of the challenges is determining the particular conventions that are at play in, and constitute the meaning of, the code for those who use it.[20]

When we learn particular codes in society such as those related to fashion, health and medicine, family, sexuality, and so on, we are learning and working with conventional images and ideas which symbolically represent things in the world. But the main function of cultural codes within a group is not to provide accurate empirical representations of, or generalizations about, reality. Their main function is primarily symbolic and social, and they don't need to accurately represent the realities of a phenomenon to effectively fulfill that function.

The mass media both rely upon and shape such codes. And as consumers of media, we come to know and use them in receiving messages, defining reality, and participating in the world. In conclusion, the investigator needs to be aware that the culture and social structure depicted in the texts being analyzed for our basic content analysis have usually been processed through the commercialized mass media and bear the influence of that.

NOTES

1. Weber, Robert Philip, *Basic Content Analysis*, Sage Publications, 1985.
2. An interesting discussion of the application of the philological term *text* to social phenomenon is provided by Clifford Geertz in his essay, "Blurred Genres: The Reconfiguration of Social Thought," in *Local Knowledge: Further Essays in Interpretive Anthropology*, Basic Books, 1983: 30–33.

3. Jacqueline Wiseman and Marcia Aron, *Field Projects for Sociology Students,* Schenkman Publishing/Harper and Row, 1970.

4. Tanya Titchkosky, "Clenched Subjectivity: Disability, Women and Medical Discourse," *Disability Studies Quarterly,* 25 (3) [Summer, 2005].

5. Aaron Cicourel, *The Social Organization of Juvenile Justice,* Wiley, 1968.

6. Arthur Asa Berger, *Media Analysis Techniques,* Revised edition, Sage, 1991: 118–126.

7. Although having literature and the arts in mind, Wolfgang Iser provides an accessible introduction to a range of theoretical approaches relevant to more qualitative approaches to analyzing texts in *How to Do Theory,* Blackwell, 2006. Also helpful are Roland Barthes, Mythologies, Noonday Press/Farrar, Straus and Giroux, 1957; Monica Morris, *An Excursion into Creative Sociology,* Columbia University Press, 1977; Ino Rossi, "Relational Structuralism," in Ino Rossi (editor), *Structural Sociology,* Columbia University Press, 1982: 3–21; and Steven Seidman and Jeffrey C. Alexander (editors), *The New Social Theory Reader,* New York: Routledge, 2001, especially the sections on new critical theory, semiotic structuralism, post-structuralism, and cultural studies.

8. Arthur Asa Berger examines some of these techniques of interpretation in a manner accessible to introductory students in *Media Analysis Techniques,* Revised edition, Sage, 1991. His chapters on semiological and Marxist analysis are particularly helpful. His section on sociological analysis, while interesting, is limited in its understanding of sociology. For Berger, sociological analysis means a sort of functionalist interpretation based on uses and the gratification of members' needs.

9. Simon Davis, "Men as Success Objects and Women as Sex Objects: A Study of Personal Advertisements," *Sex Roles,* 23 (1-2) [1990]: 43–50.

10. Herbert J. Gans, *Deciding What's News: A Study of CBS Evening News, NBC Nightly News, Newsweek and Time,* Vintage, 1979. Also see Todd Gitlin, "Prime Time Ideology: The Hegemonic Process in Television Entertainment," *Social Problems,* 26 (3) [1979] to see how such an analysis can be extended to television programming more generally.

11. Raymond Morris, "Canada as a Family: Ontarian Responses to the Quebec Independence Movement," *Canadian Review of Sociology and Anthropology,* 21 (2) [1984]: 181–201.

12. Angela Miles, "Confessions of a Harlequin Reader: Learning Romance and the Myth of Male Mothers," *Canadian Journal of Political and Social Theory,* XII (1-2) [1988]: 1–37.

13. Tami Bereska, *Deviance, Conformity and Social Control in Canada,* Pearson/Prentice Hall, 2004: 148–150.

14. Roger Rosenblatt, "Growing up on Television," *Daedalus,* Vol. 105 *[Fall, 1976].*

15. Both Harold Innis' and Marshall McLuhan's classic work on communication have demonstrated that the nature of the medium through which a message is communicated influences the overall meaning of the message. See Harold Innis, *The Bias of Communication,* University of Toronto Press, 1951; and Marshall McLuhan, *Understanding Media: The Extensions of Man,* Toronto: McGraw-Hill, 1964.

16. We will be working with this example extensively. The ideas are based on a study originally done by Craig Haney and John Manzolati, "Television Criminology: Network Illusions of Criminal Justice Realities," in Elliot Aronson (editor), *Readings About the Social Animal,* 3rd edition, W. H. Freeman, 1981: 125–136.

17. For instance, the following studies of the influence of crime and violence in the media on the audience challenge the conventional wisdom of media determinism: Monica Belkaoui, "The Mass Media in Canada," in Dennis Forcese and Stephen Richer (editors), *Social Issues: Sociological Views of Canada,* 2nd edition, Prentice-Hall, 1988: 440–468; Kenneth Dowler, "Media Consumption and Public Attitudes toward Crime and Justice: The Relationship between Fear of Crime, Punitive Attitudes, and Perceived Police Effectiveness," *Journal of Criminal Justice and Popular Culture,* 10 (2) [2003]: 109–126.; Craig Haney and John Manzolati, "Television Criminology: Network Illusions of Criminal Justice Realities," in Elliot Aronson (editor), *Readings About the Social Animal,* 3rd edition, W. H. Freeman, 1981: 125–136; Vincent F. Sacco, "Media Constructions of Crime," *Annals of the American Academy of Political and Social Science,* 539 [May, 1995]: 141–154; and Karen Sternheimer, *The Truth about Pop Culture's Influence on Children,* Westview, 2003.

18. Michael Czerny and Jamie Swift, *Getting Started on Social Analysis in Canada,* 2nd edition, Between the Lines, 1988.

19. Ian McDermid Gomme, *The Shadow Line: Deviance and Crime in Canada,* 3rd edition, Thomson Nelson, 2002: 232–260.

20. John Berger, *Ways of Seeing,* Viking Press, 1973: 45–64.

Doing the Right Thing: Ethics in Research

Lisa J. McIntyre

Doing what's right in situations is usually rather self-evident, even if it may be difficult. The rules of ethical conduct appropriate to a situation generally reflect the principle of not doing harm to others, and are oriented to the welfare of the larger society in contrast to the one's self-interest. A general exception involves the traditional professions—doctors, lawyers, and clergy. Situations involving professions are complicated by the professional's ethical and legal duty to protect the confidentiality of his/her client.

Over the last few decades the social sciences have faced an increasing concern with ethics in research. But why? What kinds of moral issues can arise in sociological research that call for ethical reflection? And what rule of thumb does McIntyre offer for dealing with ethical questions and issues in sociological research?

Ethical guides are not simply prohibitions; they also support our positive responsibilities. For example, scientists have an obligation to advance knowledge through research. They also have a responsibility to conduct research as competently as they can and to communicate their findings accurately to other scientists.
—Diener and Crandall 1978

To begin, it is important to note that the term *ethics* has both a conventional (or everyday) meaning and a technical meaning. The fact that there are two ways to use this term causes a great deal of confusion. In the

Source: Lisa J. McIntyre, "Doing the Right Thing: Ethics in Research," *The Practical Skeptic: Readings in Sociology*, 3rd Edition, McGraw-Hill, 2006. Reprinted with permission.

conventional or everyday sense, ethics is synonymous with morality, and doing the ethical thing simply means doing the moral thing. Conversely, unethical behavior is immoral behavior.

Ethics: Technical Meaning and Origins

In the technical sense, ethics and morals are different. Although in many cases there may be an overlap between ethical and moral behavior, there are no guarantees that this will occur. The following two scenarios give examples of how morals and ethics may diverge.

Scenario 1

Chris confesses to a friend to having killed someone and hidden the body under a pile of garbage near an old shack at the lake. The police can't find the missing victim, nor do they know that Chris is the killer. The friend calls the police and tells them where to find Chris and the victim's body.

Has the friend done something moral or immoral? In spite of the fact that there is a widely accepted rule in society against snitching on one's friends, most people probably would say that the friend did the moral thing. It is not right to allow murderers to go free, and the victim certainly has a right to a proper burial.

Scenario 2

Chris hires an attorney and then confesses the murder and burial to that attorney. The police can't find the missing victim, nor do they know that Chris is the killer. The lawyer isn't all that sure about Chris's story and drives up to the lake to check it out. A search through the garbage reveals the body. After taking a Polaroid of the body, the lawyer reburies it in the garbage. Returning to town, the lawyer urges Chris to go to the police but does not call the police.[1]

Because the lawyer does not turn in the murderer, he or she does not seem to be acting morally. But there is an important consideration. Whereas in scenario 1 the friend may have a moral obligation to turn Chris in to the police, the lawyer has an ethical obligation to keep the client's confidence— even when the client is a murderer! This is not a gray area, either; this ethical requirement is spelled out clearly in the legal profession's *Code of Professional Responsibility*: "The lawyer must hold in strictest confidence the disclosures made by the client in the professional relationship. The first duty of an ethical attorney is 'to keep the secrets of his client'" (Ethical Consideration 4-1). Preserving a client's secrets is such an important ethical obligation that had the lawyer turned Chris in, the attorney could have been disbarred and never again allowed to practice law.

In the technical sense, behavior is ethical insofar as it follows the rules that have been specifically oriented to the welfare of the larger society and not to the self-interests of the professional. So, ethics are designed to promote the welfare of others. You might be wondering how this lawyer could be promoting the welfare of others. The short answer is this: The legal profession is committed to the idea that people are innocent until proved guilty, that everyone accused of a crime has a right to the best defense, and that this is possible only if those accused of crimes can trust their attorneys. No client would trust an attorney who did not keep his or her secrets.

To be ethical, professionals have the burden of having to do things that others might consider to be immoral. For lawyers, in addition to keeping possibly nasty secrets, being ethical involves the duty to defend what may be unpopular cases and vicious criminals.

To become a professional, one must promise to abide by the relevant professional ethical codes. This is not a matter of personal choice—if you want to be a physician, you must follow the medical rules of ethics. To act unethically is to act unprofessionally.

It is important to be precise here: What do I mean by "professional"? In conventional language, we use the term at least three different ways. Sometimes, we apply the word to a job that is well done: "You did a very professional job of building that doghouse." Other times, we label someone a professional because he or she is paid to do something, regardless of the quality of the outcome: "John was a professional baseball player—but he never could hit a curve ball."

When sociologists use the term professional, however, they generally mean something else. Sociologically speaking, a professional is a member of a special kind of occupational group. Originally, only three occupational groups qualified as professions: lawyers, physicians, and clergy. These three groups have several things in common that set them apart from other occupations:

1. Their practitioners study for years to acquire technical knowledge and skills.
2. The knowledge they possess involves traditions and secrets that are not shared by outsiders.
3. Their knowledge is useful to outsiders and frequently means the difference between life and death.
   From the first three characteristics of the professional derives a fourth—and this is really what sets the professional apart from other workers:
4. The work of a professional cannot be judged or supervised by anyone who is not a member of the same profession.

As sociologists see it, these qualities are characteristic of doctors, lawyers, and clergy, but not of plumbers or hairstylists. Although people in any occupational group may refer to themselves as professionals ("I am a professional hairstylist" or "I am a professional plumber"), relatively few occupations really are professions.[2]

The nature of the professional's job is such that *how* it is done is as important as (if not more important than) what the outcome is. The problem is that as laypeople, we cannot judge how well the job was done. For example, if patient Q's family suspects that the surgeon wasn't competent and that this incompetence led to Q's death, they might want to sue the surgeon for malpractice. Q's family might feel righteous in claiming that if the surgeon had done the job right, Q would still be alive. But that is not necessarily true (recall the old saying, "The operation was a success, but the patient died"). The views of Q's family are legally irrelevant and will hold no water in court. According to the law, only another physician can judge whether the surgeon was truly negligent. So, Q's family will have to find another surgeon who is willing to testify that it was the poor quality of the surgeon's work that led to the patient's death.[3]

Even though they could not tell if their doctor, lawyer, priest, or rabbi was doing all that ought to be done, for a long time people trusted these professionals to do the right thing. But by the mid-nineteenth century, many people were growing increasingly suspicious of professionals and were beginning to suggest that perhaps professionals ought not to be given so much freedom and autonomy.

Professionals responded by emphasizing the fact that their actions were prompted not by self-interest, but by their concern for their clients, patients, or parishioners. According to members of the medical profession, then, doctors do not perform surgery simply to make money, but to relieve people's suffering. And according to members of the legal profession, then, lawyers do not represent clients simply for the money, but because everyone has a right to representation.

One of the ways that professionals emphasized their commitment to the public welfare over self-interest was to promulgate or announce codes of ethics. They promised that they would follow these codes of ethics and punish any member of their profession who failed to do so. As we moved into the twentieth century, however, it became increasingly clear that at least one aspect of professional work was not being regulated properly by the professionals themselves: research.

RESEARCH ATROCITIES

Gross abuses of professional power in research became public knowledge after World War II. When Nazi physicians were brought to trial at Nuremberg in the late 1940s, the tales of their "research" horrified the world:

Physicians forced people [in concentration camps] to drink seawater to find out how long a man might survive without fresh water. At Dachau, Russian prisoners of war were immersed in icy waters to see how long a pilot might survive when shot down over the English Channel and to find out what kinds of protective gear or rewarming techniques were most effective. Prisoners were placed in vacuum chambers to find out how the human body responds when pilots are forced to bail out at high altitudes.... At Auschwitz, physicians experimented with new ways to sterilize or castrate people as part of the plan to repopulate Eastern Europe with Germans. Physicians performed limb and bone transplants (on persons with no medical need) and, in at least one instance, injected prisoners' eyes with dyes to see if eye color could be permanently changed. At Buchenwald, Gerhard Rose infected prisoners with spotted fever to test experimental vaccines against the disease; at Dachau, Ernst Grawitz infected prisoners with a broad range of pathogens to test [different cures].... Hundreds of people died in these experiments; many of those who survived were forced to live with painful physical or psychological scars. (Annas and Grodin 1992, 26)

At their trials, many Nazi physicians protested that they "had only been following orders." But much evidence suggested otherwise: "Contrary to postwar apologies, doctors were never forced to perform such experiments. Physicians volunteered—and in several cases, *Nazi officials actually had to restrain overzealous physicians from pursuing even more ambitious experiments"* (Annas and Grodin 1992, 26; emphasis added).

What could have motivated these physicians, these professional healers, to misapply their professional skills so horribly? At the time, many Americans believed that there was something fundamentally wrong with the German "personality type." For one thing (or so it was thought), Germans were all too quick to follow orders without exercising independent judgment. Certainly, such things could never happen in the United States!

What many people did not know or appreciate was the long tradition among U.S. physicians of conducting questionable research. For example, in the nineteenth century, orphans, the "feeble-minded," and hospital patients frequently were made the unwilling victims of medical experiments.

In his autobiography, physician J. Marion Sims described how, between 1845 and 1849, he kept several black female slaves at his hospital to test his discovery of a repair for vesicovaginal fistula. The fistulas, allowing urine or feces to leak through the vaginal opening, caused great discomfort and distress.... Sims performed dozens of operations on the women—this in the days before anesthetics—and praised their "heroism and bravery." (Lederer 1995, 115-116)

There have even been cases in which U.S. military personnel were required to participate in surgical experiments—under threat of court-martial!

In any case, outraged at the evidence they heard, the judges of Nuremberg promulgated the Nuremberg Code. The ten principles of the code were written to protect the rights of research subjects. Never again, the judges said, would humans be placed at risk of serious harm by being used as unwilling guinea pigs. But less than 30 years later, there was another research scandal. This time, the physicians were not only Americans but were employed by the United States Public Health Service! This study, known as the Tuskegee Syphilis Experiment, began in 1932 when public health workers came to Macon County, Georgia, in search of African American men who suffered from syphilis. The physicians preyed on the poverty of the men and recruited research subjects by offering to "pay" for their participation—free medical exams, transportation to and from the medical facilities where the exams would be held, and free meals on examination days. The biggest incentive was that the families of each subject would be paid $50 to help with burial expenses.

Not one of the subjects was told that he had syphilis—though each was told that he had "bad blood." And, although a cure for syphilis was widely available throughout most of the 30-year period during which the study was conducted, not one of the men was given this medication despite the fact that it would have saved his life.[4] You see, the researchers were intent on studying the effects of *untreated* syphilis. The study continued until 1972 when its existence became public. At that point, the research was terminated because of public outrage (Jones 1981).

The [N]uremberg [C]ode

1. The voluntary consent of the human subject is absolutely essential....
2. The research should be such as to yield fruitful results for the good of society, unprocurable by other methods or means....
3. The research should be so designed ... so that the anticipated results will justify the performance of the experiment.
4. The research should be so conducted as to avoid all unnecessary physical and mental suffering and injury.
5. No research should be conducted where there is ... reason to believe that death or disabling injury will occur.

6. The degree of risk to be taken should never exceed that determined by the humanitarian importance of the problem to be solved by the research.
7. Proper preparations should be made ... to protect the research subject against even remote possibilities of injury, disability, or death.
8. The research should be conducted only by scientifically qualified persons. . . .
9. During the course of the research the human subject should be at liberty to bring the research to an end if he has reached the physical or mental state where continuation of the research seems to him to be impossible.
10. During the course of the research the scientist in charge must be prepared to terminate the research at any stage, if he has probable cause to believe ... the continuation of the research is likely to result in injury, disability, or death to the research subject.

Note: The writers of the original code emphasized the need to protect subjects in experimental research. Because social scientists use a variety of techniques, I have substituted the word "research" for "experiment."

THE CASE FOR SOCIOLOGICAL RESEARCH

You might well think that members of the general public have little to fear from sociologists. After all, what harm can a bunch of geeks with clipboards do by asking questions?

There is potential for harm in any sort of research that involves human subjects. The potential harm in sociological research frequently involves not what we do or do not do to our research subjects, but what we find out about them. Sociologists and other social scientists find out information that people often would prefer to keep private.

One of the most famous examples of social science research that many believed crossed the ethical lines was Laud Humphreys' study, which he titled *Tearoom Trade* (1970).

Technically speaking, *Tearoom Trade* was a study of impersonal sexual activity between male homosexuals. Less technically speaking, Humphreys began his research (or so he later said) by trying to find an answer to a question posed by his graduate advisor: "Where does the average guy go just to get a blow job?"

As Humphreys discovered, the answer to that particular question was "a tearoom" (that is, a restroom in a public park). In these tearooms, Humphreys did observational research. More specifically, to hide the fact that he was a researcher, he took on the role of "watch queen" (a third man who serves as a lookout for those engaged in homosexual acts and obtains voyeuristic pleasure from his observations).

From his observations, Humphreys obtained a great deal of information about how men approach each other and negotiate sex. But, given the circumstances, he could not very well find out much else. Humphreys wanted to know, Who are these men? How do they spend the rest of their time?

So, in addition to making his secret observations, Humphreys recorded each participant's license plate number. He then took this list of numbers to the police, told them he was doing "market research," and obtained the names and addresses of each man.

But, then what? He could hardly show up at the men's doorsteps and announce, "Hi, I saw you engaging in homosexual sex in the park last month, and now I would like to ask you a few questions about the rest of your life." (Sometimes, the most straightforward approach simply does not work.)

Around that time, another researcher at the same university was conducting a study on issues related to health care. Humphreys persuaded this researcher to include the names of his tearoom players on the list of subjects for the health study and schedule them for interviews. Humphreys himself would interview these men. To reduce the chances that the men would recognize him, Humphreys waited a year and changed his hairstyle. Then, no doubt armed with that ubiquitous clipboard, Humphreys visited and interviewed each of the men. This way, posing as a health-care researcher, he was able to find out all about the men's socioeconomic status (mostly middle class), their educational level (pretty high), and their family life (mostly married with children). Humphreys discovered that the only nonconventional thing about these men was that they visited tearooms for anonymous sex.

What might be ethically questionable about Humphreys' research? Although some might object that the very topic of Humphreys' research was immoral, the nature of the topic is not an *ethical* concern. What is of concern ethically is the fact that Humphreys deceived his subjects—they never knew that they were participating in research, and they didn't have the opportunity to choose to participate. Moreover, Humphreys conducted his research during a time when homosexual behavior was illegal where the research was conducted. By recording their names and addresses, Humphreys was placing his research subjects in great jeopardy. After his book was published, what if the police had demanded that Humphreys turn over his list of subjects' names and addresses? There was a great risk not only of legal prosecution but of psychological and social harm as well. And, had their names been discovered, some of the men might even have been subjected to extortion.

Humphreys defended his research by pointing out that it is important for sociologists to know about such men and their activities in order to understand them. In point of face, Humphreys' research did contradict many social myths about men who have sex in public bathrooms. Most were established members of the community with wives and children, and in practicing consensual sex, they were not hurting anyone and certainly not bothering children. Humphreys' research was published and widely cited and may well have played a role in decriminalizing some sexual acts between consenting adults.

Humphreys' research was perhaps extreme in this respect, but it is not unusual for sociologists to uncover embarrassing details. Sometimes, what we learn not only is embarrassing but may place the research subject in legal jeopardy. In such cases, we have to figure out what our duty is—do we keep the secrets only of those whom we respect as "good people"? The problem may be compounded by the fact that the people we study are often those who have little power in society: it almost seems as it sociologists are obsessed with marginalized people (the poor, the homeless, street criminals, and so on).[5]

There are few hard-and-fast rules about what is and is not ethical behavior in sociological research. As far as I am concerned, the only thing that is consistently unethical is to not think through the possible consequences of our research.

As we think through the possible consequences of our research, we need to remember that we have an obligation not only to our research subjects but to other sociologists, to the university, and to members of the community at large. Making ethical decisions involves weighing the costs and benefits of the research to all of these groups.

This takes a great deal of thought; frequently, our research may have consequences that extend beyond the obvious. For example, in the early 1960s, a woman named Kitty Genovese was raped and murdered in New York City. What set the Genovese murder apart from the many other murders that happened that year was the fact that a number of people had heard her screams for help, which lasted for many minutes, but not a single one called the police.

The Genovese murder caught the imagination of social researchers in a big way. Under what conditions would people help strangers in trouble? What followed was a multitude of so-called bystander intervention studies. Some of these were pretty benign, such as a boy on crutches dropping all of his school books to see whether anyone would stop to help. Other versions included scenes of staged violence, such as a woman yelling from the bushes, "Help, rape." Soon, people grew leery and distrustful as they walked around college campuses and nearby neigh-borhoods—there were so many researchers out and about that one never knew when one might become an involuntary research subject.

Then, the inevitable but still unthinkable happened:

At the University of Washington in Seattle in 1973, a male student accosted another student on campus and shot him. Students on their way to class did not stop to aid the victim, nor did anyone follow the assailant (who was caught anyway). When the campus reporters asked some students about their lack of concern over the murder, they said they thought it was just a psychology experiment. (Diener and Crandall 1978, 87)

In this case, the harm caused by the overdoing of bystander and other sorts of research in the field did not affect only the research subjects. It contaminated the researcher's world by making people distrust researchers. And, far worse, it may have contributed to the death of a college student.

The nature of informed consent

"Informed consent is the procedure in which individuals choose whether to participate in an investigation after being informed of facts that would be likely to influence their decision. Informed consent includes several key elements: (a) subjects learn that the research is voluntary; (b) they are informed about aspects of the research that might influence their decision to participate; and (c) they exercise a continuous free choice to participate that lasts throughout the study. The greater the possibility of danger in the study and the greater the potential harm involved, or the greater the rights relinquished, the more thorough must be the procedure of obtaining informed consent" (Diener and Crandall 1978).

INSTITUTIONAL REVIEW BOARDS: THE DAWN OF A NEW ERA

These days, before any member of the university community (student, faculty, or staff) can conduct research that involves humans, they must submit a research proposal to a university officer or committee charged with ensuring that research is done ethically. If there is any question of risk to the human subjects, a committee consisting of both faculty (from a variety of disciplines) and community members will scrutinize the proposal. These committees are commonly called Institutional Review Boards (IRBs). If the members of the IRB judge that the researcher has not created sufficient safeguards to protect the rights of the research subjects and the general public, and even the researcher him- or herself, the researcher is prohibited from going on.

Like the Nuremberg Code, contemporary ethical guidelines place a great deal of emphasis on treating research subjects with respect. In many cases, researchers must obtain not merely *consent* from potential subjects but *informed consent*. As a general rule, deception must be kept to the absolute minimum. Members of IRBs are particularly skeptical of any research that places research subjects at risk of injury (physical, psychological, emotional, or legal) greater than the risk that surrounds the routine activities of everyday life.

HOW HEROIC MUST AN ETHICAL RESEARCHER BE?

To what extremes must the sociological researcher go to fulfill his or her ethical duty? As I noted previously, one of the reasons Laud Humphreys was criticized when he published *Tearoom Trade* was that homosexual acts were prohibited by law where he did his research. In theory, the district attorney could have subpoenaed the list of names and addresses of Humphreys' subjects and prosecuted these men.[6] Would being ethical have required Humphreys to choose jail over releasing his information? In fact, this course of action apparently was contemplated, though it never materialized.

A little over a decade later, another sociologist came even closer to being forced to decide between breaching confidentiality and going to jail. Mario Brajuhas, a graduate student at the State University of New York at Stony Brook, was doing participant observation research as a waiter in a restaurant. When the restaurant burned down, the police suspected arson. Investigators knew of Brajuhas's research and of the fact that he had taken copious field notes; they suspected that those field notes might help identify the arsonist. The local prosecutor subpoenaed the notes, but Brajuhas refused to hand them over, even when threatened with jail. Finally (after 2 years), the major suspects in the fire died, and the prosecutor dropped the case.

In the early 1990s, Rik Scarce, a graduate student at Washington State University, took a vacation. He left an acquaintance of his, Rodney Coronado, behind as a housesitter. Scarce and Coronado had become acquainted when Scarce, prior to going to graduate school, had been researching a book on radical environmentalists entitled *Eco-Warriors: Understanding the Radical Environmental Movement* (1990). Coronado was involved with the Animal Liberation Front (ALF), which was adamantly opposed to the use of animals for research.

While Scarce was on vacation, the ALF raided a research laboratory at Washington State University. Several animals were set free, and the researchers' computers were destroyed. The university estimated the damage at about $100,000. Several months later, Scarce was subpoenaed and commanded to appear before a grand jury that had been convened to investigate the crime. Scarce did appear and answer several questions, but he declined to answer questions that, he said, required him to breach the confidentiality of his research subjects. Scarce quoted from the American Sociological Association's Code of Ethics, which states that "confidential information provided by research participants must be treated as such, even when this information enjoys no legal protection or privilege and legal force is applied." As Scarce later explained, "I told the judge that I feared for my ability to earn a living as a sociologist if I were compelled to testify. Research subjects might not be willing to speak with me, and institutions might not be willing to hire an unethical researcher" (Scarce 1994).

The judge was not moved by Scarce's explanation. Ultimately, after he continued to refuse to testify, Scarce was sent to jail as a "recalcitrant [unwilling] witness." He spent more than 5 months in jail before the judge, finally convinced that Scarce could not be compelled to testify, freed him.

The Scarce case sounded a warning bell to sociologists everywhere. According to their ethical code, they have an obligation to keep con-fidential information to themselves even when they have no legal right to do so. This puts sociologists in a different position than lawyers and doctors. Communication between lawyers and clients and between doctors and patients is *legally privileged*. Lawyers and physicians have not only an *ethical duty* to keep information confidential but the *legal right* to do so. Sociologists, on the other hand, have no such clear legal right, although they do have an ethical duty.

NOTES

1. A very similar case happened several years ago in Lake Pleasant, New York.

2. I do not mean to disrespect members of any occupational group by claiming that they are not professionals. I simply mean to illustrate how sociologists use the term *professional* to highlight the qualities of certain occupational groups.

3. There are exceptions. Some errors are so obvious that the law does not require an expert witness to testify—as when a surgeon cuts off the wrong leg or sews a surgical instrument into the wound. (The legal phrase for such exceptions is *res ipsa loquitur*, a Latin term meaning "the thing speaks for itself.")

4. Alexander Fleming discovered the cure (penicillin) in 1928, but it would be another decade before the drug was used by medical practitioners. In the late 1930s, just in time for World War II, two British researchers, Ernst Chain and Howard Florey, discovered a process that purified penicillin and made it safe. The drug was widely used by the military in the war and in the civilian sector after the war.

5. Part of the reason for this apparent obsession is that it is much easier to gain access to people with little power. It's easier to get permission to examine, say, prison inmates than executives of Ford Motor Company.

6. A subpoena (sa-PEE-na) is nothing to fool around with. It is a command from a legal authority to appear and give testimony. If you refuse to comply with a subpoena, you can be charged with contempt of court and sent to jail until you change your mind (or until the judge accepts the fact that nothing is going to change your mind).

References

Annas, George J., and Michael A. Grodin. 1992. *The Nazi Doctors and the Nuremberg Code.* New York: Oxford University Press.

Diener, Edward, and Rick Crandall. 1978. *Ethics in Social and Behavioral Research.* Chicago: University of Chicago Press.

Humphreys, Laud. 1970. *Tearoom Trade: Impersonal Sex in Public Places.* Chicago: Aldine.

Jones, James H. 1981. *Bad Blood: The Tuskegee Syphilis Experiment.* New York: Free Press.

Lederer, Susan E. 1995. *Subjected to Science: Human Experimentation in America Before the Second World War.* Baltimore: Johns Hopkins University.

Scarce, Rik, 1990. *Eco-Warriors: Understanding the Radical Environmental Movement.* Chicago: Noble Press.

Scarce, Rik. 1994. "(No) Trial (But) Tribulations: When Courts and Ethnography Conflict." *Journal of Contemporary Ethnography* 23: 123–149.